FILTERING AND SYSTEM IDENTIFICATION

Filtering and system identification are powerful techniques for building models of complex systems in communications, signal processing, control, and other engineering disciplines. This book discusses the design of reliable numerical methods to retrieve missing information in models derived using these techniques. Particular focus is placed on the least squares approach as applied to estimation problems of increasing complexity to retrieve missing information about a linear state-space model.

The authors start with key background topics including linear matrix algebra, signal transforms, linear system theory, and random variables. They then cover various estimation and identification methods in the state-space model. A broad range of filtering and system-identification problems are analyzed, starting with the Kalman filter and concluding with the estimation of a full model, noise statistics, and state estimator directly from the data. The final chapter on the system-identification cycle prepares the reader for tackling real-world problems.

With end-of-chapter exercises, MATLAB simulations and numerous illustrations, this book will appeal to graduate students and researchers in electrical, mechanical, and aerospace engineering. It is also a useful reference for practitioners. Additional resources for this title, including solutions for instructors, are available online at www.cambridge.org/9780521875127.

MICHEL VERHAEGEN is professor and co-director of the Delft Center for Systems and Control at the Delft University of Technology in the Netherlands. His current research involves applying new identification and controller design methodologies to industrial benchmarks, with particular focus on areas such as adaptive optics, active vibration control, and global chassis control.

VINCENT VERDULT was an assistant professor in systems and control at the Delft University of Technology in the Netherlands, from 2001 to 2005, where his research focused on system identification for nonlinear state-space systems. He is currently working in the field of information theory.

Filtering and System Identification

A Least Squares Approach

Michel Verhaegen
Technische Universiteit Delft, the Netherlands

Vincent Verdult
Technische Universiteit Delft, the Netherlands

CAMBRIDGE
UNIVERSITY PRESS

CAMBRIDGE UNIVERSITY PRESS
Cambridge, New York, Melbourne, Madrid, Cape Town,
Singapore, São Paulo, Delhi, Mexico City

Cambridge University Press
The Edinburgh Building, Cambridge CB2 8RU, UK

Published in the United States of America by Cambridge University Press, New York

www.cambridge.org
Information on this title: www.cambridge.org/9781107405028

© Cambridge University Press 2007

First published 2007
First paperback edition 2011

A catalogue record for this publication is available from the British Library

ISBN 978-0-521-87512-7 Hardback
ISBN 978-1-107-40502-8 Paperback

Contents

Preface

This book is intended as a first-year graduate course for engineering students. It stresses the role of linear algebra and the least-squares problem in the field of filtering and system identification. The experience gained with this course at the Delft University of Technology and the University of Twente in the Netherlands has shown that the review of undergraduate study material from linear algebra, statistics, and system theory makes this course an ideal start to the graduate course program. More importantly, the geometric concepts from linear algebra and the central role of the least-squares problem stimulate students to understand how filtering and identification algorithms arise and also to start developing new ones. The course gives students the opportunity to see mathematics at work in solving engineering problems of practical relevance.

The course material can be covered in seven lectures:

(i) Lecture 1: Introduction and review of linear algebra (Chapters 1 and 2)

(ii) Lecture 2: Review of system theory and probability theory (Chapters 3 and 4)

(iii) Lecture 3: Kalman filtering (Chapter 5)

(iv) Lecture 4: Estimation of frequency-response functions (Chapter 6)

(v) Lecture 5: Estimation of the parameters in a state-space model (Chapters 7 and 8)

(vi) Lecture 6: Subspace model identification (Chapter 9)

(vii) Lecture 7: From theory to practice: the system-identification cycle (Chapter 10).

The authors are of the opinion that the transfer of knowledge is greatly improved when each lecture is followed by working classes in which the students do the exercises of the corresponding classes under the supervision of a tutor. During such working classes each student has the opportunity to ask individual questions about the course material covered. At the Delft University of Technology the course is concluded by a real-life case study in which the material covered in this book has to be applied to identify a mathematical model from measured input and output data.

The authors have used this book for teaching MSc students at Delft University of Technology and the University of Twente in the Netherlands. Students attending the course were from the departments of electrical, mechanical, and aerospace engineering, and also applied physics. Currently, this book is being used for an introductory course on filtering and identification that is part of the core of the MSc program Systems and Control offered by the Delft Center for Systems and Control (http://www.dcsc.tudelft.nl). Parts of this book have been used in the graduate teaching program of the Dutch Institute of Systems and Control (DISC). Parts of this book have also been used by Bernard Hanzon when he was a guest lecturer at the Technische Universität Wien in Austria, and by Jonas Sjöberg for undergraduate teaching at Chalmers University of Technology in Sweden.

The writing of this book stems from the attempt of the authors to make their students as enthusiastic about the field of filtering and system identification as they themselves are. Though these students have played a stimulating and central role in the creation of this book, its final format and quality has been achieved only through close interaction with scientist colleagues. The authors would like to acknowledge the following persons for their constructive and helpful comments on this book or parts thereof: Dietmar Bauer (Technische Universität Wien, Austria), Bernard Hanzon (University College Cork, Ireland), Gjerrit Meinsma (University of Twente, the Netherlands), Petko Petkov (Technical University of Sofia, Bulgaria), Phillip Regalia (Institut National des Télécommunications, France), Ali Sayed (University of California, Los Angeles, USA), Johan Schoukens (Free University of Brussels, Belgium), Jonas Sjöberg (Chalmers University of Technology, Sweden), and Rufus Fraanje (TU Delft).

Special thanks go to Niek Bergboer (Maastricht University, the Netherlands) for his major contributions in developing the Matlab software and guide for the identification methods described in the book. We finally would like to thank the PhD students Paolo Massioni and Justin Rice for help in proof reading and with the solution manual.

Notation and symbols

$\mathbb{Z}$	the set of integers
$\mathbb{N}$	the set of positive integers
$\mathbb{C}$	the set of complex numbers
$\mathbb{R}$	the set of real numbers
$\mathbb{R}^n$	the set of real-valued n-dimensional vectors
$\mathbb{R}^{m \times n}$	the set of real-valued m by n matrices
∞	infinity
Re	real part
Im	imaginary part
$\in$	belongs to
$=$	equal
$\approx$	approximately equal
$\square$	end of proof
$\otimes$	Kronecker product
I_n	the $n \times n$ identity matrix
$[A]_{i,j}$	the (i,j)th entry of the matrix A
$A(i,:)$	the ith row of the matrix A
$A(:,i)$	the ith column of the matrix A
A^{T}	the transpose of the matrix A
A^{-1}	the inverse of the matrix A
$A^{1/2}$	the symmetric positive-definite square root of the matrix A
$\mathrm{diag}(a_1, a_2, \ldots, a_n)$	an $n \times n$ diagonal matrix whose (i,i)th entry is a_i
$\det(A)$	the determinant of the matrix A
$\mathrm{range}(A)$	the column space of the matrix A
$\mathrm{rank}(A)$	the rank of the matrix A
$\mathrm{trace}(A)$	the trace of the matrix A

$\text{vec}(A)$	a vector constructed by stacking the columns of the matrix A on top of each other
$\|A\|_2$	the 2-norm of the matrix A
$\|A\|_F$	the Frobenius norm of the matrix A
$[x]_i$	the ith entry of the vector x
$\|x\|_2$	the 2-norm of the vector x
$\lim$	limit
$\min$	minimum
$\max$	maximum
$\sup$	supremum (least upper bound)
$E[\,\cdot\,]$	statistical expected value
$\delta(t)$	Dirac delta function (Definition 3.8 on page 53)
$\Delta(k)$	unit pulse function (Definition 3.3 on page 44)
$s(k)$	unit step function (Definition 3.4 on page 44)
$X \sim (m, \sigma^2)$	Gaussian random variable X with mean m and variance σ^2

List of abbreviations

ARX	Auto-Regressive with eXogeneous input
ARMAX	Auto-Regressive Moving Average with eXogeneous input
BIBO	Bounded Input, Bounded Output
BJ	Box–Jenkins
CDF	Cumulative Distribution Function
DARE	Discrete Algebraic Ricatti Equation
DFT	Discrete Fourier Transform
DTFT	Discrete-Time Fourier Transform
ETFE	Empirical Transfer-Function Estimate
FFT	Fast Fourier Transform
FIR	Finite Impulse Response
FRF	Frequency-Response Function
IID	Independent, Identically Distributed
IIR	Infinite Impulse Response
LTI	Linear Time-Invariant
LTV	Linear Time-Varying
MIMO	Multiple Input, Multiple Output
MOESP	Multivariable Output-Error State-sPace
N4SID	Numerical algorithm for Subspace IDentification
PDF	Probability Density Function
PEM	Prediction-Error Method
PI	Past Inputs
PO	Past Outputs
OE	Output-Error
RMS	Root Mean Square
SISO	Single Input, Single Output
SRCF	Square-Root Covariance Filter
SVD	Singular-Value Decomposition
WSS	Wide-Sense Stationary

1
Introduction

Making observations through the senses of the environment around us is a natural activity of living species. The information acquired is diverse, consisting for example of sound signals and images. The information is processed and used to make a particular model of the environment that is applicable to the situation at hand. This act of model building based on observations is embedded in our human nature and plays an important role in daily decision making.

Model building through observations also plays a very important role in many branches of science. Despite the importance of making observations through our senses, scientific observations are often made via measurement instruments or sensors. The measurement data that these sensors acquire often need to be processed to judge or validate the experiment, or to obtain more information on conducting the experiment. Data are often used to build a mathematical model that describes the dynamical properties of the experiment. System-identification methods are systematic methods that can be used to build mathematical models from measured data. One important use of such mathematical models is in predicting model quantities by filtering acquired measurements.

A milestone in the history of filtering and system identification is the method of least squares developed just before 1800 by Johann Carl Friedrich Gauss (1777–1855). The use of least squares in filtering and identification is a recurring theme in this book. What follows is a brief sketch of the historical context that characterized the early development of the least-squares method. It is based on an overview given by Bühler (1981).

At the time Gauss first developed the least-squares method, he did not consider it very important. The first publication on the least-squares

1

method was published by Adrien-Marie Legendre (1752–1833) in 1806,
when Gauss had already clearly and frequently used the method much
earlier. Gauss motivated and derived the method of least squares sub-
stantially in the papers *Theoria combinationis observationum erroribus
minimis obnoxiae* I and II of 1821 and 1823. Part I is devoted to the theory and Part II contains applications, mostly to problems from astron-
omy. In Part I he developed a probability theory for accidental errors
(*Zufallsfehler*). Here Gauss defined a (probability distribution) function
$\phi(x)$ for the error in the observation x. On the basis of this function, the
product $\phi(x)\mathrm{d}x$ is the probability that the error falls within the interval
between x and $x+\mathrm{d}x$. The function $\phi(x)$ had to satisfy the normalization
condition

$$\int_{-\infty}^{\infty} \phi(x)\mathrm{d}x = 1.$$

The decisive requirement postulated by Gauss is that the integral

$$\int_{-\infty}^{\infty} x^2 \phi(x)\mathrm{d}x$$

attains a minimum. The selection of the square of the error as the most
suitable weight is why this method is called the method of least squares.
This selection was doubted by Pierre-Simon Laplace (1749–1827), who
had earlier tried to use the absolute value of the error. Computationally
the choice of the square is superior to Laplace's original method.

After the development of the basic theory of the least-squares method,
Gauss had to find a suitable function $\phi(x)$. At this point Gauss introduced, after some heuristics, the Gaussian distribution

$$\phi(x) = \frac{1}{\pi}\mathrm{e}^{-x^2}$$

as a "natural" way in which errors of observation occur. Gauss never
mentioned in his papers statistical distribution functions different from
the Gaussian one. He was caught in his own success; the applications
to which he applied his theory did not stimulate him to look for other
distribution functions. The least-squares method was, at the beginning of
the nineteenth century, his indispensable theoretical tool in experimental
research; and he saw it as the most important witness to the connection
between mathematics and Nature.

Still today, the ramifications of the least-squares method in mathematical modeling are tremendous and any book on this topic has to narrow

itself down to a restrictive class of problems. In this introductory textbook on system identification we focus mainly on the identification of
linear state-space models from measured data sequences of inputs and
outputs of the engineering system that we want to model. Though this
focused approach may at first seem to rule out major contributions in the
field of system identification, the contrary is the case. It will be shown
in the book that the state-space approach chosen is capable of treating
many existing identification methods for estimating the parameters in a
difference equation as special cases. Examples are given for the widely
used ARX and ARMAX models (Ljung, 1999).

The central goal of the book is to help the reader discover how the
linear least-squares method can solve, or help in solving, different variants of the linear state-space model-identification problem. The linear
least-squares method can be formulated as a deterministic parameter-
optimization problem of the form

$$\min_x \mu^{\mathrm{T}} \mu \quad \text{subject to } y = Fx + \mu, \tag{1.1}$$

with the vector $y \in \mathbb{R}^N$ and the matrix $F \in \mathbb{R}^{N \times n}$ given and with
$x \in \mathbb{R}^n$ the vector of unknown parameters to be determined. The solution of this optimization problem is the subject of a large number of
textbooks. Although its analytic solution can be given in a proof of only
a few lines, these textbooks analyze the least-squares solution from different perspectives. Examples are the statistical interpretation of the
solution under various assumptions on the entries of the matrix F and
the perturbation vector μ, or the numerical solution in a computationally efficient manner by exploiting structure in the matrix F. For an
advanced study of the least-squares problem and its applications in
many signal-processing problems, we refer to the book of Kailath *et al.*
(2000).

The main course of this book is preceded by three introductory chapters. In Chapter 2 a refreshment survey of matrix linear algebra is given.
Chapter 3 gives a brief overview of signal transforms and linear system
theory for deterministic signals and systems. Chapter 4 treats random
variables and random signals. Understanding the system-identification
methods discussed in this book depends on a profound mastering of the
background material presented in these three chapters.

Often, the starting point of identifying a dynamical model is the determination of a predictor. Therefore, in Chapter 5, we first study the

prediction of the state of a linear state-space model. The state-prediction or state-observation problem requires, in addition to the inputs and outputs, knowledge of the dynamic model (in state-space form) and the mean and covariance matrix of the stochastic perturbations. The goal is to reconstruct the state sequence from this knowledge. The optimality of the state-reconstruction problem can be defined in a least-squares sense. In Chapter 5, it is shown that the optimal predictor or Kalman-filter problem can be formulated and solved as a (weighted) linear least-squares problem. This formulation and solution of the Kalman-filter problem was first proposed by Paige (1985). The main advantage of this formulation is that a (recursive) solution can simply be derived from elementary linear-algebra concepts, such as Gaussian elimination for solving an overdetermined set of equations. We will briefly discuss the application of Kalman filtering for estimating unknown inputs of a dynamical system.

Chapter 6 discusses the estimation of input–output descriptions of linear state-space models in the frequency domain. The estimation of such descriptions, like the frequency response function (FRF) is based on the (discrete) Fourier transform of time sequences. The study in this chapter includes the effect that the practical constraint of the finite duration of the experiment has on the accuracy of the FRF estimate. A brief exposition on the use of the fast Fourier transform (FFT) in deriving fast algorithmic implementations is given. The availability of fast algorithms is one of the main advantages of frequency-domain methods when dealing with large amounts of data. In major parts of industry, such as the automobile and aircraft industry, it is therefore still the main tool for retrieving information about dynamic systems.

Chapter 7 discusses the estimation of the entries of the system matrices of a state-space model, under the assumptions that the output observations are corrupted by additive white noise and the state vector of the model has a fixed and known dimension. This problem gives rise to the so-called output-error methods (Ljung, 1999). This elementary estimation problem reveals a number of issues that are at the heart of a wide variety of identification approaches. The key problem to start with is that of how to express the entries of the system matrices as functions of an unknown parameter vector. The choice of this parameter vector is referred to in this textbook as the parameterization problem. Various alternatives for parameterizing multivariable state-space models are proposed. Once a parameterization has been chosen, the output-error problem can also be formulated as the following least-squares problem:

$$\min_{x_1, x_2} \mu^{\mathrm{T}} \mu \quad \text{subject to} \quad y = F(x_1)x_2 + \mu, \tag{1.2}$$

where the unknown parameter vectors x_1 and x_2 need to be determined. This type of least-squares problem is much harder to tackle than its linear variant (1.1), because the matrix F depends on the unknown parameters x_1. It is usually solved iteratively and therefore requires starting values of the parameter vectors x_1 and x_2. Furthermore, in general there is no guarantee that such an iterative numerical procedure converges to the global optimum of the cost function $\mu^{\mathrm{T}} \mu$. In this chapter special attention is paid to the numerical implementation of iterative procedures for output-error optimization. After having obtained an estimate of the unknown parameter vector, the problem of assessing the accuracy of the obtained estimates is addressed via the evaluation of the covariance matrices of the estimates under the assumption that the estimates are unbiased. We end this chapter by discussing how to avoid biased solutions when the additive noise is no longer white.

Chapter 8 presents the classical prediction-error method (PEM) (Ljung, 1999) for the identification of a predictor model (Kalman filter) with a fixed and known state dimension from measured input and output data. The problem boils down to estimating the parameters of a predictor model given by the innovation representation of the Kalman filter. The problems and solutions presented in Chapter 7 for the output-error case are adapted for these predictor models. In addition to the presentation of the prediction-error method for general multivariable state-space models, special attention is given to single-input, single-output (SISO) systems. This is done, first, to show that well-known model structures such as the ARMAX model can be treated as a particular canonical parameterization of a state-space model. Second, it enables a qualitative analysis of the bias when identifying a model that has a state dimension or a noise model different from the system that generated the data.

Chapter 9 treats the recently developed class of subspace identification methods. These methods are capable of providing accurate estimates of multivariable state-space models under general noise perturbations by just solving a linear least-squares problem of the form (1.1). The interest in subspace methods, both in academia and in industry, stems partly from the fact that no model parameterization is necessary to estimate a model and its order. This is achieved by relating key subspaces defined from matrices of the model to structured matrices constructed from the available observations. The central role the sub-space plays explains the name given to these methods. A distinction is made among different types of subspace methods, depending on how they use the concept of instrumental variables to cope consistently with various noise scenarios. Although, in practice, it has been demonstrated that subspace methods immediately provide accurate models, they do not optimize the prediction-error criterion as the prediction-error method does. To achieve this statistical optimum, we could use the estimates obtained with subspace methods as starting values for the prediction-error method. This concept has been proposed by Ljung (MathWorks, 2000a), for example.

Chapter 10 establishes the link between model estimation algorithms and their use in a real-life identification experiment. To set up, analyze, and improve an identification experiment, a cyclic procedure such as that outlined by Ljung (1999) is discussed. The cyclic procedure aims at a systematic treatment of many choices that need to be made in system identification. These choices include the selection of the experimental circumstances (for example sampling frequency, experiment duration, and type of input signal), the treatment of the recorded time sequences (detrending, removing outliers, and filtering) and the selection of a model structure (model order and delay) for the parameter-estimation algorithms. Here we include a brief discussion on how the subspace methods of Chapter 9 and the parametric methods of Chapters 7 and 8 can work together in assisting the system-identification practitioner to make choices regarding the model structure. It is this merging of subspace and prediction-error methods that makes the overall identification cycle feasible for multivariable systems. When using the prediction-error method in isolation, finding the appropriate model structure would require the testing of an extremely large amount of possibilities. This is infeasible in practice, since often not just one model needs to be identified, but a series of models for different experimental conditions.

At the end of each chapter dedicated exercises are included to let the reader experiment with the development and application of new algorithms. To facilitate the use of the methods described, the authors have developed a Matlab toolbox containing the identification methods described, together with a comprehensive software guide (Verhaegen *et al.*, 2003).

Filtering and system identification are excellent examples of multidisciplinary science, not only because of their versatility of application in many different fields, but also because they bring together fundamental knowledge from a wide number of (mathematical) disciplines. The authors are convinced that the current outline of the textbook should be considered as just an introduction to the fascinating field of system identification. System identification is a branch of science that illustrates very well the saying that the proof of the pudding is in the eating. Study and master the material in this textbook, but, most importantly, use it!

2
Linear algebra

After studying this chapter you will be able to

- apply basic operations to vectors and matrices;
- define a vector space;
- define a subspace of a vector space;
- compute the rank of a matrix;
- list the four fundamental subspaces defined by a linear transformation;
- compute the inverse, determinant, eigenvalues, and eigenvectors of a square matrix.
- describe what positive-definite matrices are;
- compute some important matrix decompositions, such as the eigenvalue decomposition, the singular-value decomposition and the QR factorization;
- solve linear equations using techniques from linear algebra;
- describe the deterministic least-squares problem; and
- solve the deterministic least-squares problem in numerically sound ways.

2.1 Introduction

In this chapter we review some basic topics from linear algebra. The material presented is frequently used in the subsequent chapters.

Since the 1960s linear algebra has gained a prominent role in engineering as a contributing factor to the success of technological breakthroughs.

Linear algebra provides tools for numerically solving system-theoretic problems, such as filtering and control problems. The widespread use of linear algebra tools in engineering has in its turn stimulated the development of the field of linear algebra, especially the numerical analysis of algorithms. A boost to the prominent role of linear algebra in engineering has certainly been provided by the introduction and widespread use of computer-aided-design packages such as Matlab (MathWorks, 2000b) and SciLab (Gomez, 1999). The user-friendliness of these packages allow us to program solutions for complex system-theoretic problems in just a few lines of code. Thus the prototyping of new algorithms is greatly speeded-up. However, on the other hand, there is also need for a word of caution: The coding in Matlab may give the user the impression that one successful Matlab run is equivalent to a full proof of a new theory. In order to avoid the cultivation of such a "proven-by-Matlab" attitude, the refreshment survey in this chapter and the use of linear algebra in later chapters concern primarily the derivation of the algorithms rather than their use. The use of Matlab routines for the class of filtering and identification problems analyzed in this book is described in detail in the comprehensive software guide (Verhaegen *et al.*, 2003).

We start this chapter with a review of two basic elements of linear algebra: vectors and matrices. Vectors are described in Section 2.2, matrices in Section 2.3. For a special class of matrices, square matrices, several important concepts exist, and these are described in Section 2.4. Section 2.5 describes some matrix decompositions that have proven to be useful in the context of filtering and estimation. Finally, in Sections 2.6 and 2.7 we focus on least-squares problems in which an overdetermined set of linear equations needs to be solved. These problems are of particular interest, since a lot of filtering, estimation, and even control problems can be written as linear (weighted) least-squares problems.

2.2 Vectors

A *vector* is an array of real or complex numbers. Throughout this book we use $\mathbb{R}$ to denote the set of real numbers and $\mathbb{C}$ to denote the set of complex numbers. Vectors come in two flavors, *column vectors* and *row vectors*. The column vector that consists of the elements $x_1, x_2, \ldots, x_n$

with $x_i \in \mathbb{C}$ will be denoted by x, that is,

$$x = \begin{bmatrix} x_1 \\ x_2 \\ \vdots \\ x_n \end{bmatrix}.$$

In this book a vector denoted by a lower-case character will always be a column vector. Row vectors are denoted by x^{T}, that is,

$$x^{\mathrm{T}} = \begin{bmatrix} x_1 & x_2 & \cdots & x_n \end{bmatrix}.$$

The row vector x^{T} is also called the *transpose* of the column vector x. The number of elements in a vector is called the *dimension* of the vector. A vector having n elements is referred to as an n-dimensional vector. We use the notation $x \in \mathbb{C}^n$ to denote an n-dimensional vector that has complex-valued elements. Obviously, an n-dimensional vector with real-valued elements is denoted by $x \in \mathbb{R}^n$. In this book, most vectors will be real-valued; therefore, in the remaining part of this chapter we will restrict ourselves to real-valued vectors. However, most results can readily be extended to complex-valued vectors.

The multiplication of a vector $x \in \mathbb{R}^n$ by a scalar $\alpha \in \mathbb{R}$ is defined as

$$\alpha x = \begin{bmatrix} \alpha x_1 \\ \alpha x_2 \\ \vdots \\ \alpha x_n \end{bmatrix}.$$

The sum of two vectors $x, y \in \mathbb{R}^n$ is defined as

$$x + y = \begin{bmatrix} x_1 \\ x_2 \\ \vdots \\ x_n \end{bmatrix} + \begin{bmatrix} y_1 \\ y_2 \\ \vdots \\ y_n \end{bmatrix} = \begin{bmatrix} x_1 + y_1 \\ x_2 + y_2 \\ \vdots \\ x_n + y_n \end{bmatrix}.$$

The standard *inner product* of two vectors $x, y \in \mathbb{R}^n$ is equal to

$$x^{\mathrm{T}}y = x_1 y_1 + x_2 y_2 + \cdots + x_n y_n.$$

The *2-norm* of a vector x, denoted by $||x||_2$, is the square root of the inner product of this vector with itself, that is,

$$||x||_2 = \sqrt{x^{\mathrm{T}}x}.$$

Two vectors are called *orthogonal* if their inner product equals zero; they are called *orthonormal* if they are orthogonal and have unit 2-norms.

Any two vectors $x, y \in \mathbb{R}^n$ satisfy the *Cauchy–Schwartz inequality*

$$|x^T y| \leq ||x||_2 ||y||_2,$$

where equality holds if and only if $x = \alpha y$ for some scalar $\alpha \in \mathbb{R}$, or $y = 0$.

The m vectors $x_i \in \mathbb{R}^n$, $i = 1, 2, \ldots, m$ are *linearly independent* if any linear combination of these vectors,

$$\alpha_1 x_1 + \alpha_2 x_2 + \cdots + \alpha_m x_m, \tag{2.1}$$

with $\alpha_i \in \mathbb{R}$, $i = 1, 2, \ldots, m$ is zero only if $\alpha_i = 0$, for all $i = 1, 2, \ldots, m$. If (2.1) equals zero for some of the coefficients α_i different from zero, then the vectors x_i, for $i = 1, 2, \ldots, m$, are *linearly dependent or collinear*. In the latter case, at least one of them, say x_m, can be expressed as a linear combination of the others, that is,

$$x_m = \beta_1 x_1 + \beta_2 x_2 + \cdots + \beta_{m-1} x_{m-1},$$

for some scalars $\beta_i \in \mathbb{R}$, $i = 1, 2, \ldots, m-1$. Note that a set of m vectors in $\mathbb{R}^n$ must be linearly dependent if $m > n$. Let us illustrate linear independence with an example.

Example 2.1 (Linear independence) Consider the vectors

$$x_1 = \begin{bmatrix} 1 \\ 1 \\ 1 \end{bmatrix}, \qquad x_2 = \begin{bmatrix} 0 \\ 1 \\ 0 \end{bmatrix}, \qquad x_3 = \begin{bmatrix} 2 \\ 3 \\ 2 \end{bmatrix}.$$

The vectors x_1 and x_2 are linearly independent, because

$$\alpha_1 x_1 + \alpha_2 x_2 = 0$$

implies $\alpha_1 = 0$ and $\alpha_1 + \alpha_2 = 0$ and thus $\alpha_1 = \alpha_2 = 0$. The vectors x_1, x_2, and x_3 are linearly dependent, because $x_3 = 2x_1 + x_2$.

A *vector space* $\mathcal{V}$ is a set of vectors, together with rules for vector addition and multiplication by real numbers, that has the following properties:

(i) $x + y \in \mathcal{V}$ for all $x, y \in \mathcal{V}$;
(ii) $\alpha x \in \mathcal{V}$ for all $\alpha \in \mathbb{R}$ and $x \in \mathcal{V}$;
(iii) $x + y = y + x$ for all $x, y \in \mathcal{V}$;
(iv) $x + (y + z) = (x + y) + z$ for all $x, y, z \in \mathcal{V}$;

(v) $x + 0 = x$ for all $x \in \mathcal{V}$;

(vi) $x + (-x) = 0$ for all $x \in \mathcal{V}$;

(vii) $1 \cdot x = x$ for all $x \in \mathcal{V}$;

(viii) $(\alpha_1 \alpha_2)x = \alpha_1(\alpha_2 x)$ for all $\alpha_1, \alpha_2 \in \mathbb{R}$ and $x \in \mathcal{V}$;

(ix) $\alpha(x + y) = \alpha x + \alpha y$ for all $\alpha \in \mathbb{R}$ and $x, y \in \mathcal{V}$; and

(x) $(\alpha_1 + \alpha_2)x = \alpha_1 x + \alpha_2 x$ for all $\alpha_1, \alpha_2 \in \mathbb{R}$ and $x \in \mathcal{V}$.

For example, the collection of all n-dimensional vectors, $\mathbb{R}^n$, forms a vector space. If every vector in a vector space $\mathcal{V}$ can be expressed as a linear combination of some vectors $x_i \in \mathcal{V}$, $i = 1, 2, \ldots, \ell$, these vectors x_i *span* the space $\mathcal{V}$. In other words, every vector $y \in \mathcal{V}$ can be written as

$$y = \sum_{i=1}^{\ell} \alpha_i x_i,$$

with $\alpha_i \in \mathbb{R}$, $i = 1, 2, \ldots, \ell$. A *basis* for a vector space is a set of vectors that span the space and are linearly independent. An *orthogonal basis* is a basis in which every vector is orthogonal to all the other vectors contained in the basis. The number of vectors in a basis is referred to as the dimension of the space. Therefore, to span an n-dimensional vector space, we need at least n vectors.

A *subspace* $\mathcal{U}$ of a vector space $\mathcal{V}$, denoted by $\mathcal{U} \subset \mathcal{V}$, is a nonempty subset that satisfies two requirements:

(i) $x + y \in \mathcal{U}$ for all $x, y \in \mathcal{U}$; and

(ii) $\alpha x \in \mathcal{U}$ for all $\alpha \in \mathbb{R}$ and $x \in \mathcal{U}$.

In other words, adding two vectors or multiplying a vector by a scalar produces again a vector that lies in the subspace. By taking $\alpha = 0$ in the second rule, it follows that the zero vector belongs to every subspace. Every subspace is by itself a vector space, because the rules for vector addition and scalar multiplication inherit the properties of the host vector space $\mathcal{V}$. Two subspaces $\mathcal{U}$ and $\mathcal{W}$ of the same space $\mathcal{V}$ are said to be orthogonal if every vector in $\mathcal{U}$ is orthogonal to every vector in $\mathcal{W}$. Given a subspace $\mathcal{U}$ of $\mathcal{V}$, the space of all vectors in $\mathcal{V}$ that are orthogonal to $\mathcal{U}$ is called the *orthogonal complement* of $\mathcal{U}$, denoted by $\mathcal{U}^{\perp}$.

Example 2.2 (Vector spaces) The vectors

$$x_1 = \begin{bmatrix} 0 \\ 1 \\ 0 \end{bmatrix}, \qquad x_2 = \begin{bmatrix} 2 \\ 0 \\ 0 \end{bmatrix}, \qquad x_3 = \begin{bmatrix} -1 \\ 0 \\ -1 \end{bmatrix}$$

are linearly independent, and form a basis for the vector space $\mathbb{R}^3$. They do not form an orthogonal basis, since x_2 is not orthogonal to x_3 The vector x_1 spans a one-dimensional subspace of $\mathbb{R}^3$ or a line. This subspace is the orthogonal complement of the two-dimensional subspace of $\mathbb{R}^3$, or a plane, spanned by x_2 and x_3, since x_1 is orthogonal both to x_2 and to x_3.

2.3 Matrices

A vector is an array that has either one column or one row. An array that has m rows and n columns is called an m- by n-dimensional *matrix*, or, briefly, an $m \times n$ matrix. An $m \times n$ matrix is denoted by

$$
A = \begin{bmatrix} a_{11} & a_{12} & \cdots & a_{1n} \\ a_{21} & a_{22} & \cdots & a_{2n} \\ \vdots & \vdots & \ddots & \vdots \\ a_{m1} & a_{m2} & \cdots & a_{mn} \end{bmatrix}.
$$

The scalar a_{ij} is referred to as the (i,j)th entry or element of the matrix. In this book, matrices will be denoted by upper-case letters. We use the notation $A \in \mathbb{C}^{m \times n}$ to denote an $m \times n$ matrix with complex-valued entries and $A \in \mathbb{R}^{m \times n}$ to denote an $m \times n$ matrix with real-valued entries. Most matrices that we encounter will be real-valued, so in this chapter we will mainly discuss real-valued matrices.

An n-dimensional column vector can, of course, be viewed as an $n \times 1$ matrix and an n-dimensional row vector can be viewed as a $1 \times n$ matrix. A matrix that has the same number of columns as rows is called a square matrix. Square matrices have some special properties, which are described in Section 2.4.

A matrix $A \in \mathbb{R}^{m \times n}$ can be viewed as a collection of n column vectors of dimension m,

$$
A = \begin{bmatrix} \begin{bmatrix} a_{11} \\ a_{21} \\ \vdots \\ a_{m1} \end{bmatrix} & \begin{bmatrix} a_{12} \\ a_{22} \\ \vdots \\ a_{m2} \end{bmatrix} & \cdots & \begin{bmatrix} a_{1n} \\ a_{2n} \\ \vdots \\ a_{mn} \end{bmatrix} \end{bmatrix},
$$

or as a collection of m row vectors of dimension n,

$$
A = \begin{bmatrix} \begin{bmatrix} a_{11} & a_{12} & a_{13} & \cdots & a_{1n} \end{bmatrix} \\ \begin{bmatrix} a_{21} & a_{22} & a_{23} & \cdots & a_{2n} \end{bmatrix} \\ \vdots \\ \begin{bmatrix} a_{m1} & a_{m2} & a_{m3} & \cdots & a_{mn} \end{bmatrix} \end{bmatrix}.
$$

This shows that we can partition a matrix into column vectors or row vectors. A more general partitioning is a partitioning into *submatrices*, for example

$$A = \begin{bmatrix} A_{11} & A_{12} \\ A_{21} & A_{22} \end{bmatrix} \in \mathbb{R}^{m \times n},$$

with matrices $A_{11} \in \mathbb{R}^{p \times q}$, $A_{12} \in \mathbb{R}^{p \times (n-q)}$, $A_{21} \in \mathbb{R}^{(m-p) \times q}$, and $A_{22} \in \mathbb{R}^{(m-p) \times (n-q)}$ for certain $p < m$ and $q < n$.

The *transpose* of a matrix A, denoted by A^{T}, is the $n \times m$ matrix that is obtained by interchanging the rows and columns of A, that is,

$$A^{\mathrm{T}} = \begin{bmatrix} a_{11} & a_{21} & \cdots & a_{m1} \\ a_{12} & a_{22} & \cdots & a_{m2} \\ \vdots & \vdots & \ddots & \vdots \\ a_{1n} & a_{2n} & \cdots & a_{nm} \end{bmatrix}.$$

It immediately follows that $(A^{\mathrm{T}})^{\mathrm{T}} = A$.

A special matrix is the *identity matrix*, denoted by I. This is a square matrix that has only nonzero entries along its diagonal, and these entries are all equal to unity, that is,

$$I = \begin{bmatrix} 1 & 0 & \cdots & 0 & 0 \\ 0 & 1 & \cdots & 0 & 0 \\ \vdots & \vdots & \ddots & \vdots & \vdots \\ 0 & 0 & \cdots & 1 & 0 \\ 0 & 0 & \cdots & 0 & 1 \end{bmatrix}. \tag{2.2}$$

If we multiply the matrix $A \in \mathbb{R}^{m \times n}$ by a scalar $\alpha \in \mathbb{R}$, we get a matrix with the (i,j)th entry given by αa_{ij} (for $i = 1, \ldots, m, j = 1, \ldots, n$). The sum of two matrices A and B of equal dimensions yields a matrix of the same dimensions with the (i,j)th entry given by $a_{ij} + b_{ij}$. Obviously, $(A + B)^{\mathrm{T}} = A^{\mathrm{T}} + B^{\mathrm{T}}$. The product of the matrices $A \in \mathbb{R}^{m \times n}$ and $B \in \mathbb{R}^{n \times p}$ is defined as an $m \times p$ matrix with the (i,j)th entry given by $\sum_{k=1}^{n} a_{ik} b_{kj}$. It is important to note that in general we have $AB \neq BA$. We also have $(AB)^{\mathrm{T}} = B^{\mathrm{T}} A^{\mathrm{T}}$.

Another kind of matrix product is the *Kronecker product*. The Kronecker product of two matrices $A \in \mathbb{R}^{m \times n}$ and $B \in \mathbb{R}^{p \times q}$, denoted by

$A \otimes B$, is the $mp \times nq$ matrix given by

$$A \otimes B = \begin{bmatrix} a_{11}B & a_{12}B & \cdots & a_{1n}B \\ a_{21}B & a_{22}B & \cdots & a_{2n}B \\ \vdots & \vdots & \ddots & \vdots \\ a_{m1}B & a_{m2}B & \cdots & a_{mn}B \end{bmatrix}.$$

The Kronecker product and the *vec operator* in combination are often useful for rewriting matrix products. The vec operator stacks all the columns of a matrix on top of each other in one big vector. For a matrix $A \in \mathbb{R}^{m \times n}$ given as $A = \begin{bmatrix} a_1 & a_2 & \cdots & a_n \end{bmatrix}$, with $a_i \in \mathbb{R}^m$, the mn-dimensional vector $\text{vec}(A)$ is defined as

$$\text{vec}(A) = \begin{bmatrix} a_1 \\ a_2 \\ \vdots \\ a_n \end{bmatrix}.$$

Given matrices $A \in \mathbb{R}^{m \times n}$, $B \in \mathbb{R}^{n \times p}$, and $C \in \mathbb{R}^{p \times q}$, we have the following relation:

$$\text{vec}(ABC) = (C^{\mathrm{T}} \otimes A)\text{vec}(B). \qquad (2.3)$$

The 2-norm of the vector $\text{vec}(A)$ can be viewed as a norm for the matrix A. This norm is called the *Frobenius norm*. The Frobenius norm of a matrix $A \in \mathbb{R}^{m \times n}$, denoted by $||A||_{\mathrm{F}}$, is defined as

$$||A||_{\mathrm{F}} = ||\text{vec}(A)||_2 = \left(\sum_{i=1}^{m} \sum_{j=1}^{n} a_{ij}^2 \right)^{1/2}.$$

A matrix that has the property that $A^{\mathrm{T}}A = I$ is called an *orthogonal matrix*; the column vectors a_i of an orthogonal matrix are orthogonal vectors that in addition satisfy $a_i^{\mathrm{T}}a_i = 1$. Hence, an orthogonal matrix has *orthonormal* columns!

The *rank* of the matrix A, denoted by $\text{rank}(A)$, is defined as the number of linearly independent columns of A. It is easy to see that the number of linearly independent columns must equal the number of linearly independent rows. Hence, $\text{rank}(A) = \text{rank}(A^{\mathrm{T}})$ and $\text{rank}(A) \le \min(m, n)$ if A is $m \times n$. A matrix A has *full rank* if $\text{rank}(A) = \min(m, n)$. An important property of the rank of the product of two matrices is stated in the following lemma.

Lemma 2.1 (Sylvester's inequality) *(Kailath, 1980) Consider matrices $A \in \mathbb{R}^{m \times n}$ and $B \in \mathbb{R}^{n \times p}$, then*

$$\text{rank}(A) + \text{rank}(B) - n \le \text{rank}(AB) \le \min\Big(\text{rank}(A), \text{rank}(B)\Big).$$

This inequality is often used to determine the rank of the matrix AB when both A and B have full rank n with $n \le p$ and $n \le m$. In this case Sylvester's inequality becomes $n + n - n \le \text{rank}(AB) \le \min(n, n)$, and it follows that $\text{rank}(AB) = n$. The next example illustrates that it is not always possible to determine the rank of the matrix AB using Sylvester's inequality.

Example 2.3 (Rank of the product of two matrices) This example demonstrates that, on taking the product of two matrices of rank n, the resulting matrix can have a rank less than n. Consider the following matrices:

$$A = \begin{bmatrix} 1 & 0 & 2 \\ 0 & 1 & 0 \end{bmatrix}, \qquad B = \begin{bmatrix} 1 & 3 \\ 1 & 1 \\ 1 & 0 \end{bmatrix}.$$

Both matrices have rank two. Their product is given by

$$AB = \begin{bmatrix} 3 & 3 \\ 1 & 1 \end{bmatrix}.$$

This matrix is of rank one. For this example, Sylvester's inequality becomes $2 + 2 - 3 \le \text{rank}(AB) \le 2$, or $1 \le \text{rank}(AB) \le 2$.

With the definition of the rank of a matrix we can now define the *four fundamental subspaces* related to a matrix. A matrix $A \in \mathbb{R}^{m \times n}$ defines a linear transformation from the vector space $\mathbb{R}^n$ to the vector space $\mathbb{R}^m$. In these vector spaces four subspaces are defined. The vector space spanned by the columns of the matrix A is a subspace of $\mathbb{R}^m$. It is called the *column space* of the matrix A and is denoted by $\text{range}(A)$. Let $\text{rank}(A) = r$, then the column space of A has dimension r and any selection of r linearly independent columns of A forms a basis for its column space. Similarly, the rows of A span a vector space. This subspace is called the *row space* of the matrix A and is a subspace of $\mathbb{R}^n$. The row space is denoted by $\text{range}(A^{\mathrm{T}})$. Any selection of r linearly independent rows of A forms a basis for its row space. Two other important subspaces associated with the matrix A are the *null space* or *kernel* and the *left null space*. The null space, denoted by $\text{ker}(A)$, consists of all vectors $x \in \mathbb{R}^n$ that satisfy $Ax = 0$. The left null space, denoted by $\text{ker}(A^{\mathrm{T}})$, consists of all vectors $y \in \mathbb{R}^m$ that satisfy $A^T y = 0$. These results are summarized in the box below.

The four fundamental subspaces of a matrix $A \in \mathbb{R}^{m \times n}$ are

$$\text{range}(A) = \{y \in \mathbb{R}^m : y = Ax \text{ for some } x \in \mathbb{R}^n\},$$
$$\text{range}(A^T) = \{x \in \mathbb{R}^n : x = A^T y \text{ for some } y \in \mathbb{R}^m\},$$
$$\ker(A) = \{x \in \mathbb{R}^n : Ax = 0\},$$
$$\ker(A^T) = \{y \in \mathbb{R}^m : A^T y = 0\}.$$

We have the following important relation between these subspaces.

Theorem 2.1 *(Strang, 1988) Given a matrix $A \in \mathbb{R}^{m \times n}$, the null space of A is the orthogonal complement of the row space of A in $\mathbb{R}^n$, and the left null space of A is the orthogonal complement of the column space of A in $\mathbb{R}^m$.*

From this we see that the null space of a matrix $A \in \mathbb{R}^{m \times n}$ with rank r has dimension $n - r$, and the left null space has dimension $m - r$.

Given a matrix $A \in \mathbb{R}^{m \times n}$ and a vector $x \in \mathbb{R}^n$, we can write

$$Ax = A(x_r + x_n) = Ax_r = y_r,$$

with

$$x_r \in \text{range}(A^T) \subseteq \mathbb{R}^n,$$
$$x_n \in \ker(A) \subseteq \mathbb{R}^n,$$
$$y_r \in \text{range}(A) \subseteq \mathbb{R}^m,$$

and also, for a vector $y \in \mathbb{R}^m$,

$$A^T y = A^T(y_r + y_n) = A^T y_r = x_r,$$

with

$$y_r \in \text{range}(A) \subseteq \mathbb{R}^m,$$
$$y_n \in \ker(A^T) \subseteq \mathbb{R}^m,$$
$$x_r \in \text{range}(A^T) \subseteq \mathbb{R}^n.$$

Example 2.4 (Fundamental subspaces) Given the matrix

$$A = \begin{bmatrix} 1 & 1 & 2 & 1 \\ 0 & 1 & 2 & 1 \\ 0 & 1 & 2 & 1 \end{bmatrix}.$$

It is easy to see that the first and second columns of this matrix are linearly independent. The third and fourth columns are scaled versions of the second column, and thus $\text{rank}(A) = 2$. The first two columns are a basis for the two-dimensional column space of A. This column space is a subspace of $\mathbb{R}^3$. The left null space of A is the orthogonal complement of the column space in $\mathbb{R}^3$. Hence, it is a one-dimensional space, a basis for this space is the vector

$$y = \begin{bmatrix} 0 \\ 1 \\ -1 \end{bmatrix},$$

because it is orthogonal to the columns of the matrix A. It is easy to see that indeed $A^{\mathrm{T}}y = 0$. The row space of A is spanned by the first two rows. This two-dimensional space is a subspace of $\mathbb{R}^4$. The two-dimensional null space is the orthogonal complement of this space. The null space can be spanned by the vectors

$$x_1 = \begin{bmatrix} 0 \\ -1 \\ 0 \\ 1 \end{bmatrix}, \qquad x_2 = \begin{bmatrix} 0 \\ 0 \\ -1 \\ 2 \end{bmatrix},$$

because they are linearly independent and are orthogonal to the rows of the matrix A. It is easy to see that indeed $Ax_1 = 0$ and $Ax_2 = 0$.

2.4 Square matrices

Square matrices have some special properties, of which some important ones are described in this section. If a square matrix A has full rank, there exists a unique matrix A^{-1}, called the inverse of A, such that

$$AA^{-1} = A^{-1}A = I.$$

A square matrix that has full rank is called an *invertible* or *nonsingular* matrix. We have $(A^{\mathrm{T}})^{-1} = (A^{-1})^{\mathrm{T}}$ and also $A^{\mathrm{T}}(A^{-1})^{\mathrm{T}} = (A^{-1})^{\mathrm{T}}A^{\mathrm{T}} = I$. If A is a square orthogonal matrix, $A^{\mathrm{T}}A = I$, and thus $A^{-1} = A^{\mathrm{T}}$ and also $AA^{\mathrm{T}} = I$. If A and B are invertible matrices then $(AB)^{-1} = B^{-1}A^{-1}$. An important formula for inverting matrices is given in the following lemma.

Lemma 2.2 (Matrix-inversion lemma) *(Golub and Van Loan, 1996) Consider the matrices $A \in \mathbb{R}^{n \times n}$, $B \in \mathbb{R}^{n \times m}$, $C \in \mathbb{R}^{m \times m}$, and $D \in \mathbb{R}^{m \times n}$. If A, C, and $A + BCD$ are invertible, then,*

$$(A + BCD)^{-1} = A^{-1} - A^{-1}B(C^{-1} + DA^{-1}B)^{-1}DA^{-1}.$$

In Exercise 2.2 on page 38 you are asked to prove this lemma.

The rank of a square matrix can be related to the rank of its submatrices, using the notion of Schur complements.

Lemma 2.3 (Schur complement) *Let $S \in \mathbb{R}^{(n+m) \times (n+m)}$ be a matrix partitioned as*

$$S = \begin{bmatrix} A & B \\ C & D \end{bmatrix},$$

with $A \in \mathbb{R}^{n \times n}$, $B \in \mathbb{R}^{n \times m}$, $C \in \mathbb{R}^{m \times n}$, and $D \in \mathbb{R}^{m \times m}$.

(i) *If $\mathrm{rank}(A) = n$, then $\mathrm{rank}(S) = n + \mathrm{rank}(D - CA^{-1}B)$, where $D - CA^{-1}B$ is called the Schur complement of A.*

(ii) *If $\mathrm{rank}(D) = m$, then $\mathrm{rank}(S) = m + \mathrm{rank}(A - BD^{-1}C)$, where $A - BD^{-1}C$ is called the Schur complement of D.*

Proof Statement (i) is easily proven using the following identity and Sylvester's inequality (Lemma 2.1):

$$\begin{bmatrix} I_n & 0 \\ -CA^{-1} & I_m \end{bmatrix} \begin{bmatrix} A & B \\ C & D \end{bmatrix} \begin{bmatrix} I_n & -A^{-1}B \\ 0 & I_m \end{bmatrix} = \begin{bmatrix} A & 0 \\ 0 & D - CA^{-1}B \end{bmatrix}.$$

Statement (ii) is easily proven from the following identity:

$$\begin{bmatrix} I_n & -BD^{-1} \\ 0 & I_m \end{bmatrix} \begin{bmatrix} A & B \\ C & D \end{bmatrix} \begin{bmatrix} I_n & 0 \\ -D^{-1}C & I_m \end{bmatrix} = \begin{bmatrix} A - BD^{-1}C & 0 \\ 0 & D \end{bmatrix}.$$

$\square$

The *determinant* of a square $n \times n$ matrix, denoted by $\det(A)$, is a scalar that is defined recursively as

$$\det(A) = \sum_{i=1}^{n} (-1)^{i+j} a_{ij} \det(A_{ij}),$$

where A_{ij} is the $(n-1) \times (n-1)$ matrix that is formed by deleting the ith row and the jth column of A. The determinant of a 1×1 matrix A (a scalar) is equal to the matrix itself, that is, $\det(A) = A$. Some important properties of the determinant are the following:

(i) $\det(A^T) = \det(A)$ for all $A \in \mathbb{R}^{n \times n}$;
(ii) $\det(\alpha A) = \alpha^n \det(A)$ for all $A \in \mathbb{R}^{n \times n}$ and $\alpha \in \mathbb{R}$;
(iii) $\det(AB) = \det(A)\det(B)$ for all $A, B \in \mathbb{R}^{n \times n}$;
(iv) $\det(A) \neq 0$ if and only if $A \in \mathbb{R}^{n \times n}$ is invertible; and
(v) $\det(A^{-1}) = 1/\det(A)$ for all invertible matrices $A \in \mathbb{R}^{n \times n}$.

The determinant and the inverse of a matrix are related as follows:

$$A^{-1} = \frac{\text{adj}(A)}{\det(A)}, \tag{2.4}$$

where the *adjugate* of A, denoted by $\text{adj}(A)$, is an $n \times n$ matrix with its (i,j)th entry given by $(-1)^{i+j} \det(A_{ji})$ and A_{ji} is the $(n-1) \times (n-1)$ matrix that is formed by deleting the jth row and ith column of A.

Example 2.5 (Determinant and inverse of a 2×2 matrix) Consider a matrix $A \in \mathbb{R}^{2 \times 2}$ given by

$$A = \begin{bmatrix} a_{11} & a_{12} \\ a_{21} & a_{22} \end{bmatrix}.$$

The determinant of this matrix is given by

$$\det(A) = (-1)^{1+1} a_{11} \det(a_{22}) + (-1)^{2+1} a_{21} \det(a_{12}) = a_{11}a_{22} - a_{21}a_{12}.$$

The adjugate of A is given by

$$\text{adj}(A) = \begin{bmatrix} (-1)^2 a_{22} & (-1)^3 a_{12} \\ (-1)^3 a_{21} & (-1)^4 a_{11} \end{bmatrix}.$$

Hence, the inverse of A is

$$A^{-1} = \frac{\text{adj}(A)}{\det(A)} = \frac{1}{a_{11}a_{22} - a_{21}a_{12}} \begin{bmatrix} a_{22} & -a_{12} \\ -a_{21} & a_{11} \end{bmatrix}.$$

It is easily verified that indeed $AA^{-1} = I$.

An *eigenvalue* of a square matrix $A \in \mathbb{R}^{n \times n}$ is a scalar $\lambda \in \mathbb{C}$ that satisfies

$$Ax = \lambda x, \tag{2.5}$$

for some nonzero vector $x \in \mathbb{C}^n$. All nonzero vectors x that satisfy (2.5) for some λ are called the *eigenvectors* of A corresponding to the eigenvalue λ. An $n \times n$ matrix can have at most n different eigenvalues. Note that, even if the matrix A is real-valued, the eigenvalues and eigenvectors can be complex-valued. Since (2.5) is equivalent to $(\lambda I - A)x = 0$, the eigenvectors of A are in the null space of the polynomial matrix $(\lambda I - A)$. To find the (nonzero) eigenvectors, λ should be such that the polynomial matrix $(\lambda I - A)$ is singular. This is equivalent to the condition

$$\det(\lambda I - A) = 0.$$

The left-hand side of this equation is a polynomial in λ of degree n and is often referred to as the *characteristic polynomial* of the matrix A. Hence, the eigenvalues of a matrix A are equal to the zeros of its characteristic polynomial. The eigenvalues of a full-rank matrix are all nonzero. This follows from the fact that a zero eigenvalue turns Equation (2.5) on page 20 into $Ax = 0$, which has no nonzero solution x if A has full rank. From this it also follows that a singular matrix A has rank(A) nonzero eigenvalues and $n - \text{rank}(A)$ eigenvalues that are equal to zero. With the eigenvalues of the matrix A, given as λ_i, $i = 1, 2, \ldots, n$, we can express its *determinant* as

$$\det(A) = \lambda_1 \cdot \lambda_2 \cdots \lambda_n = \prod_{i=1}^{n} \lambda_i.$$

The *trace* of a square matrix $A \in \mathbb{R}^{n \times n}$, denoted by $\text{tr}(A)$, is the sum of its diagonal entries, that is,

$$\text{tr}(A) = \sum_{i=1}^{n} a_{ii}.$$

Then, in terms of the eigenvalues of A,

$$\text{tr}(A) = \lambda_1 + \lambda_2 + \cdots + \lambda_n = \sum_{i=1}^{n} \lambda_i.$$

The characteristic polynomial $\det(\lambda I - A)$ has the following important property, which is often used to write A^{n+k} for some integer $k \geq 0$ as a linear combination of $I, A, \ldots, A^{n-1}$.

Theorem 2.2 (Cayley–Hamilton) *(Strang, 1988) If the characteristic polynomial of the matrix $A \in \mathbb{R}^{n \times n}$ is given by*

$$\det(\lambda I - A) = \lambda^n + a_{n-1}\lambda^{n-1} + \cdots + a_1\lambda + a_0$$

for some $a_i \in \mathbb{R}$, $i = 0, 1, \ldots, n - 1$, then

$$A^n + a_{n-1}A^{n-1} + \cdots + a_1A + a_0I = 0.$$

Example 2.6 (Eigenvalues and eigenvectors) Consider the matrix

$$\begin{bmatrix} 4 & -5 \\ 2 & -3 \end{bmatrix}.$$

We have $\det(A) = -2$ and $\mathrm{tr}(A) = 1$. The eigenvalues of this matrix are computed as the zeros of

$$\det(\lambda I - A) = \det\left(\begin{bmatrix} \lambda - 4 & 5 \\ -2 & \lambda + 3 \end{bmatrix}\right) = \lambda^2 - \lambda - 2 = (\lambda + 1)(\lambda - 2).$$

Hence, the eigenvalues are -1 and 2. We see that indeed $\det(A) = -1 \cdot 2$ and $\mathrm{tr}(A) = -1 + 2$. The eigenvector x corresponding to the eigenvalue of -1 is found by solving $Ax = -x$. Let

$$x = \begin{bmatrix} x_1 \\ x_2 \end{bmatrix}.$$

Then it follows that

$$4x_1 - 5x_2 = -x_1,$$
$$2x_1 - 3x_2 = -x_2$$

or, equivalently, $x_1 = x_2$;, thus an eigenvector corresponding to the eigenvalue -1 is given by

$$\begin{bmatrix} 1 \\ 1 \end{bmatrix}.$$

Similarly, it can be found that

$$\begin{bmatrix} 5 \\ 2 \end{bmatrix}$$

is an eigenvector corresponding to the eigenvalue 2.

An important property of the eigenvalues of a matrix is summarized in the following lemma.

Lemma 2.4 *(Strang, 1988) Suppose that a matrix $A \in \mathbb{R}^{n \times n}$ has a complex-valued eigenvalue λ, with corresponding eigenvector x. Then the complex conjugate of λ is also an eigenvalue of A and the complex conjugate of x is a corresponding eigenvector.*

An important question is the following: when are the eigenvectors of a matrix linearly independent? The next lemma provides a partial answer to this question.

Lemma 2.5 *(Strang, 1988) Eigenvectors $x_1, x_2, \ldots, x_k$, corresponding to distinct eigenvalues $\lambda_1, \lambda_2, \ldots, \lambda_k$, $k \le n$ of a matrix $A \in \mathbb{R}^{n \times n}$ are linearly independent.*

Eigenvectors corresponding to repetitive eigenvalues can be linearly independent, but this is not necessary. This is illustrated in the following example.

Example 2.7 (Linearly dependent eigenvectors) Consider the matrix

$$\begin{bmatrix} 3 & 1 \\ 0 & 3 \end{bmatrix}.$$

The two eigenvalues of this matrix are both equal to 3. Solving $Ax = 3x$ to determine the eigenvectors shows that all eigenvectors are multiples of

$$\begin{bmatrix} 1 \\ 0 \end{bmatrix}.$$

Thus, the eigenvectors are linearly dependent.

Now we will look at some special square matrices. A *diagonal matrix* is a matrix in which nonzero entries occur only along its diagonal. An example of a diagonal matrix is the identity matrix, see Equation (2.2) on page 14. A diagonal matrix $A \in \mathbb{R}^{n \times n}$ is often denoted by $A = \text{diag}(a_{11}, a_{22}, \ldots, a_{nn})$. Hence, $I = \text{diag}(1, 1, \ldots, 1)$.

An *upper-triangular matrix* is a square matrix in which all the entries below the diagonal are equal to zero. A *lower-triangular matrix* is a square matrix in which all the entries above the diagonal are equal to zero. Examples of such matrices are:

$$\begin{bmatrix} a_{11} & a_{12} & a_{13} \\ 0 & a_{22} & a_{23} \\ 0 & 0 & a_{33} \end{bmatrix}, \quad \begin{bmatrix} a_{11} & 0 & 0 \\ a_{21} & a_{22} & 0 \\ a_{31} & a_{32} & a_{33} \end{bmatrix}.$$

Triangular matrices have some interesting properties. The transpose of an upper-triangular matrix is a lower-triangular matrix and vice versa. The product of two upper (lower)-triangular matrices is again upper (lower)-triangular. The inverse of an upper (lower)-triangular matrix is also upper (lower)-triangular. Finally, the determinant of a triangular matrix $A \in \mathbb{R}^{n \times n}$ equals the product of its diagonal entries, that is,

$$\det(A) = \prod_{i=1}^{n} a_{ii}.$$

A square matrix that satisfies $A^{\mathrm{T}} = A$ is called a *symmetric matrix*. The inverse matrix of a symmetric matrix is again symmetric.

Lemma 2.6 *(Strang, 1988) The eigenvalues of a symmetric matrix $A \in \mathbb{R}^{n \times n}$ are real and its eigenvectors corresponding to distinct eigenvalues are orthogonal.*

A square symmetric matrix $A \in \mathbb{R}^{n \times n}$ is called *positive-definite* if for all nonzero vectors $x \in \mathbb{R}^n$ it satisfies $x^{\mathrm{T}} A x > 0$. We write $A > 0$ to denote that the matrix A is positive-definite. A square symmetric matrix $A \in \mathbb{R}^{n \times n}$ is called *positive-semidefinite* if for all nonzero vectors $x \in \mathbb{R}^n$ it satisfies $x^{\mathrm{T}} A x \geq 0$. This is denoted by $A \geq 0$. By reversing the inequality signs we can define *negative-definite* and *negative-semidefinite* matrices in a similar way. A positive-semidefinite matrix has eigenvalues λ_i that satisfy $\lambda_i \geq 0$. A positive-definite matrix has only positive eigenvalues. An important relation for partitioned positive-definite matrices is the following.

Lemma 2.7 *Let $S \in \mathbb{R}^{(n+m) \times (n+m)}$ be a symmetric matrix partitioned as*

$$S = \begin{bmatrix} A & B \\ B^{\mathrm{T}} & C \end{bmatrix},$$

with $A \in \mathbb{R}^{n \times n}$, $B \in \mathbb{R}^{n \times m}$, and $C \in \mathbb{R}^{m \times m}$.

  (i) *If A is positive-definite, then S is positive-(semi)definite if and only if the Schur complement of A given by $C - B^{\mathrm{T}} A^{-1} B$ is positive-(semi)definite.*
  (ii) *If C is positive-definite, then S is positive-(semi)definite if and only if the Schur complement of C given by $A - B C^{-1} B^{\mathrm{T}}$ is positive-(semi)definite.*

In Exercise 2.5 on page 38 you are asked to prove this lemma.

Example 2.8 (Positive-definite matrix) Consider the symmetric matrix

$$A = \begin{bmatrix} 3 & 1 \\ 1 & 3 \end{bmatrix}.$$

The eigenvalues of this matrix are 2 and 4. Therefore this matrix is positive-definite. We could also arrive at this conclusion by using the Schur complement, which gives $(3 - 1 \cdot \frac{1}{3} \cdot 1) > 0$.

2.5 Matrix decompositions

In this section we look at some useful matrix decompositions. The first one is called the *eigenvalue decomposition*. Suppose that the matrix $A \in \mathbb{R}^{n \times n}$ has n linearly independent eigenvectors. Let these eigenvectors be the columns of the matrix V, then it follows that $AV = V\Lambda$, where Λ is a diagonal matrix with the eigenvalues of the matrix A along its diagonal. When the eigenvectors are assumed to be linearly independent, the matrix V is invertible and we have the following important result.

Theorem 2.3 (Eigenvalue decomposition) *(Strang, 1988) Any matrix $A \in \mathbb{R}^{n \times n}$ that has n linearly independent eigenvectors can be decomposed as*

$$A = V\Lambda V^{-1},$$

where $\Lambda \in \mathbb{R}^{n \times n}$ is a diagonal matrix containing the eigenvalues of the matrix A, and the columns of the matrix $V \in \mathbb{R}^{n \times n}$ are the corresponding eigenvectors.

Not all $n \times n$ matrices have n linearly independent eigenvectors. The eigenvalue decomposition can be performed only for matrices with n independent eigenvectors, and these matrices are called *diagonalizable*, since $V^{-1}AV = \Lambda$. With Lemma 2.5 on page 23 it follows that any matrix with distinct eigenvalues is diagonalizable. The converse is not true: there exist matrices with repeated eigenvalues that are diagonalizable. An example is the identity matrix I: it has all eigenvalues equal to 1, but does have n linearly independent eigenvectors if we take $V = I$.

For symmetric matrices we have a theorem that is somewhat stronger than Theorem 2.3 and is related to Lemma 2.6 on page 24.

Theorem 2.4 (Spectral theorem) *(Strang, 1988) Any symmetric matrix $A \in \mathbb{R}^{n \times n}$ can be diagonalized by an orthogonal matrix*

$$Q^{\mathrm{T}} A Q = \Lambda,$$

where $\Lambda \in \mathbb{R}^{n \times n}$ is a diagonal matrix containing the eigenvalues of the matrix A and the columns of the matrix $Q \in \mathbb{R}^{n \times n}$ form a complete set of orthonormal eigenvectors, such that $Q^{\mathrm{T}} Q = I$.

Since a positive-definite matrix A has real eigenvalues that are all positive, the spectral theorem shows that such a positive-definite matrix can be decomposed as

$$A = Q\Lambda Q^{\mathrm{T}} = RR^{\mathrm{T}},$$

where $R = Q\Lambda^{1/2}$ and $\Lambda^{1/2}$ is a diagonal matrix containing the positive square roots of the eigenvalues of A along the diagonal. The matrix R is called a *square root* of A, denoted by $R = A^{1/2}$. The square root of a matrix is not unique. One other way to obtain a matrix square root is by the *Cholesky factorization*.

Theorem 2.5 (Cholesky factorization) *(Golub and Van Loan, 1996) Any symmetric positive-definite matrix $A \in \mathbb{R}^{n \times n}$ can be decomposed as*

$$A = RR^{\mathrm{T}},$$

with R a unique lower-triangular matrix with positive diagonal entries.

The eigenvalue decomposition exists only for square matrices that have a complete set of linearly independent eigenvectors. The decomposition that we describe next, the *singular-value decomposition* (SVD) exists for any, even nonsquare, matrices.

Theorem 2.6 (Singular-value decomposition) *(Strang, 1988) Any matrix $A \in \mathbb{R}^{m \times n}$ can be decomposed as*

$$A = U\Sigma V^{\mathrm{T}},$$

where $U \in \mathbb{R}^{m \times m}$ and $V \in \mathbb{R}^{n \times n}$ are orthogonal matrices and $\Sigma \in \mathbb{R}^{m \times n}$ has its only nonzero elements along the diagonal. These elements σ_i are ordered such that

$$\sigma_1 \geq \sigma_2 \geq \cdots \geq \sigma_r > \sigma_{r+1} = \cdots = \sigma_k = 0,$$

where $r = \mathrm{rank}(A)$ and $k = \min(m, n)$.

The diagonal elements σ_i of the matrix Σ are called the *singular values* of A, the columns of the matrix U are called the *left singular vectors*, and the columns of the matrix V are called the *right singular vectors*. The SVD of the matrix A can be related to the eigenvalue decompositions of the symmetric matrices AA^{T} and $A^{\mathrm{T}}A$, since

$$AA^{\mathrm{T}} = U\Sigma V^{\mathrm{T}} V\Sigma^{\mathrm{T}} U^{\mathrm{T}} = U\Sigma\Sigma^{\mathrm{T}} U^{\mathrm{T}},$$
$$A^{\mathrm{T}}A = V\Sigma^{\mathrm{T}} U^{\mathrm{T}} U\Sigma V^{\mathrm{T}} = V\Sigma^{\mathrm{T}}\Sigma V^{\mathrm{T}}.$$

Hence, the matrix U contains all the eigenvectors of the matrix AA^{T}, the $m \times m$ matrix $\Sigma\Sigma^{\mathrm{T}}$ contains the corresponding eigenvalues, with r nonzero eigenvalues σ_i^2, $i = 1, 2, \ldots, r$; the matrix V contains the eigenvectors of the matrix $A^{\mathrm{T}}A$, the $n \times n$ matrix $\Sigma^{\mathrm{T}}\Sigma$ contains the corresponding eigenvalues with r nonzero eigenvalues σ_i^2, $i = 1, 2, \ldots,$ r. Note that, if A is a positive-definite symmetric matrix, the spectral theorem (Theorem 2.4 on page 25) shows that the SVD of A is in fact an eigenvalue decomposition of A and thus $U = V$.

When a matrix $A \in \mathbb{R}^{m \times n}$ has rank r, such that $r < m$ and $r < n$, the SVD can be partitioned as follows:

$$A = \begin{bmatrix} U_1 & U_2 \end{bmatrix} \begin{bmatrix} \Sigma_1 & 0 \\ 0 & 0 \end{bmatrix} \begin{bmatrix} V_1^T \\ V_2^T \end{bmatrix},$$

where $U_1 \in \mathbb{R}^{m \times r}$, $U_2 \in \mathbb{R}^{m \times (m-r)}$, $\Sigma_1 \in \mathbb{R}^{r \times r}$, $V_1 \in \mathbb{R}^{n \times r}$, and $V_2 \in \mathbb{R}^{n \times (n-r)}$. From this relation we immediately see that the columns of the matrices U_1, U_2, V_1, and V_2 provide orthogonal bases for all four fundamental subspaces of the matrix A, that is,

$$\text{range}(A) = \text{range}(U_1),$$
$$\ker(A^T) = \text{range}(U_2),$$
$$\text{range}(A^T) = \text{range}(V_1),$$
$$\ker(A) = \text{range}(V_2).$$

The SVD is a numerically reliable factorization. The computation of singular values is not sensitive to (rounding) errors in the computations. The numerical reliability of the SVD can be established by performing an error analysis as explained in the book by Golub and Van Loan (1996).

Another numerically attractive factorization is the *QR factorization.*

Theorem 2.7 *(Strang, 1988) Any matrix $A \in \mathbb{R}^{m \times n}$ can be decomposed as*

$$A = QR,$$

where $Q \in \mathbb{R}^{m \times m}$ is an orthogonal matrix and $R \in \mathbb{R}^{m \times n}$ is upper-triangular, augmented with columns on the right for $n > m$ or augmented with zero rows at the bottom for $m > n$.

When a matrix $A \in \mathbb{R}^{m \times n}$ has rank r, such that $r < m$ and $r < n$, the QR factorization can be partitioned as follows:

$$A = \begin{bmatrix} Q_1 & Q_2 \end{bmatrix} \begin{bmatrix} R_1 & R_2 \\ 0 & 0 \end{bmatrix},$$

where $Q_1 \in \mathbb{R}^{m \times r}$, $Q_2 \in \mathbb{R}^{m \times (m-r)}$, $R_1 \in \mathbb{R}^{r \times r}$, and $R_2 \in \mathbb{R}^{r \times (n-r)}$. This relation immediately shows that

$$\begin{aligned} \text{range}(A) &= \text{range}(Q_1), \\ \ker(A^{\mathrm{T}}) &= \text{range}(Q_2), \\ \text{range}(A^{\mathrm{T}}) &= \text{range}(R_1^{\mathrm{T}}). \end{aligned}$$

The *RQ factorization* of the matrix $A \in \mathbb{R}^{m \times n}$ is given by

$$A = RQ, \tag{2.6}$$

where $Q \in \mathbb{R}^{n \times n}$ is an orthogonal matrix and $R \in \mathbb{R}^{m \times n}$ is lower-triangular, augmented with rows at the bottom for $m > n$, or augmented with zero columns on the right for $n > m$. The RQ factorization of A is, of course, related to the QR factorization of A^{T}. Let the QR factorization of A^{T} be given by $A^{\mathrm{T}} = \overline{Q}\,\overline{R}$, then Equation (2.6) shows that $\overline{Q} = Q^{\mathrm{T}}$ and $\overline{R} = R^{\mathrm{T}}$.

The RQ factorization can be performed in a numerically reliable way. However, it cannot be used to determine the rank of a matrix in a reliable way. To reliably determine the rank, the SVD should be used.

2.6 Linear least-squares problems

The mathematical concepts of vectors, matrices, and matrix decompositions are now illustrated in the analysis of linear least-squares problems. Least-squares problems have a long history in science, as has been discussed in Chapter 1. They arise in solving an overdetermined set of equations. With the definition of a data matrix $F \in \mathbb{R}^{m \times n}$ and an observation vector $y \in \mathbb{R}^m$, a set of equations in an unknown parameter vector $x \in \mathbb{R}^n$ can be denoted by

$$Fx = y. \tag{2.7}$$

This is a compact matrix notation for the set of linear equations

$$\begin{aligned} f_{11}x_1 + f_{12}x_2 + \cdots + f_{1n}x_n &= y_1, \\ f_{21}x_1 + f_{22}x_2 + \cdots + f_{2n}x_n &= y_2, \end{aligned}$$

$$\vdots$$

$$f_{m1}x_1 + f_{m2}x_2 + \cdots + f_{mn}x_n = y_m.$$

The matrix–vector product Fx geometrically means that we seek linear combinations of the columns of the matrix F that equal the vector y. Therefore, a solution x to (2.7) exists only provided that the vector y

lies in the column space of the matrix F. When the vector y satisfies this condition, we say that the set of equations in (2.7) is consistent, otherwise, it is called *inconsistent*.

To find a solution to an inconsistent set of equations, we pose the following least-squares problem:

$$\min_{x} \epsilon^{T}\epsilon \quad \text{subject to } y = Fx + \epsilon. \tag{2.8}$$

That is, we seek a vector x that minimizes the norm of the residual vector ϵ, such that $y \approx Fx$. The residual vector $\epsilon \in \mathbb{R}^{m}$ can be eliminated from the above problem formulation, thus reducing it to the more standard form

$$\min_{x} \|Fx - y\|_{2}^{2}. \tag{2.9}$$

Of course, it would be possible to minimize the distance between y and Fx using some other cost function. This would lead to solution methods completely different from the ones discussed below. In this book we stick to the least-squares cost function, because it is widely used in science and engineering.

The cost function of the least-squares minimization problem (2.9) can be expanded as

$$\begin{aligned} f(x) &= \|Fx - y\|_{2}^{2} \\ &= (Fx - y)^{T}(Fx - y) \\ &= x^{T}F^{T}Fx - x^{T}F^{T}y - y^{T}Fx + y^{T}y. \end{aligned}$$

From this expression it is easy to compute the gradient (see also Exercise 2.9 on page 38)

$$\frac{\partial}{\partial x}f(x) = \begin{bmatrix} \dfrac{\partial f(x)}{\partial x_1} \\ \vdots \\ \dfrac{\partial f(x)}{\partial x_n} \end{bmatrix} = 2F^{T}Fx - 2F^{T}y.$$

The solution $\widehat{x}$ to the least-squares problem (2.9) is found by setting the gradient $\partial f(x)/\partial x$ equal to 0. This yields the so-called *normal equations*,

$$F^{T}F\widehat{x} = F^{T}y, \tag{2.10}$$

or, alternatively,

$$F^{T}\left(F\widehat{x} - y\right) = 0. \tag{2.11}$$

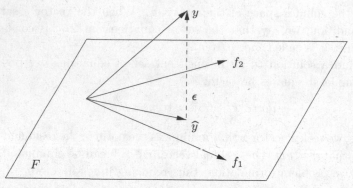

Fig. 2.1. The solution to the least-squares problem. The matrix F is 3×2; its two column vectors f_1 and f_2 span a two-dimensional plane in a three-dimensional space. The vector y does not lie in this plane. The projection of y onto the plane spanned by f_1 and f_2 is the vector $\widehat{y} = F\widehat{x}$. The residual $\epsilon = y - \widehat{y}$ is orthogonal to the plane spanned by f_1 and f_2.

Without computing the actual solution $\widehat{x}$ from the normal equations, this relationship shows that the residual, $\epsilon = y - F\widehat{x}$, is orthogonal to the range space of F. This is illustrated in a three-dimensional Euclidean space in Figure 2.1.

Using Equation (2.11), the minimal value of the cost function $\|Fx - y\|_2^2$ in (2.9) is

$$-y^{\mathrm{T}}(F\widehat{x} - y).$$

The calculation of the derivative of the quadratic form $f(x)$ requires a lot of book-keeping when relying on standard first- or second-year calculus. Furthermore, to prove that the value of x at which the gradient is equal to zero is a minimum, it must also be shown that the Hessian matrix (containing second-order derivatives) evaluated at this value of x is positive-definite; see for example Strang (1988). A more elegant way to solve the minimization of (2.9) is by applying the "completion-of-squares" argument (Kailath *et al.*, 2000). Before applying this principle to the solution of (2.9), let us first illustrate the use of this argument in the following example.

Example 2.9 ("Completion-of-squares" argument) Consider the following minimization:

$$\min_x f(x) = \min_x (1 - 2x + 2x^2).$$

Then the completion-of-squares argument first rewrites the functional $f(x)$ as

$$1 - 2x + 2x^2 = 2\left(x - \frac{1}{2}\right)^2 + \left(1 - \frac{1}{2}\right).$$

Second, one observes that the first term of the rewritten functional is the only term that depends on the argument x and is always nonnegative thanks to the square and the positive value of the coefficient 2. The second term of the rewritten functional is independent of the argument x. Hence, the minimum of the functional is obtained by setting the term $2(x - \frac{1}{2})^2$ equal to zero, that is, for the following value of the argument of the functional:

$$\widehat{x} = \frac{1}{2}.$$

For this argument the value of the functional equals the second term of the rewritten functional, that is,

$$f(\widehat{x}) = \left(1 - \frac{1}{2}\right).$$

To apply the completion-of-squares argument to solve 2.9, we rewrite the functional $f(x) = \|Fx - y\|_2^2$ as

$$f(x) = \begin{bmatrix} 1 & x^{\mathrm{T}} \end{bmatrix} \underbrace{\begin{bmatrix} y^{\mathrm{T}}y & -y^{\mathrm{T}}F \\ -F^{\mathrm{T}}y & F^{\mathrm{T}}F \end{bmatrix}}_{M} \begin{bmatrix} 1 \\ x \end{bmatrix},$$

and factorize the matrix M with the help of the Schur complement (Lemma 2.3 on page 19),

$$M = \begin{bmatrix} I & -\widehat{x}^{\mathrm{T}} \\ 0 & I \end{bmatrix} \begin{bmatrix} y^{\mathrm{T}}y - y^{\mathrm{T}}F\widehat{x} & 0 \\ 0 & F^{\mathrm{T}}F \end{bmatrix} \begin{bmatrix} I & 0 \\ -\widehat{x} & I \end{bmatrix},$$

for $\widehat{x}$ satisfying

$$F^{\mathrm{T}}F\widehat{x} = F^{\mathrm{T}}y.$$

Using this factorization, the above cost function $f(x)$ can now be formulated in a suitable form to apply the completion-of-squares principle, namely

$$f(x) = (y^{\mathrm{T}}y - y^{\mathrm{T}}F\widehat{x}) + (x - \widehat{x})^{\mathrm{T}}F^{\mathrm{T}}F(x - \widehat{x}).$$

Since the matrix $F^{\mathrm{T}}F$ is nonnegative, the second term above is always nonnegative and depends only on the argument x. Therefore the minimum is obtained for $x = \widehat{x}$ and the value of the cost function at the minimum is (again)

$$f(\widehat{x}) = y^{\mathrm{T}}(y - F\widehat{x}).$$

2.6.1 Solution if the matrix F has full column rank

From Equation (2.10), it becomes clear that the argument $\widehat{x}$ that minimizes (2.9) is *unique* if the matrix F has full column rank n. In that case, the matrix $F^{\mathrm{T}}F$ is square and invertible and the solution $\widehat{x}$ is

$$\widehat{x} = (F^{\mathrm{T}}F)^{-1}F^{\mathrm{T}}y. \tag{2.12}$$

The corresponding value of the cost function $\|F\widehat{x} - y\|_2^2$ in (2.9) is

$$\|F\widehat{x} - y\|_2^2 = \|(F(F^{\mathrm{T}}F)^{-1}F^{\mathrm{T}} - I_m)y\|_2^2$$
$$= y^{\mathrm{T}}(I_m - F(F^{\mathrm{T}}F)^{-1}F^{\mathrm{T}})y.$$

The matrix $(F^{\mathrm{T}}F)^{-1}F^{\mathrm{T}}$ is called the *pseudo-inverse* of the matrix F, because $((F^{\mathrm{T}}F)^{-1}F^{\mathrm{T}})F = I_n$. The matrix $F(F^{\mathrm{T}}F)^{-1}F^{\mathrm{T}} \in \mathbb{R}^{m\times m}$ determines the *orthogonal projection* of a vector in $\mathbb{R}^m$ onto the space spanned by the columns of the matrix F. We denote this projection by Π_F. It has the following properties.

(i) $\Pi_F \cdot \Pi_F = \Pi_F$.

(ii) $\Pi_F F = F$.

(iii) If the QR factorization of the matrix F is denoted by

$$F = \begin{bmatrix} Q_1 & Q_2 \end{bmatrix} \begin{bmatrix} R \\ 0 \end{bmatrix},$$

with $R \in \mathbb{R}^{m\times m}$ square and invertible, then $\Pi_F = Q_1 Q_1^{\mathrm{T}}$.

(iv) $\|y\|_2^2 = \|\Pi_F y\|_2^2 + \|(I - \Pi_F)y\|_2^2$ for all $y \in \mathbb{R}^m$.

(v) $\|\Pi_F y\|_2 \le \|y\|_2$ for all $y \in \mathbb{R}^m$.

Example 2.10 (Overdetermined set of linear equations) Given

$$F = \begin{bmatrix} 1 & 1 \\ 2 & 1 \\ 1 & 1 \end{bmatrix}, \qquad y = \begin{bmatrix} 1 \\ 0 \\ 0 \end{bmatrix},$$

consider the set of three equations in two unknowns, $Fx = y$. The least-squares solution is given by

$$\widehat{x} = (F^{\mathrm{T}}F)^{-1}F^{\mathrm{T}}y$$

$$= \begin{bmatrix} \frac{3}{2} & -2 \\ -2 & 3 \end{bmatrix} \begin{bmatrix} 1 & 2 & 1 \\ 1 & 1 & 1 \end{bmatrix} \begin{bmatrix} 1 \\ 0 \\ 0 \end{bmatrix}$$

$$= \begin{bmatrix} -\frac{1}{2} \\ 1 \end{bmatrix}.$$

The least-squares residual is $\|\epsilon\|_2^2 = \|F\widehat{x} - y\|_2^2 = \frac{1}{2}$.

2.6.2 Solutions if the matrix F does not have full column rank

When the columns of the matrix $F \in \mathbb{R}^{m \times n}$ are not independent, multiple solutions to Equation (2.9) exist. We illustrate this claim for the special in which when the matrix F is of rank r and partitioned as

$$F = \begin{bmatrix} F_1 & F_2 \end{bmatrix}, \tag{2.13}$$

with $F_1 \in \mathbb{R}^{m \times r}$ of full column rank and $F_2 \in \mathbb{R}^{m \times (n-r)}$. Since F and F_1 are of rank r, the columns of the matrix F_2 can be expressed as linear combinations of the columns of the matrix F_1. By introducing a nonzero matrix $X \in \mathbb{R}^{r \times (n-r)}$ we can write

$$F_2 = F_1 X.$$

Then, if we partition x according to the matrix F into

$$\begin{bmatrix} x_1 \\ x_2 \end{bmatrix},$$

with $x_1 \in \mathbb{R}^r$ and $x_2 \in \mathbb{R}^{n-r}$, the minimization problem (2.9) can be written as

$$\min_{x_1, x_2} \left\| \begin{bmatrix} F_1 & F_2 \end{bmatrix} \begin{bmatrix} x_1 \\ x_2 \end{bmatrix} - y \right\|_2^2 = \min_{x_1, x_2} \| F_1(x_1 + X x_2) - y \|_2^2.$$

We introduce the auxiliary variable $\eta = x_1 + X x_2$. Since the matrix F_1 has full column rank, an optimal solution $\hat{\eta}$ satisfies

$$\hat{\eta} = \hat{x}_1 + X\hat{x}_2 = (F_1^T F_1)^{-1} F_1^T y. \tag{2.14}$$

Any choice for $\hat{x}_2$ fixes the estimate $\hat{x}_1$ at $\hat{x}_1 = \hat{\eta} - X\hat{x}_2$. This shows that $\hat{x}_2$ can be chosen independently of $\hat{x}_1$ and thus there exists an infinite number of solutions $(\hat{x}_1, \hat{x}_2)$ that yield the minimum value of the cost function, which is given by

$$\| F_1 \hat{\eta} - y \|_2^2 = \left\| F_1 (F_1^T F_1)^{-1} F_1^T y - y \right\|_2^2 = \| (\Pi_{F_1} - I_m) y \|_2^2.$$

To yield a unique solution, additional constraints on the vector x are needed. One commonly used constraint is that the solution x should have minimal 2-norm. In that case, the optimization problem (2.9) can be formulated as

$$\min_{x \in \mathcal{X}} \| x \|_2^2, \quad \text{with } \mathcal{X} = \left\{ x : x = \arg\min_z \| Fz - y \|_2^2 \right\}. \tag{2.15}$$

Now we derive the general solution to problem (2.15), where no specific partitioning of F is assumed. This solution can be found through the use of the SVD of the matrix F (see Section 2.5). Let the SVD of the matrix F be given by

$$F = \begin{bmatrix} U_1 & U_2 \end{bmatrix} \begin{bmatrix} \Sigma & 0 \\ 0 & 0 \end{bmatrix} \begin{bmatrix} V_1^{\mathrm{T}} \\ V_2^{\mathrm{T}} \end{bmatrix} = U_1 \Sigma V_1^{\mathrm{T}},$$

with $\Sigma \in \mathbb{R}^{r \times r}$ nonsingular, then the minimization problem in the definition of the set $\mathcal{X}$ can be written as

$$\min_z \|Fz - y\|_2^2 = \min_z \|U_1 \Sigma V_1^{\mathrm{T}} z - y\|_2^2.$$

We define the partitioned vector

$$\begin{bmatrix} \xi_1 \\ \xi_2 \end{bmatrix} = \begin{bmatrix} V_1^{\mathrm{T}} \\ V_2^{\mathrm{T}} \end{bmatrix} z.$$

With this vector, the problem becomes

$$\min_z \|Fz - y\|_2^2 = \min_{\xi_1} \|U_1 \Sigma \xi_1 - y\|_2^2.$$

We find that $\widehat{\xi}_1 = \Sigma^{-1} U_1^{\mathrm{T}} y$ and that ξ_2 does not change the value of $\|Fz - y\|_2^2$. Therefore, ξ_2 can be chosen arbitrarily. The solutions become

$$\widehat{z} = \begin{bmatrix} V_1 & V_2 \end{bmatrix} \begin{bmatrix} \widehat{\xi}_1 \\ \widehat{\xi}_2 \end{bmatrix} = V_1 \Sigma^{-1} U_1^{\mathrm{T}} y + V_2 \widehat{\xi}_2,$$

with $\widehat{\xi}_2 \in \mathbb{R}^{n-r}$. Thus

$$\mathcal{X} = \left\{ x : x = V_1 \Sigma^{-1} U_1^{\mathrm{T}} y + V_2 \widehat{\xi}_2, \ \widehat{\xi}_2 \in \mathbb{R}^{n-r} \right\}.$$

The solution to Equation (2.15) is obtained by selecting the vector x from the set $\mathcal{X}$ that has the smallest 2-norm. Since $V_1^{\mathrm{T}} V_2 = 0$, we have

$$\|x\|_2^2 = \|V_1 \Sigma^{-1} U_1^{\mathrm{T}} y\|_2^2 + \|V_2 \widehat{\xi}_2\|_2^2,$$

and the minimal 2-norm solution is obtained by taking $\widehat{\xi}_2 = 0$. Thus, the solution to (2.14) in terms of the SVD of the matrix F is

$$\widehat{x} = V_1 \Sigma^{-1} U_1^{\mathrm{T}} y. \tag{2.16}$$

Example 2.11 (Underdetermined set of linear equations) Given

$$F = \begin{bmatrix} 1 & 2 & 1 \\ 1 & 1 & 1 \end{bmatrix}, \qquad y = \begin{bmatrix} 1 \\ 0 \end{bmatrix},$$

consider the set of two equations in three unknowns $Fx = y$. It is easy to see that the solution is not unique, Namely the set of vectors $\{x(a) : a \in \mathbb{R}\}$ contains all solutions to $Fx(a) = y$, with $x(a)$ given as

$$x(a) = \begin{bmatrix} -1 - a \\ 1 \\ a \end{bmatrix}.$$

The minimum-norm solution, derived by solving Exercise 2.10 on page 40, is unique and is given by

$$\widehat{x} = F^{\mathrm{T}}(FF^{\mathrm{T}})^{-1}y$$

$$= \begin{bmatrix} 1 & 1 \\ 2 & 1 \\ 1 & 1 \end{bmatrix} \begin{bmatrix} \frac{3}{2} & -2 \\ -2 & 3 \end{bmatrix} \begin{bmatrix} 1 \\ 0 \end{bmatrix}$$

$$= \begin{bmatrix} -\frac{1}{2} \\ 1 \\ -\frac{1}{2} \end{bmatrix}.$$

The 2-norm of this solution is $\|\widehat{x}\|_2 = \sqrt{3/2}$. It is easy to verify that this solution equals $x(\widehat{a})$ with

$$\widehat{a} = \arg\min_a \|x(a)\|_2^2 = -\frac{1}{2}.$$

2.7 Weighted linear least-squares problems

A generalization of the least-squares problem (2.8) on page 29 is obtained by introducing a matrix $L \in \mathbb{R}^{m \times m}$ as follows:

$$\min_x \epsilon^{\mathrm{T}} \epsilon \quad \text{subject to } y = Fx + L\epsilon. \tag{2.17}$$

The data matrices $F \in \mathbb{R}^{m \times n}$ ($m \geq n$) and $L \in \mathbb{R}^{m \times m}$ and the data vector $y \in \mathbb{R}^m$ are known quantities. The unknown parameter vector $x \in \mathbb{R}^n$ is to be determined. We observe that the solution is not changed by the application of a nonsingular transformation matrix $T_\ell \in \mathbb{R}^{m \times m}$ and an orthogonal transformation matrix $T_r \in \mathbb{R}^{m \times m}$ as follows:

$$\min_x \epsilon^{\mathrm{T}} T_r^{\mathrm{T}} T_r \epsilon \quad \text{subject to } T_\ell y = T_\ell Fx + T_\ell LT_r^{\mathrm{T}} T_r \epsilon. \tag{2.18}$$

Let $T_r \epsilon$ be denoted by $\widetilde{\epsilon}$, then we can denote the above problem more compactly as

$$\min_x \widetilde{\epsilon}^{\mathrm{T}} \widetilde{\epsilon} \quad \text{subject to } T_\ell y = T_\ell Fx + T_\ell LT_r^{\mathrm{T}} \widetilde{\epsilon}.$$

The transformations T_ℓ and T_r may introduce a particular pattern of zeros into the matrices $T_\ell F$ and $T_\ell L T_r^T$ such that the solution vector x can be determined, for example by back substitution. An illustration of this is given in Chapter 5 in the solution of the Kalman-filter problem. An outline of a numerically reliable solution for general F and L matrices was given by Paige (Kourouklis and Paige, 1981; Paige, 1979).

If the matrix L is invertible, we can select the allowable transformations T_ℓ and T_r equal to L^{-1} and I_m, respectively. Then the weighted least-squares problem (2.17) has the same solution as the following linear least-squares problem:

$$\min_x \epsilon^T \epsilon \quad \text{subject to } L^{-1} y = L^{-1} F x + \epsilon. \tag{2.19}$$

Let the matrix W be defined as $W = (LL^T)^{-1}$, so that we may denote this problem by the standard formulation of the weighted least-squares problem

$$\min_x (Fx - y)^T W (Fx - y). \tag{2.20}$$

Note that this problem formulation is less general than (2.17), because it requires L to be invertible. The necessary condition for the solution x of this problem follows on setting the gradient of the cost function $(Fx - y)^T W (Fx - y)$ equal to zero. This yields the following normal equations for the optimal value $\hat{x}$:

$$(F^T W F)\hat{x} = F^T W y.$$

Through the formulation of the weighted least-squares problem as the standard least-squares problem (2.19), we could obtain the above normal equations immediately from (2.10) by replacing therein the matrix F and the vector y by $L^{-1}F$ and $L^{-1}y$, respectively, and invoking the definition of the matrix W. If we express the normal equations as

$$F^T W (F\hat{x} - y) = 0,$$

then we observe that the residual vector $\epsilon = y - F\hat{x}$ is orthogonal to the column space of the matrix WF.

The explicit calculation of the solution $\hat{x}$ depends further on the invertibility of the matrix $F^T W F$. If this matrix is invertible, the solution $\hat{x}$ becomes

$$\hat{x} = (F^T W F)^{-1} F^T W y,$$

and the minimum value of the cost function in (2.20) is

$$(F\widehat{x} - y)^{\mathrm{T}} W (F\widehat{x} - y) = y^{\mathrm{T}} \left(W F (F^{\mathrm{T}} W F)^{-1} F^{\mathrm{T}} - I_m \right) W$$
$$\times \left(F (F^{\mathrm{T}} W F)^{-1} F^{\mathrm{T}} W - I_m \right) y,$$
$$= y^{\mathrm{T}} \left(W - W F (F^{\mathrm{T}} W F)^{-1} F^{\mathrm{T}} W \right) y.$$

2.8 Summary

We have reviewed some important concepts from linear algebra. First, we focused on vectors. We defined the inner product, the 2-norm, and orthogonality. Then we talked about linear independence, dimension, vector spaces, span, bases, and subspaces. We observed that a linear transformation between vector spaces is characterized by a matrix. The review continued with the definition of the matrix product, the Kronecker product, the vec operator, the Frobenius norm, and the rank of a matrix. The rank equals the number of independent columns and independent rows. This led to the definition of the four fundamental subspaces defined by a matrix: these are its column space, its row space, its null space, and its left null space.

Next, we looked at some important concepts for square matrices. The inverse and the determinant of a matrix were defined. We stated the matrix-inversion lemma, described the Schur complement, and introduced eigenvalues and eigenvectors. For symmetric matrices the concepts of positive and negative definiteness were discussed.

Matrix decomposition was the next topic in this chapter. The decompositions described include the eigenvalue decomposition, the Cholesky factorization, the singular-value decomposition (SVD), and the QR and RQ factorization.

Finally, we analyzed a problem that we will encounter frequently in the remaining part of this book, namely the least-squares problem. A generalization of the least-squares problem, the weighted least-squares problem, was also discussed.

Exercises

2.1   Given a matrix $A \in \mathbb{R}^{m \times n}$, $m \geq n$, and a vector $x \in \mathbb{R}^n$,

(a) show that there exists a decomposition $x = x_r + x_n$ such that $x_r \in \mathrm{range}(A^{\mathrm{T}})$ and $x_n \in \mathrm{ker}(A)$.

(b) Show that for this decomposition $x_r^{\mathrm{T}} x_n = 0$.

2.2   Prove the matrix-inversion lemma (Lemma 2.2 on page 19).

2.3   Given a square, symmetric, and invertible matrix $B \in \mathbb{R}^{(n+1) \times (n+1)}$ that can be partitioned as

$$B = \begin{bmatrix} A & v \\ v^T & \sigma \end{bmatrix},$$

express the inverse of the matrix B in terms of the inverse of the matrix $A \in \mathbb{R}^{n \times n}$.

2.4   Compute the determinant of the matrix

$$\begin{bmatrix} a_{11} & a_{12} & a_{13} \\ a_{21} & a_{22} & a_{23} \\ a_{31} & a_{32} & a_{33} \end{bmatrix}.$$

2.5   Prove Lemma 2.7 on page 24.

2.6   Given a matrix $A \in \mathbb{R}^{n \times n}$, if the eigenvalue decomposition of this matrix exists, and all eigenvalues have a magnitude strictly smaller than 1, show that

$$(I - A)^{-1} = \sum_{i=0}^{\infty} A^i = I + A + A^2 + A^3 + \cdots.$$

2.7   Given a symmetric matrix $A \in \mathbb{R}^{n \times n}$ with an eigenvalue decomposition equal to

$$A = U \begin{bmatrix} \lambda_1 & 0 & \cdots & 0 \\ 0 & \lambda_2 & \cdots & 0 \\ \vdots & & \ddots & \\ 0 & \cdots & & \lambda_n \end{bmatrix} U^T,$$

(a) show that the matrix $U \in \mathbb{R}^{n \times n}$ is an orthogonal matrix.

(b) Show that the eigenvalues of the matrix $(A + \mu I)$, for some scalar $\mu \in \mathbb{C}$, equal $\lambda_i + \mu$, $i = 1, 2, \ldots, n$.

2.8   Explain how the SVD can be used to compute the inverse of a nonsingular matrix.

2.9   Given the scalar function

$$f(x) = \|Ax - b\|_2^2,$$

with $A \in \mathbb{R}^{N \times n}$ $(N \gg n)$, $b \in \mathbb{R}^N$, and $x = \begin{bmatrix} x_1 & x_2 & \cdots & x_n \end{bmatrix}^T$,

(a) prove that

$$\frac{\mathrm{d}}{\mathrm{d}x} f(x) = 2A^T A x - 2A^T b.$$

(b) Prove that a solution to the minimization problem

$$\min_x f(x) \qquad \text{(E2.1)}$$

satisfies the linear set of equations (which are called the normal equations)

$$A^{\mathrm{T}} A x = A^{\mathrm{T}} b. \qquad \text{(E2.2)}$$

(c) When is the solution to Equation (E2.1) unique?

(d) Let $Q \in \mathbb{R}^{N \times N}$ be an orthogonal matrix, that is, $Q^{\mathrm{T}} Q = Q Q^{\mathrm{T}} = I_N$. Show that

$$\|Ax - b\|_2^2 = \|Q^{\mathrm{T}}(Ax - b)\|_2^2.$$

(e) Assume that the matrix A has full column rank, and that its QR factorization is given by

$$A = Q \begin{bmatrix} R \\ 0 \end{bmatrix}.$$

Show, without using Equation (E2.2), that the solution to Equation (E2.1) is

$$x = R^{-1} b_1, \qquad \text{(E2.3)}$$

where $b_1 \in \mathbb{R}^n$ comes from the partitioning

$$Q^{\mathrm{T}} b = \begin{bmatrix} b_1 \\ b_2 \end{bmatrix},$$

with $b_2 \in \mathbb{R}^{N-n}$. Show also that

$$\min \|Ax - b\|_2^2 = \|b_2\|_2^2.$$

(f) Assume that the matrix A has full column rank and that its SVD is given by $A = U \Sigma V^{\mathrm{T}}$ with

$$\Sigma = \begin{bmatrix} \Sigma_n \\ 0 \end{bmatrix}, \quad \Sigma_n = \begin{bmatrix} \sigma_1 & 0 & \cdots & 0 \\ 0 & \sigma_2 & & \\ & & \ddots & \\ 0 & \cdots & 0 & \sigma_n \end{bmatrix}.$$

Show, without using Equation (E2.2), that the solution to Equation (E2.1) is

$$x = \sum_{i=1}^{n} \frac{u_i^{\mathrm{T}} b}{\sigma_i} v_i, \qquad \text{(E2.4)}$$

where u_i and v_i represent the ith column vector of the matrix U and the matrix V, respectively. Show also that

$$\min\|Ax - b\|_2^2 = \sum_{i=n+1}^{N} (u_i^{\mathrm{T}}b)^2.$$

(g) Show that, if A has full column rank, the solutions (E2.3) and (E2.4) are equivalent to the solution obtained by solving Equation (E2.2).

2.10   Let an underdetermined set of equations be denoted by

$$b = Ax, \tag{E2.5}$$

with $b \in \mathbb{R}^m$ and $A \in \mathbb{R}^{m \times n}$ $(m < n)$ of full rank m. Let $e \in \mathbb{R}^n$ be a vector in the kernel of A, that is, $Ae = 0$.

(a) Prove that any solution to Equation (E2.5) is given by

$$x = A^{\mathrm{T}}(AA^{\mathrm{T}})^{-1}b + e.$$

(b) Prove that the solution to Equation (E2.5) that has minimal norm is given by the above equation for $e = 0$.

2.11   Consider a polynomial approximation of the exponential function e^t of order n,

$$\mathrm{e}^t = \begin{bmatrix} 1 & t & t^2 & \cdots & t^{n-1} \end{bmatrix}\theta + v(t),$$

for $t \in \{0, 0.1, 0.2, \ldots, 3\}$, where $\theta \in \mathbb{R}^n$ contains the polynomial coefficients and $v(t)$ is the approximation error.

(a) Formulate the determination of the unknown parameter vector θ as a linear least-squares problem.

(b) Check numerically whether it is possible to determine a unique solution to this least-squares problem for $n = 2, 3, \ldots, 6$ using the Equations (E2.2), (E2.3), and (E2.4) derived in Exercise 2.9 on page 38. Summarize the numerical results for each method in tabular form, as illustrated below.

Order n	θ_1	θ_2	θ_3	θ_4	θ_5	θ_6
2						
3						
4						
5						
6						

(c) Explain the differences between the numerical estimates of all three methods obtained in part (b) and the coefficients of the Taylor-series expansion

$$e^t = 1 + t + \frac{t^2}{2} + \frac{t^3}{3!} + \cdots \quad \text{for } |t| < \infty.$$

2.12 (Exercise 5.2.6 from Golub and Van Loan (1996)). Given a partial QR factorization of a full-column-rank matrix $A \in \mathbb{R}^{N \times n}$ $(N \gg n)$

$$Q^{\mathrm{T}} A = \begin{bmatrix} R & w \\ 0 & v \end{bmatrix}, \qquad Q^{\mathrm{T}} b = \begin{bmatrix} c \\ d \end{bmatrix},$$

with $R \in \mathbb{R}^{(n-1) \times (n-1)}$, $w \in \mathbb{R}^{n-1}$, $v \in \mathbb{R}^{N-n+1}$, $c \in \mathbb{R}^{n-1}$, and $d \in \mathbb{R}^{N-n+1}$, show that

$$\min_x \|Ax - b\|_2^2 = \|d\|_2^2 - \left(\frac{v^{\mathrm{T}} d}{\|v\|_2} \right)^2.$$

3

Discrete-time signals and systems

3.1 Introduction

This chapter deals with two important topics: signals and systems. A *signal* is basically a value that changes over time. For example, the outside temperature as a function of the time of the day is a signal. More specifically, this is a continuous-time signal; the signal value is defined at every time instant. If we are interested in measuring the outside temperature, we will seldom do this continuously. A more practical approach is to measure the temperature only at certain time instants, for example every minute. The signal that is obtained in that way is a sequence of numbers; its values correspond to certain time instants. Such an ordered sequence is called a discrete-time signal. In our example the temperature is known only every minute, not every second. A *system* relates different signals to each other. For example, a mercury thermometer is a system that relates the outside temperature to the expansion of the mercury it contains. Naturally, we can speak of continuous-time systems and discrete-time systems, depending on what signals we are considering. Although most real-life systems operate in continuous time, it is often desirable to have a discrete-time system that accurately describes this continuous-time system, because often the signals related to this system are measured in discrete time.

This chapter provides an overview of important topics related to discrete-time signals and systems. It is not meant as a first introduction to the theory of discrete-time signals and systems, but rather as a review of the concepts that are used in the remaining chapters of this book. We have chosen to review discrete-time signals and systems only, not their continuous-time counterparts, because this book focuses on discrete-time systems. This does not mean, however, that we do not encounter any continuous-time systems. Especially in Chapter 10, which deals with setting up identification experiments for real-life (continuous-time) systems, the relation between the discrete-time system and the corresponding continuous-time system cannot be neglected. We assume that the reader has some basic knowledge about discrete-time and continuous-time signals and systems. Many good books that provide a more in-depth introduction to the theory of signals and systems have been written (see for example Oppenheim and Willsky (1997)).

This chapter is organized as follows. Section 3.2 provides a very short introduction to discrete-time signals. Section 3.3 reviews the z-transform and the discrete-time Fourier transform. Section 3.4 introduces discrete-time linear systems and discusses their properties. Finally, Section 3.5 deals with the interaction between linear systems.

3.2 Signals

We start by giving the formal definition of a discrete-time signal. We use $\mathbb{Z}$ to denote the set of all integers.

Definition 3.1 *A discrete-time signal x is a function $x : T \to R$ with the time axis $T \subset \mathbb{Z}$ and signal range $R \subset \mathbb{C}$.*

If R is a subset of $\mathbb{R}$ then the signal is said to be real-valued, otherwise it is complex-valued. Most signals we encounter will be real-valued.

The definition states that a discrete-time signal is a function of time; it can, however, also be viewed as an ordered sequence, i.e.

$$\ldots, x(-2), x(-1), x(0), x(1), x(2), \ldots$$

Both of these interpretations are useful.

Often, a discrete-time signal is obtained by uniformly *sampling* a continuous-time signal. Before we explain what this means we define a continuous-time signal.

Definition 3.2 *A continuous-time signal x is a function $x : T \to R$ with the time axis T an interval of $\mathbb{R}$ and signal range $R \subset \mathbb{C}$*

Let $x_{\mathrm{c}}(t)$, $t \in \mathbb{R}$ be a continuous-time signal. We construct a discrete-time signal $x_{\mathrm{d}}(k)$, $k \in \mathbb{Z}$, by observing the signal $x_{\mathrm{c}}(t)$ only at time instants that are multiples of $T \in \mathbb{R}$, that is, $x_{\mathrm{d}}(k) = x_{\mathrm{c}}(kT)$. This process is called equidistant *sampling* and T is called the *sampling interval*. Note that the signal $x_{\mathrm{d}}(k)$ is defined only at the sampling instances, not in between.

Two special discrete-time signals that we often use are the unit pulse and the unit step. These signals are defined below, and shown in Figure 3.1.

Definition 3.3 *The unit pulse $\Delta(k)$ is a discrete-time signal that satisfies*

$$\Delta(k) = \begin{cases} 1, & \text{for } k = 0, \\ 0, & \text{for } k \neq 0. \end{cases}$$

Definition 3.4 *The unit step $s(k)$ is a discrete-time signal that satisfies*

$$s(k) = \begin{cases} 1, & \text{for } k \geq 0, \\ 0, & \text{for } k < 0. \end{cases}$$

We can use $\Delta(k)$ to express an arbitrary signal $x(k)$ as

$$x(k) = \sum_{i=-\infty}^{\infty} x(i)\Delta(k-i).$$

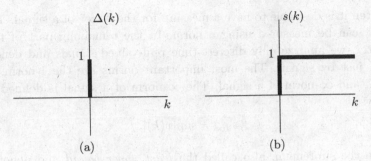

Fig. 3.1. The unit pulse (a) and unit-step signal (b).

For $s(k)$ this expression becomes

$$s(k) = \sum_{i=-\infty}^{\infty} s(i)\Delta(k-i) = \sum_{i=0}^{\infty} \Delta(k-i).$$

Definition 3.5 *A signal x is periodic if there exists an integer $P \in \mathbb{Z}$ with $P > 0$ such that*

$$x(k+P) = x(k),$$

for all $k \in \mathbb{Z}$. The smallest $P > 0$ for which this holds is called the period of the signal.

The following example shows that a discrete-time signal obtained by sampling a periodic continuous-time signal does not have to be periodic.

Example 3.1 (Discrete-time sine wave) Consider the discrete-time sine wave given by

$$x(k) = \sin(k\omega).$$

For $\omega = \pi$ this signal is periodic with period 2, because

$$\sin(k\pi) = \sin(k\pi + 2\pi) = \sin[(k+2)\pi].$$

However, for $\omega = 1$ the signal is nonperiodic, because there does not exist an integer P such that

$$\sin(k) = \sin(k+P).$$

Note that the signals with $\omega = \omega_0 + 2\pi\ell$, $\ell \in \mathbb{Z}$, are indistinguishable, because

$$\sin(k\omega) = \sin(k\omega_0 + 2\pi k\ell) = \sin(k\omega_0).$$

Often it is desirable to have a measure for the "size" of a signal. The "size" can be measured using a norm. In the remaining part of this chapter, we analyze only discrete-time real-valued signals and denote them just by signals. The most important norms are the 1-norm, 2-norm, and ∞-norm of a signal. The ∞-norm of a signal is defined as follows:

$$||x||_\infty = \sup_{k \in \mathbb{Z}} |x(k)|,$$

where the supremum, also called the *least upper bound* and denoted by "sup," is the smallest number $\alpha \in \mathbb{R} \cup \{\infty\}$ such that $|x(k)| \leq \alpha$ for all $k \in \mathbb{Z}$. If $\alpha = \infty$, the signal is called unbounded. The ∞-norm is also called the *amplitude* of the signal. Note that there are signals, for example $x(k) = e^{kT}$ for T a positive real number, for which the amplitude is infinite. The 2-norm of a signal is defined as follows:

$$||x||_2 = \left(\sum_{k=-\infty}^{\infty} |x(k)|^2 \right)^{1/2}.$$

The square of the 2-norm, $||x||_2^2$, is called the *energy* of the signal. Note that there are signals for which the 2-norm is infinite; these signals do not have finite energy. The *power* of a signal is defined as follows:

$$\lim_{N \to \infty} \frac{1}{2N} \sum_{k=-N}^{N} |x(n)|^2.$$

The square root of the power is called the *root-mean-square* (RMS) value. The 1-norm of a signal is defined as follows:

$$||x||_1 = \sum_{k=-\infty}^{\infty} |x(k)|.$$

Signals that have a finite 1-norm are called *absolutely summable*.

Example 3.2 (Norms) Consider the signal $x(k) = a^k s(k)$ with $a \in \mathbb{R}$, $|a| < 1$. We have

$$||x||_\infty = 1,$$

$$||x||_2 = \left(\sum_{k=0}^{\infty} a^{2k} \right)^{1/2} = \frac{1}{\sqrt{1-a^2}},$$

$$||x||_1 = \sum_{k=0}^{\infty} |a^k| = \frac{1}{1-|a|}.$$

The norms do not need to be finite; this depends on the convergence properties of the sums involved. For the signal $x(k) = as(k)$, $a \in \mathbb{R}$ the ∞-norm equals $|a|$, but the 2-norm and the 1-norm are infinite; the power of this signal is, however, bounded and equals

$$\lim_{N \to \infty} \frac{1}{2N} \sum_{k=0}^{N} |a|^2 = \lim_{N \to \infty} \frac{1}{2N}(N+1)|a|^2 = \frac{|a|^2}{2}.$$

Sometimes we have to consider several signals at once. A convenient way to deal with this is to define a *vector signal*. A vector signal is simply a vector with each entry a signal.

An often-used operation on two signals is the *convolution*. The convolution of two vector signals x and y, with both $x(k)$ and $y(k) \in \mathbb{R}^n$, results in the signal z with samples $z(k) \in \mathbb{R}$ given by

$$z(k) = \sum_{\ell=-\infty}^{\infty} x(k-\ell)^{\mathrm{T}} y(\ell) = \sum_{\ell=-\infty}^{\infty} x(\ell)^{\mathrm{T}} y(k-\ell).$$

This is often denoted by

$$z(k) = x(k)^{\mathrm{T}} * y(k).$$

The convolution of two scalar signals x and y can of course be written as $x(k) * y(k)$.

3.3 Signal transforms

Discrete-time signals are defined in the time domain. It can, however, be useful to transform these signals to another domain, in which the signals have certain properties that simplify the calculations or make the interpretation of signal features easier. Two examples of transforms that have proven to be useful are the z-transform and the discrete-time Fourier transform. Both these transforms convert the time signal to a domain in which the convolution of two signals becomes a simple product. These transforms possess many nice properties, besides this one. This section describes the z-transform and the discrete-time Fourier transform and some of their properties.

3.3.1 *The z-transform*

The z-transform converts a time signal, which is a function of time, into a function of the complex variable z. It is defined as follows.

Definition 3.6 (z-Transform) *The z-transform of a signal $x(k)$ is given by*

$$X(z) = \sum_{k=-\infty}^{\infty} x(k)z^{-k}, \quad z \in E_x.$$

The existence region E_x consists of all $z \in \mathbb{C}$ for which the sum converges.

The z-transform of an n-dimensional vector signal $x(k)$ is an n-dimensional function $X(z)$, whose ith entry equals the z-transform of the ith entry of $x(k)$.

Example 3.3 (z-Transform of a unit step signal) The z-transform of a unit step signal $x(k) = s(k)$ is given by

$$X(z) = \sum_{k=-\infty}^{\infty} s(k)z^{-k} = \sum_{k=0}^{\infty} z^{-k}$$

$$= \frac{1}{1 - z^{-1}} = \frac{z}{z - 1},$$

where the sum converges only for $|z| > 1$. Hence, the existence region contains all complex numbers z that satisfy $|z| > 1$.

The existence region is crucial in order for one to be able to carry out the inverse operation of the z-transform. Given a function $X(z)$, the corresponding signal $x(k)$ is uniquely determined only if the existence region of $X(z)$ is given. The next example shows that without the existence region it is not possible to determine $x(k)$ uniquely from $X(z)$.

Example 3.4 (Existence region of the z-transform) The z-transform of the signal $x(k) = -s(-k-1)$ is given by

$$X(z) = \sum_{k=-\infty}^{\infty} -s(-k-1)z^{-k} = -\sum_{k=-\infty}^{-1} z^{-k}$$

$$= 1 - \sum_{\ell=0}^{\infty} z^{\ell} = 1 - \frac{1}{1 - z}$$

$$= \frac{z}{z - 1},$$

where the existence region equals $|z| < 1$. Comparing this result with Example 3.3 above shows that, without the existence region given, the function $X(x) = z/(z-1)$ can correspond both to $s(k)$ and to $-s(-k-1)$.

Table 3.1. *Properties of the z-transform*

Property	Time signal	z-Transform	Conditions
Linearity	$ax(k) + by(k)$	$aX(z) + bY(z)$	$a, b \in \mathbb{C}$
Time reversal	$x(-k)$	$X(z^{-1})$	
Multiple shift	$x(k - \ell)$	$z^{-\ell}X(z)$	$\ell \in \mathbb{Z}$
Multiplication by a^k	$a^k x(k)$	$X\left(\dfrac{z}{a}\right)$	$a \in \mathbb{C}$
Multiplication by k	$kx(k)$	$-z\dfrac{\mathrm{d}}{\mathrm{d}z}X(z)$	
Convolution	$x(k) * y(k)$	$X(z)Y(z)$	

Given the existence region of $X(z)$, we can derive the inverse z-transform as follows.

Theorem 3.1 (Inverse z-transform) *The inverse z-transform of a function $X(z)$ defined for $z \in E_x$ is given by*

$$x(k) = \frac{1}{2\pi j} \oint_C X(z)z^{k-1}\,\mathrm{d}z, \tag{3.1}$$

where the contour integral is taken counterclockwise along an arbitrary closed path C that encircles the origin and lies entirely within the existence region E_x.

This is a rather formal formulation. For the actual calculation of the inverse z-transform, complex-function theory is used. The inverse z-transform is rarely used. Instead, the inverse is often calculated using the properties of the z-transform listed in Table 3.1 and the z-transform pairs listed in Table 3.2. This is illustrated in the next example.

Example 3.5 (Inverse z-transform) We want to compute the inverse z-transform of

$$X(z) = \frac{z}{(z-a)(z-b)}, \tag{3.2}$$

with $a, b \in \mathbb{C}$, $|a| > |b|$, and existence region $|z| > |a|$. Instead of computing the inverse with (3.1), we will use Tables 3.1 and 3.2. First, we expand $X(z)$ into partial fractions, that is, we write $X(z)$ as

$$X(z) = \frac{cz}{z-a} + \frac{dz}{z-b},$$

Table 3.2. *Standard z-transform pairs*

Time signal	z-Transform	Existence region	Conditions
$\Delta(k)$	1	$z \in \mathbb{C}$	
$\Delta(k - \ell)$	$z^{-\ell}$	$z \neq 0$	$\ell \in \mathbb{Z}, \ell > 0$
$\Delta(k + \ell)$	z^{ℓ}	$z \in \mathbb{C}$	$\ell \in \mathbb{Z}, \ell \geq 0$
$a^k s(k)$	$\dfrac{z}{z - a}$	$\|z\| > \|a\|$	$a \in \mathbb{C}$
$a^k s(k - 1)$	$\dfrac{a}{z - a}$	$\|z\| > \|a\|$	$a \in \mathbb{C}$
$k a^k s(k - 1)$	$\dfrac{az}{(z - a)^2}$	$\|z\| > \|a\|$	$a \in \mathbb{C}$
$-a^k s(-k - 1)$	$\dfrac{z}{z - a}$	$\|z\| < \|a\|$	$a \in \mathbb{C}$
$-a^k s(-k)$	$\dfrac{a}{z - a}$	$\|z\| < \|a\|$	$a \in \mathbb{C}$

for some $c, d \in \mathbb{C}$. This expansion can also be written as

$$X(z) = \frac{(c + d)z^2 - (cb + ad)z}{(z - a)(z - b)}.$$

Comparing this with Equation (3.2) yields $c + d = 0$ and $-(cb + ad) = 1$, and, therefore,

$$c = -d = \frac{1}{a - b}.$$

Now we can write (3.2) as

$$X(z) = \frac{1}{a - b}\left(\frac{z}{z - a} - \frac{z}{z - b}\right).$$

With the help of Tables 3.1 and 3.2, we get

$$x(k) = \frac{1}{a - b}\left(a^k - b^k\right)s(k).$$

3.3.2 The discrete-time Fourier transform

Another useful signal transform is the discrete-time Fourier transform. This transform can be obtained from the z-transform by taking $z = e^{j\omega T}$ with $-\pi/T \leq \omega \leq \pi/T$. Hence, ωT is the arc length along the unit circle in the complex-z plane, with $\omega T = 0$ corresponding to $z = 1$, and $\omega T = \pm\pi$ corresponding to $z = -1$. This is illustrated in Figure 3.2. Hence, given the z-transform $X(z)$ of a signal $x(k)$, and assuming that

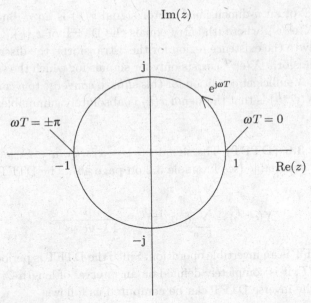

Fig. 3.2. The relation between the z-transform and the DTFT. The z-transform is defined in the existence region E_z, while the DTFT is defined only on the unit circle $z = e^{j\omega T}$.

the unit circle in the complex plane belongs to the existence region, the discrete-time Fourier transform of $x(k)$ is equal to

$$X(z)|_{z=e^{j\omega T}} = X(e^{j\omega T}).$$

Note that the z-transform is defined on its entire existence region, while the discrete-time Fourier transform is defined only on the unit circle.

Definition 3.7 (Discrete-time Fourier transform) *The discrete-time Fourier transform (DTFT) of a signal $x(k)$ is given by*

$$X(e^{j\omega T}) = \sum_{k=-\infty}^{\infty} x(k)e^{-j\omega kT}, \tag{3.3}$$

with frequency $\omega \in \mathbb{R}$ and sampling time $T \in \mathbb{R}$.

The variable ω is called the *frequency* and is expressed in radians per second. The DTFT is said to transform the signal $x(k)$ from the time-domain to the frequency domain. The DTFT is a continuous function of ω that is periodic with period $2\pi/T$, that is,

$$X\left(e^{j(\omega+2\pi/T)T}\right) = X(e^{j\omega T}).$$

The DTFT of an n-dimensional vector signal $x(k)$ is an n-dimensional function $X(\mathrm{e}^{\mathrm{j}\omega T})$, whose ith entry equals the DTFT of $x_i(k)$.

Similarly to the existence region for the z-transform, the discrete-time Fourier transform $X(\mathrm{e}^{\mathrm{j}\omega T})$ exists only for signals for which the sum (3.3) converges. A sufficient condition for this sum to converge to a continuous function $X(\mathrm{e}^{\mathrm{j}\omega T})$ is that the signal $x(k)$ is absolutely summable, that is, $||x||_1 < \infty$.

Example 3.6 (DTFT) The signal $x(k) = a^k s(k)$, $a \in \mathbb{C}$, $|a| < 1$, is absolutely summable (see Example 3.2 on page 46). The DTFT of $x(k)$ is given by

$$X(\mathrm{e}^{\mathrm{j}\omega T}) = \sum_{k=0}^{\infty} a^k \mathrm{e}^{-\mathrm{j}\omega kT} = \frac{1}{1 - a\mathrm{e}^{-\mathrm{j}\omega T}}.$$

The DTFT is an invertible operation. Since the DTFT is periodic with period $2\pi/T$, it is completely defined on an interval of length $2\pi/T$, and therefore the inverse DTFT can be computed as follows.

Theorem 3.2 (Inverse discrete-time Fourier transform) *The inverse discrete-time Fourier transform of a periodic function $X(\mathrm{e}^{\mathrm{j}\omega T})$ with period $2\pi/T$ is given by*

$$x(k) = \frac{T}{2\pi} \int_{-\frac{\pi}{T}}^{\frac{\pi}{T}} X(\mathrm{e}^{\mathrm{j}\omega T}) \mathrm{e}^{\mathrm{j}\omega kT} \, \mathrm{d}\omega,$$

with time $k \in \mathbb{Z}$ and sampling time $T \in \mathbb{R}$.

An important relation for the DTFT is Plancherel's identity.

Theorem 3.3 (Plancherel's identity) *Given the signals $x(k)$ and $y(k)$, with corresponding DTFTs $X(\mathrm{e}^{\mathrm{j}\omega T})$ and $Y(\mathrm{e}^{\mathrm{j}\omega T})$, it holds that*

$$\sum_{k=-\infty}^{\infty} x(k) y^*(k) = \frac{T}{2\pi} \int_{-\frac{\pi}{T}}^{\frac{\pi}{T}} X(\mathrm{e}^{\mathrm{j}\omega T}) Y^*(\mathrm{e}^{\mathrm{j}\omega T}) \mathrm{d}\omega, \qquad (3.4)$$

where $Y^*(\mathrm{e}^{\mathrm{j}\omega T})$ is the complex conjugate of $Y(\mathrm{e}^{\mathrm{j}\omega T})$.

In Exercise 3.4 on page 83 you are asked to prove this relation. A special case of Theorem 3.3, in which $y(k) = x(k)$, shows that the energy of $x(k)$ can also be calculated using the DTFT of $x(k)$, that is,

$$||x||_2^2 = \sum_{k=-\infty}^{\infty} |x(k)|^2 = \frac{T}{2\pi} \int_{-\frac{\pi}{T}}^{\frac{\pi}{T}} |X(\mathrm{e}^{\mathrm{j}\omega T})|^2 \, \mathrm{d}\omega. \qquad (3.5)$$

Table 3.3. *Properties of the discrete-time Fourier transform for scalar signals*

Property	Time signal	DTFT	Conditions
Linearity	$ax(k) + by(k)$	$aX(e^{j\omega T}) + bY(e^{j\omega T})$	$a, b \in \mathbb{C}$
Time reversal	$x(-k)$	$X(e^{-j\omega T})$	
Multiple shift	$x(k - \ell)$	$e^{-j\ell\omega T} X(e^{j\omega T})$	$\ell \in \mathbb{Z}$
Multiplication by k	$kx(k)$	$\dfrac{j}{T}\dfrac{d}{d\omega} X(e^{j\omega T})$	
Convolution	$x(k) * y(k)$	$X(e^{j\omega T})Y(e^{j\omega T})$	

Table 3.4. *Properties of the Dirac delta function*

Property	Expression	Conditions
Symmetry	$\delta(-t) = \delta(t)$	
Scaling	$\delta(at) = \dfrac{1}{|a|}\delta(t)$	$a \in \mathbb{R},\ a \neq 0$
Derivative of unit step	$\dfrac{d}{dt}s(t) = \delta(t)$	
Multiplication	$x(t)\delta(t - \tau) = x(\tau)\delta(t - \tau)$	$\tau \in \mathbb{R}$
Sifting	$\displaystyle\int_{-\infty}^{\infty} \delta(t - \tau)x(t)dt = x(\tau)$	$\tau \in \mathbb{R}$

This relation is know as Parseval's identity. Table 3.3 lists some other important properties of the DTFT.

For signals that are not absolutely summable, we can still define the DTFT, if we use generalized functions. For this purpose, we introduce the *Dirac delta function*.

Definition 3.8 *The Dirac delta function $\delta(t)$ is the generalized function such that $\delta(t) = 0$ for all $t \in \mathbb{R}$ except for $t = 0$ and*

$$\int_{-\infty}^{\infty} \delta(t)dt = 1.$$

The Dirac delta function can be thought of as an impulse having infinite magnitude at time instant zero. Some important properties of the Dirac delta function are listed in Table 3.4.

Table 3.5. *Standard discrete-time Fourier-transform pairs*

Time signal	DTFT	Conditions
1	$\dfrac{2\pi}{T}\delta(\omega)$	
$\Delta(k)$	1	
$\Delta(k-\ell)$	$\mathrm{e}^{-\mathrm{j}\omega T\ell}$	$\ell \in \mathbb{Z}$
$a^k s(k)$	$\dfrac{1}{1 - a\mathrm{e}^{-\mathrm{j}\omega T}}$	$a \in \mathbb{C},\ \|a\| < 1$
$\sin(ak)$	$\mathrm{j}\dfrac{\pi}{T}\delta\left(\omega + \dfrac{a}{T}\right) - \mathrm{j}\dfrac{\pi}{T}\delta\left(\omega - \dfrac{a}{T}\right)$	$a \in \mathbb{R}$
$\cos(ak)$	$\dfrac{\pi}{T}\delta\left(\omega + \dfrac{a}{T}\right) + \dfrac{\pi}{T}\delta\left(\omega - \dfrac{a}{T}\right)$	$a \in \mathbb{R}$

With the Dirac delta function it is now possible to define the DTFT of a signal for which the 1-norm is infinite. This is illustrated in the example below.

Example 3.7 (DTFT) The DTFT of the signal $x(k) = 1$, which is clearly not absolutely summable, is given by

$$X(\mathrm{e}^{\mathrm{j}\omega T}) = \frac{2\pi}{T}\delta(\omega).$$

This is easily verified using the inverse DTFT and the properties of the Dirac delta function:

$$x(k) = \int_{-\frac{\pi}{T}}^{\frac{\pi}{T}} \delta(\omega)\mathrm{e}^{\mathrm{j}\omega kT}\,\mathrm{d}\omega = 1.$$

We conclude this section by listing a number of standard DTFT pairs in Table 3.5 and giving a final example.

Example 3.8 (DTFT of a rectangular window) We compute the DTFT of a rectangular window $x(k)$ defined as

$$x(k) = \begin{cases} 1, & k = 0, 1, 2, \ldots, N - 1, \\ 0, & \text{otherwise}, \end{cases}$$

for $N \in \mathbb{Z}$, $N > 0$. We have

$$X(\mathrm{e}^{\mathrm{j}\omega T}) = \sum_{k=0}^{N-1} \mathrm{e}^{-\mathrm{j}\omega kT}.$$

To evaluate this sum, we first prove that, for $z \in \mathbb{C}$, the following sum holds:

$$\sum_{k=0}^{N-1} z^k = \begin{cases} N, & z = 1, \\ \dfrac{1 - z^N}{1 - z}, & z \neq 1. \end{cases} \qquad (3.6)$$

The relation for $z = 1$ is obvious; the relation for $z \neq 1$ is easily proven by induction. Clearly, for $N = 1$ Equation (3.6) holds. Now assume that for $N = n$ Equation (3.6) holds. By showing that with this assumption it also holds for $N = n + 1$, we in fact show that it holds for all n:

$$\sum_{k=0}^{n} z^k = z^n + \sum_{k=0}^{n-1} z^k = z^n + \frac{1 - z^n}{1 - z} = \frac{1 - z^{n+1}}{1 - z}.$$

Now we can write $X(\mathrm{e}^{j\omega T})$ as follows:

$$X(\mathrm{e}^{j\omega T}) = \begin{cases} N, & \omega = n\dfrac{2\pi}{T}, n \in \mathbb{Z}, \\ \dfrac{1 - \mathrm{e}^{-j\omega T N}}{1 - \mathrm{e}^{-j\omega T}}, & \text{otherwise.} \end{cases}$$

This can be simplified using

$$\frac{1 - \mathrm{e}^{-j\omega T N}}{1 - \mathrm{e}^{-j\omega T}} = \frac{\mathrm{e}^{jN\omega T/2} - \mathrm{e}^{-jN\omega T/2}}{\mathrm{e}^{j\omega T/2} - \mathrm{e}^{-j\omega T/2}} \mathrm{e}^{-j(N-1)\omega T/2}$$

$$= \frac{\sin(N\omega T/2)}{\sin(\frac{1}{2}\omega T)} \mathrm{e}^{-j(N-1)\omega T/2}.$$

Since l'Hôpital's rule shows that

$$\lim_{\omega \to 0} \frac{\sin(N\omega T/2)}{\sin(\frac{1}{2}\omega T)} = \lim_{\omega \to 0} N \frac{\cos(N\omega T/2)}{\cos(\frac{1}{2}\omega T)} = N,$$

the DTFT $X(\mathrm{e}^{j\omega T})$ can be denoted compactly by

$$X(\mathrm{e}^{j\omega T}) = \frac{\sin(N\omega T/2)}{\sin(\frac{1}{2}\omega T)} \mathrm{e}^{-j(N-1)\omega T/2},$$

for all $\omega \in \mathbb{R}$.

3.4 Linear systems

A system converts a certain set of signals, called input signals, into another set of signals, called output signals. Most systems are *dynamical systems*. The word dynamical refers to the fact that the system has some memory, so that the current output signal is influenced by its own time history and the time history of the input signal. In this book we will look at a particular class of dynamical systems, called discrete-time state-space systems. In addition to the input and output signals, these systems have a third signal that is called the *state*. This is an internal signal of the system, which can be thought of as the memory of the system. We formally define a discrete-time state-space system as follows.

Definition 3.9 *Let $u(k) \in \mathcal{U} \subset \mathbb{R}^m$ be a vector of input signals, let $y(k) \in \mathcal{Y} \subset \mathbb{R}^\ell$ be a vector of output signals, and let $x(k) \in \mathcal{X} \subset \mathbb{R}^n$ be a vector of the states of the system. $\mathcal{U}$ is called the input space, $\mathcal{Y}$ the output space, and $\mathcal{X}$ the state space. A discrete-time state-space system can be represented as follows:*

$$x(k+1) = f\Big(k, x(k), u(k)\Big), \tag{3.7}$$

$$y(k) = h\Big(k, x(k), u(k)\Big), \tag{3.8}$$

with $k \in \mathcal{T} \subset \mathbb{Z}$, $f : \mathcal{T} \times \mathcal{X} \times \mathcal{U} \to \mathcal{X}$ and $h : \mathcal{T} \times \mathcal{X} \times \mathcal{U} \to \mathcal{Y}$.

Equation (3.7), the *state equation*, is a difference equation that describes the time evolution of the state $x(k)$ of the system. It describes how to determine the state at time instant $k + 1$, given the state and the input at time instant k. This equation shows that the system is *causal*. For a causal system the output at a certain time instant k does not depend on the input signal at time instants later than k. A simple example of a noncausal system is $y(k) = x(k) = u(k + 1)$. Equation (3.8), the *output equation*, describes how the value of the output at a certain time instant depends on the values of the state and the input at that particular time instant.

Note that in Definition 3.9 the input and output are both vector signals. Such a system is called a multiple-input, multiple-output system, or *MIMO system* for short. If the input and output are scalar signals, we often refer to the system as a *SISO system*, where SISO stands for single-input, single-output.

An important class of system is the class of *time-invariant systems*.

Definition 3.10 (Time-invariance) *The system (3.7)–(3.8) is time-invariant if, for every triple of signals $(u(k), x(k), y(k))$ that satisfies (3.7) and (3.8) for all $k \in \mathbb{Z}$, also the time-shifted signals $(u(k-\ell), x(k-\ell), y(k-\ell))$ satisfy (3.7) and (3.8) for any $\ell \in \mathbb{Z}$.*

In other words, a time-invariant system is described by functions f and h that do not change over time:

$$x(k+1) = f\Big(x(k), u(k)\Big), \tag{3.9}$$

$$y(k) = h\Big(x(k), u(k)\Big). \tag{3.10}$$

Another important class of systems is the class of linear systems. A linear system possesses a lot of interesting properties. To define a linear system, we first need the definition of a linear function.

Definition 3.11 *The function $f : \mathbb{R}^m \to \mathbb{R}^n$ is linear if, for any two vectors $x_1, x_2 \in \mathbb{R}^m$ and any $\alpha, \beta \in \mathbb{R}$,*

$$f(\alpha x_1 + \beta x_2) = \alpha f(x_1) + \beta f(x_2).$$

Definition 3.12 (Linear system) *The state-space system (3.7)–(3.8) is a linear system if the functions f and h are linear functions with respect to $x(k)$ and $u(k)$.*

A linear state-space system that is time varying can be represented as follows:

$$x(k+1) = A(k)x(k) + B(k)u(k), \tag{3.11}$$
$$y(k) = C(k)x(k) + D(k)u(k), \tag{3.12}$$

with $A(k) \in \mathbb{R}^{n \times n}$, $B(k) \in \mathbb{R}^{n \times m}$, $C(k) \in \mathbb{R}^{\ell \times n}$, and $D(k) \in \mathbb{R}^{\ell \times m}$. Note the linear dependence on $x(k)$ and $u(k)$ both in the state equation and in the output equation. Such a system is often called an LTV (*linear time-varying*) state-space system. If the system matrices $A(k)$, $B(k)$, $C(k)$, and $D(k)$ do not depend on the time k, the system is time-invariant. A *linear time-invariant* (LTI) state-space system can thus be represented as

$$x(k+1) = Ax(k) + Bu(k), \tag{3.13}$$
$$y(k) = Cx(k) + Du(k). \tag{3.14}$$

In this book our attention will be mainly focused on LTI systems. Therefore, in the remaining part of this section we give some properties of LTI systems.

3.4.1 Linearization

An LTI system can be used to approximate a time-invariant nonlinear system like (3.9)–(3.10) in the vicinity of a certain operating point. Let the operating point be given by $(\overline{u}, \overline{x}, \overline{y})$, such that $\overline{x} = f(\overline{x}, \overline{u})$ and $\overline{y} = g(\overline{x}, \overline{u})$. Then we can approximate the function $f(x(k), u(k))$ for $x(k)$ close to $\overline{x}$ and $u(k)$ close to $\overline{u}$ with a first-order Taylor-series expansion as follows:

$$f\Big(x(k), u(k)\Big) \approx f(\overline{x}, \overline{u}) + A\Big(x(k) - \overline{x}\Big) + B\Big(u(k) - \overline{u}\Big),$$

where

$$A = \frac{\partial f}{\partial x^{\mathrm{T}}}(\overline{x}, \overline{u}), \qquad B = \frac{\partial f}{\partial u^{\mathrm{T}}}(\overline{x}, \overline{u}).$$

A similar thing can be done for the output equation. We can approximate the function $h(x(k), u(k))$ for $x(k)$ close to $\overline{x}$ and $u(k)$ close to $\overline{u}$ as

$$h\Big(x(k), u(k)\Big) \approx h(\overline{x}, \overline{u}) + C\Big(x(k) - \overline{x}\Big) + D\Big(u(k) - \overline{u}\Big),$$

where

$$C = \frac{\partial h}{\partial x^{\mathrm{T}}}(\overline{x}, \overline{u}), \qquad D = \frac{\partial h}{\partial u^{\mathrm{T}}}(\overline{x}, \overline{u}).$$

Taking $\widetilde{x}(k) = x(k) - \overline{x}$, we see that the system (3.9)–(3.10) can be approximated by the system

$$\widetilde{x}(k+1) = A\widetilde{x}(k) + B\Big(u(k) - \overline{u}\Big),$$

$$y(k) = C\widetilde{x}(k) + D\Big(u(k) - \overline{u}\Big) + \overline{y}.$$

If we define the signals $\widetilde{u}(k) = u(k) - \overline{u}$ and $\widetilde{y}(k) = y(k) - \overline{y}$ which describe the differences between the actual values of the input and output and their corresponding values $\overline{u}$ and $\overline{y}$ at the operating point, we get

$$\widetilde{x}(k+1) = A\widetilde{x}(k) + B\widetilde{u}(k),$$

$$\widetilde{y}(k) = C\widetilde{x}(k) + D\widetilde{u}(k).$$

This system describes the input–output behavior of the system (3.9)–(3.10) as the difference from the nominal behavior at the point $(\overline{u}, \overline{x}, \overline{y})$.

Example 3.9 (Linearization) Consider the nonlinear system

$$x(k+1) = f\Big(x(k)\Big),$$

$$y(k) = x(k),$$

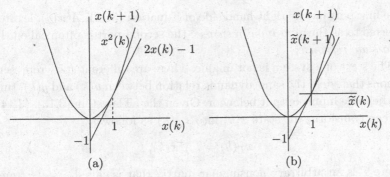

Fig. 3.3. (a) Linearization of $f(x) = x^2$ in the operating point $(1, 1)$. (b) Local coordinates $\widetilde{x}(k)$ with respect to the operating point $(1, 1)$.

where $x(k)$ is a scalar signal and $f(x) = x^2$. For $\overline{x} = 1$ we have $\overline{x} = f(\overline{x}) = 1$. It is easy to see that

$$x(k + 1) \approx f(\overline{x}) + \frac{\partial f}{\partial x}(\overline{x}) \left(x(k) - \overline{x} \right) = 1 + 2\left(x(k) - 1 \right) = 2x(k) - 1.$$

This is illustrated in Figure 3.3(a). Taking $\widetilde{x}(k) = x(k) - \overline{x}$ yields the system $\widetilde{x}(k + 1) \approx 2\widetilde{x}(k)$. This system describes the behavior of $x(k)$ relative to the point $\overline{x}$. This is illustrated in Figure 3.3(b).

3.4.2 System response and stability

The response of an LTI system at time instant k to an initial state $x(0)$ and an input signal from time 0 to k can be found from the state equation by recursion:

$$x(1) = Ax(0) + Bu(0),$$
$$x(2) = A^2x(0) + ABu(0) + Bu(1),$$
$$x(3) = A^3x(0) + A^2Bu(0) + ABu(1) + Bu(2),$$

$$\vdots$$

$$x(k) = A^kx(0) + A^{k-1}Bu(0) + \cdots + ABu(k - 2) + Bu(k - 1).$$

Or, equivalently, the response from time instant k to time instant $k + j$ is given by

$$x(k + j) = A^kx(j) + \sum_{i=0}^{k-1} A^{k-i-1}Bu(i + j). \qquad (3.15)$$

The first part on the right-hand side of Equation (3.15), $A^k x(j)$, is often referred to as the *zero-input response*; the second part is often called the *zero-state response*.

The state of a system is not unique. There are different state representations that yield the same dynamic relation between $u(k)$ and $y(k)$, that is, the same input–output behavior. Given the LTI system (3.13)–(3.14), we can transform the state $x(k)$ into $x_T(k)$ as follows:

$$x_T(k) = T^{-1}x(k),$$

where T is an arbitrary nonsingular matrix that is called a *state transformation* or a *similarity transformation*. The system that corresponds to the transformed state is given by

$$x_T(k+1) = A_T x_T(k) + B_T u(k), \tag{3.16}$$
$$y(k) = C_T x_T(k) + D_T u(k), \tag{3.17}$$

where

$$A_T = T^{-1}AT, \qquad B_T = T^{-1}B,$$
$$C_T = CT, \qquad\qquad D_T = D.$$

If the matrix A has n linearly independent eigenvectors, we can use these as the columns of the similarity transformation T. In this case the transformed A matrix will be diagonal with the eigenvalues of the matrix A as its entries. This special representation of the system is called the *modal form*. Systems that have an A matrix that is not diagonalizable can also be put into a special form similar to the *modal form* called the Jordan normal form (Kwakernaak and Sivan, 1991). We will, however, not discuss this form in this book.

Using an eigenvalue decomposition of the matrix A, the zero-input response of the system (3.13)–(3.14) can be decomposed as

$$x(k) = A^k x(0) = V\Lambda^k V^{-1} x(0) = \sum_{i=1}^{n} \alpha_i \lambda_i^k v_i, \tag{3.18}$$

where the matrix V consists of column vectors v_i that are the eigenvectors of the matrix A that correspond to the ith eigenvector of A and α_i is a scalar obtained by multiplying the ith row of the matrix V^{-1} by $x(0)$. The scalars α_i are the coefficients of a linear expansion of $x(0)$ in the vectors $v_1, v_2, \ldots, v_n$. The quantities $\lambda_i^k v_i$ are called the *modes* of the system. For this reason, the decomposition (3.18) is called the *modal form*; it is a linear combination of the modes of the system. A mode $\lambda_i^k v_i$ is said to be *excited* if the corresponding coefficient α_i is nonzero.

Note that the eigenvalues λ_i can be complex numbers, and that the corresponding modes are then also complex-valued. If the A matrix of the system is real-valued, the complex-valued eigenvalues always come as conjugate pairs (see Lemma 2.4 on page 22). We can combine two complex-valued modes into a real-valued expression as follows. Let λ be a complex-valued eigenvalue with corresponding eigenvector v and expansion coefficient α. The complex conjugate of λ, denoted by $\overline{\lambda}$, is also an eigenvalue of the system; its corresponding eigenvector is $\overline{v}$ and its expansion coefficient is $\overline{\alpha}$. Let $\alpha = |\alpha|e^{j\psi}$, $\lambda = |\lambda|e^{j\phi}$, and $v = r + js$. Then it follows that

$$\alpha\lambda^k v + \overline{\alpha}\overline{\lambda}^k\overline{v} = |\alpha||\lambda|^k e^{j(\psi+k\phi)}(r+js) + |\alpha||\lambda|^k e^{-j(\psi+k\phi)}(r-js)$$
$$= 2|\alpha||\lambda|^k \Big(r\cos(k\phi+\psi) - s\sin(k\phi+\psi)\Big).$$

The last expression is clearly real-valued. Because of the presence of the sine and cosine functions, it follows that the complex-valued eigenvalues give rise to oscillatory behavior.

The modal expansion (3.18) tells us something about the nature of the response of the system. There are three important cases:

(i) $|\lambda_i| < 1$, the corresponding mode decreases exponentially over time;

(ii) $|\lambda_i| > 1$, the corresponding mode increases exponentially over time without bound; and

(iii) $|\lambda_i| = 1$, the corresponding mode is bounded or increases over time.

Thus, the eigenvalues of the A matrix determine whether the state of the system is bounded when time increases. If the state is bounded, the system is called *stable*. Stability of an LTI system without inputs

$$x(k+1) = Ax(k), \tag{3.19}$$
$$y(k) = Cx(k), \tag{3.20}$$

is formally defined as follows.

Definition 3.13 (Stability) *The system (3.19)–(3.20) is stable for $k \geq k_0$ if, to each value of $\epsilon > 0$, however small, there corresponds a value of $\delta > 0$ such that $\|x(k_0)\|_2 < \epsilon$ implies that $\|x(k_1)\|_2 < \delta$ for all $k_1 \geq k_0$.*

Obviously, the modes of the system with eigenvalues of magnitude larger than unity keep growing over time and hence make the state unbounded and the system unstable. Modes with eigenvalues of magnitude smaller than unity go to zero and make the system stable. What about the modes with eigenvalues of magnitude exactly equal to unity? It turns out that these modes result in a bounded state only when they have a complete set of linearly independent eigenvectors. This will be illustrated by an example.

Example 3.10 (Stability) Given the matrix

$$A = \begin{bmatrix} \lambda & 1 \\ 0 & \lambda \end{bmatrix},$$

with $\lambda \in \mathbb{R}$, it is easy to see that the two eigenvalues of this matrix are both equal to λ. Solving

$$\begin{bmatrix} \lambda & 1 \\ 0 & \lambda \end{bmatrix} \begin{bmatrix} v_1 \\ v_2 \end{bmatrix} = \lambda \begin{bmatrix} v_1 \\ v_2 \end{bmatrix}$$

shows that $v_2 = 0$ defines the only eigenvector. As a consequence, the matrix A cannot be decomposed into the product of a matrix V containing all the eigenvectors of A, a diagonal matrix containing the eigenvalues on its diagonal, and the inverse of V as in Equation (3.18). To determine the response of the system $x(k+1) = Ax(k)$ to a nonzero initial state vector, we need to find an expression for A^k. Observe that

$$A^2 = \begin{bmatrix} \lambda & 1 \\ 0 & \lambda \end{bmatrix} \begin{bmatrix} \lambda & 1 \\ 0 & \lambda \end{bmatrix} = \begin{bmatrix} \lambda^2 & 2\lambda \\ 0 & \lambda^2 \end{bmatrix},$$

$$A^3 = \begin{bmatrix} \lambda^2 & 2\lambda \\ 0 & \lambda^2 \end{bmatrix} \begin{bmatrix} \lambda & 1 \\ 0 & \lambda \end{bmatrix} = \begin{bmatrix} \lambda^3 & 3\lambda^2 \\ 0 & \lambda^3 \end{bmatrix}.$$

Continuing this exercise shows that

$$A^k = \begin{bmatrix} \lambda^k & k\lambda^{k-1} \\ 0 & \lambda^k \end{bmatrix},$$

and therefore

$$x_1(k+1) = \lambda^k x_1(0) + k\lambda^{k-1} x_2(0),$$
$$x_2(k+1) = \lambda^k x_2(0).$$

We see that $x_1(k)$ is bounded only when $|\lambda| < 1$.

We can conclude that the system (3.19)–(3.20) is stable if and only if all the eigenvalues of the matrix A have magnitudes smaller than or equal to unity, and the number of independent eigenvectors corresponding to the eigenvalues of magnitude unity must equal the number of the latter eigenvalues (Kwakernaak and Sivan, 1991).

A stronger notion of stability is *asymptotic stability*, defined as follows.

Definition 3.14 (Asymptotic stability) *The system (3.19)–(3.20) is asymptotically stable if it is stable for $k \geq k_0$ and in addition there exists an $\eta > 0$ such that $\|x(k_0)\|_2 < \eta$ implies $\lim_{k \to \infty} \|x(k)\|_2 = 0$.*

If the system (3.19)–(3.20) is asymptotically stable, its state is bounded, and goes to zero with increasing time, regardless of the initial conditions. Now, the following theorem should not come as a surprise.

Theorem 3.4 *(Rugh, 1996) The system (3.19)–(3.20) is asymptotically stable if and only if all the eigenvalues of the matrix A have magnitudes strictly smaller than unity.*

It is important to note that the stability properties of an LTI system are not changed by applying a similarity transformation to the state, since such a similarity transformation does not change the eigenvalues of the A matrix.

Regarding asymptotic stability, we also have the following important result, which is called the Lyapunov stability test.

Theorem 3.5 (Lyapunov stability test) *(Rugh, 1996) The system (3.19) is asymptotically stable if and only if, for every positive-definite matrix $Q \in \mathbb{R}^{n \times n}$, there exists a unique positive-definite matrix $P \in \mathbb{R}^{n \times n}$ such that $P - APA^{\mathrm{T}} = Q$.*

The proof of this theorem can be found in Kailath (1980), for example. To get a feeling for this theorem, consider the scalar case: $p - apa = (1 - a^2)p = q$. If the system is asymptotically stable, $|a| < 1$, then $(1 - a^2)$ is positive, so there exist positive scalars p and q, such that $(1 - a^2)p = q$. Furthermore, given q, the scalar p is uniquely determined. If the system is not asymptotically stable, $|a| \geq 1$, then $(1 - a^2)$ is either zero or negative. For a negative $(1 - a^2)$, it is not possible to have two positive scalars p and q such that $(1 - a^2)p = q$ holds. If $(1 - a^2) = 0$ then $(1 - a^2)p = q$ holds for any p, positive or negative, and $q = 0$.

The previous theorems deal solely with systems without inputs. Adding an input to the system can change the response of the system drastically. To be able to determine whether the output of an LTI system with inputs remains bounded over time, the concept of bounded-input, bounded-output (BIBO) stability is introduced.

Definition 3.15 (Bounded-input, bounded-output stability)
(Rugh, 1996) The system (3.13)–(3.14) is bounded-input, bounded-output stable if there exists a finite constant η such that for any j and any input signal $u(k)$ the corresponding response $y(k)$ with $x(j) = 0$ satisfies

$$\sup_{k \geq j} \|y(k)\|_2 \leq \eta \sup_{k \geq j} \|u(k)\|_2.$$

We have the following result.

Theorem 3.6 *(Rugh, 1996) The system (3.13)–(3.14) is bounded-input, bounded-output stable if the eigenvalues of the matrix A have magnitudes smaller than unity.*

For the system (3.13)–(3.14) to be bounded-input, bounded-output stable, we need the zero-state response to be bounded. Taking $x(j) = 0$ in Equation (3.15) on page 59, we see that the zero-state response consists of a sum of terms of the form $A^j Bu(k)$. Each of these terms can be expressed as a linear combination of the modes of the system:

$$A^j Bu(k) = \sum_{i=1}^{n} \beta_i(k) \lambda_i^j v_i,$$

where v_i are the eigenvectors of the matrix A and β_i are the coefficients of a linear expansion of $Bu(k)$ in the vectors $v_1, v_2, \ldots, v_n$. Therefore, $\|A^j Bu(k)\|_2 < \infty$ if all the eigenvalues of the matrix A have magnitudes smaller than unity.

3.4.3 Controllability and observability

Depending on the nature of the system, some of the components of the state vector $x(k)$ might not be influenced by the input vector $u(k)$ over time (see Equation (3.13) on page 57). If the input can be used to steer any state of the system to the zero state within a finite time interval, the system is said to be *controllable*. The formal definition of a controllable system is as follows.

Definition 3.16 (Controllability) *The LTI system (3.13)–(3.14) is controllable if, given any initial state $x(k_a)$, there exists an input signal $u(k)$ for $k_a \leq k \leq k_b$ such that $x(k_b) = 0$ for some k_b.*

The problem with the concept of controllability in discrete time is that certain systems are controllable while the input cannot actually be used to steer the state. A simple example is a system that has $A = 0$ and $B = 0$; any initial state will go to zero, but the input does not influence the state at all. Therefore, a stronger notion of controllability, called *reachability*, is often used.

Definition 3.17 (Reachability) *The LTI system (3.13)–(3.14) is reachable if for any two states x_a and x_b there exists an input signal $u(k)$ for $k_a \leq k \leq k_b$ that will transfer the system from the state $x(k_a) = x_a$ to $x(k_b) = x_b$.*

A trivial example of a nonreachable system is a system that has a state equation that is not influenced by the input at all, that is, $x(k + 1) = Ax(k)$. Reachability implies controllability, but not necessarily vice versa. Only if the matrix A is invertible does controllability imply reachability.

The reachability of the LTI system (3.13)–(3.14) can be determined from the rank of the matrix

$$\mathcal{C}_n = \begin{bmatrix} B & AB & \cdots & A^{n-1}B \end{bmatrix}.$$

In the literature this matrix is called the *controllability matrix*. Although "reachability matrix" would seem to be a more appropriate name, we adopt the more commonly used name "controllability matrix."

Lemma 3.1 (Reachability rank condition) *(Rugh, 1996) The LTI system (3.13)–(3.14) is reachable if and only if*

$$\text{rank}(\mathcal{C}_n) = n. \tag{3.21}$$

Proof Using Equation (3.15) on page 59, we can write

$$\underbrace{\begin{bmatrix} B & AB & \cdots & A^{n-1}B \end{bmatrix}}_{\mathcal{C}_n} \begin{bmatrix} u(j + n - 1) \\ u(j + n - 2) \\ \vdots \\ u(j) \end{bmatrix} = x(n + j) - A^n x(j). \tag{3.22}$$

If the rank condition for $\mathcal{C}_n$ is satisfied, we can determine an input sequence that takes the system from state $x(j)$ to state $x(n + j)$ as follows:

$$\begin{bmatrix} u(j+n-1) \\ u(j+n-2) \\ \vdots \\ u(j) \end{bmatrix} = \mathcal{C}_n^{\mathrm{T}}(\mathcal{C}_n\mathcal{C}_n^{\mathrm{T}})^{-1}\Big(x(n+j) - A^n x(j)\Big).$$

Note that in this argument we have taken n steps between the initial state $x(j)$ and the final state $x(n+j)$. For a system with one input ($m = 1$), it is clear that the rank condition cannot hold for fewer than n steps; for multiple inputs ($m > 1$) it might be possible that it holds. Thus, in general we cannot take fewer than n steps. Consideration of more than n steps is superfluous, because Theorem 2.2 on page 21 (Cayley–Hamilton) can be used to express $A^{n+k}B$ for $k \geq 0$ as a linear combination of $B, AB, \ldots, A^{n-1}B$.

The "only" part of the proof is established by a contradiction argument. If the rank condition does not hold, there exists a nonzero vector $\bar{x}$ such that $\bar{x}^{\mathrm{T}}\mathcal{C}_n = 0$. Now, suppose that the system is reachable, then there is an input sequence $u(j), u(j+1), \ldots, u(j+n-1)$ in Equation (3.22) that steers the state from $x(j)$ to $x(n+j)$. If we take $\bar{x} = x(n+j) - A^k x(j)$ different from zero and multiply both sides of Equation (3.22) on the left by $\bar{x}^{\mathrm{T}}$, we get $\bar{x}^{\mathrm{T}}\bar{x} = 0$. This implies $\bar{x} = 0$; which is a contradiction to the fact that $\bar{x}$ should be nonzero. $\qquad\square$

Example 3.11 (Reachability) Consider the system

$$\begin{bmatrix} x_1(k+1) \\ x_2(k+1) \end{bmatrix} = \begin{bmatrix} a_1 & 0 \\ 0 & a_2 \end{bmatrix} \begin{bmatrix} x_1(k) \\ x_2(k) \end{bmatrix} + \begin{bmatrix} b \\ 0 \end{bmatrix} u(k), \tag{3.23}$$

$$y(k) = \begin{bmatrix} c & 0 \end{bmatrix} \begin{bmatrix} x_1(k) \\ x_2(k) \end{bmatrix}. \tag{3.24}$$

with a_1, a_2, b and c real scalars. Since the input $u(k)$ influences only the state $x_1(k)$ and since there is no coupling between the states $x_1(k)$ and $x_2(k)$, the state $x_2(k)$ cannot be steered and hence the system is not reachable. The controllability matrix is given by

$$\mathcal{C}_n = \begin{bmatrix} b & a_1 b \\ 0 & 0 \end{bmatrix}.$$

It is clear that this matrix does not have full rank.

Another useful test for reachability is the following.

Lemma 3.2 (Popov–Belevitch–Hautus reachability test) *(Kailath, 1980) The LTI system (3.13)–(3.14) is reachable if and only if, for all $\lambda \in \mathbb{C}$ and $x \in \mathbb{C}^n$, $x \neq 0$, such that $A^T x = \lambda x$, it holds that $B^T x \neq 0$.*

With the definition of reachability, we can specialize Lyapunov's stability test in Theorem 3.5 to the following.

Theorem 3.7 (Lyapunov stability for reachable systems)*(Kailath, 1980) If the system (3.13)–(3.14) is asymptotically stable and reachable, then there exists a unique positive-definite matrix $P \in \mathbb{R}^{n \times n}$ such that $P - APA^T = BB^T$.*

Equation (3.14) on page 57 shows that the output of the system is related to the state of the system, but the state is not directly observed. The dependence of the time evolution of the state on the properties of the output equation can be observed from the time evolution of the output. If there is a unique relation, in time, between state and output, the system is called observable. The formal definition of observability is given below.

Definition 3.18 (Observability) *The LTI system (3.13)–(3.14) is observable if any initial state $x(k_a)$ is uniquely determined by the corresponding zero-input response $y(k)$ for $k_a \leq k \leq k_b$ with k_b finite.*

A trivial example of a nonobservable system is a system that has an output equation that is not influenced by the state at all, that is, $y(k) = Du(k)$.

To test observability of an LTI system, the *observability matrix*, given by

$$\mathcal{O}_n = \begin{bmatrix} C \\ CA \\ \vdots \\ CA^{n-1} \end{bmatrix}, \tag{3.25}$$

is used.

Lemma 3.3 (Observability rank condition) *(Rugh, 1996) The LTI system (3.13)–(3.14) is observable if and only if*

$$\text{rank}(\mathcal{O}_n) = n. \tag{3.26}$$

Proof We can use Equation (3.15) on page 59 to derive for the zero-input response $y(k)$ the relation

$$
\mathcal{O}_n x(j) = \begin{bmatrix} C \\ CA \\ CA^2 \\ \vdots \\ CA^{n-1} \end{bmatrix} x(j) = \begin{bmatrix} y(j) \\ y(j+1) \\ y(j+2) \\ \vdots \\ y(j+n-1) \end{bmatrix}. \tag{3.27}
$$

If the rank condition holds, we can determine the initial state $x(j)$ uniquely from $y(j), y(j+1), \ldots, y(j+n-1)$ as follows:

$$
x(j) = (\mathcal{O}_n^{\mathrm{T}} \mathcal{O}_n)^{-1} \mathcal{O}_n^{\mathrm{T}} \begin{bmatrix} y(j) \\ y(j+1) \\ y(j+2) \\ \vdots \\ y(j+n-1) \end{bmatrix}.
$$

We have taken n output samples. For a system with one output ($\ell = 1$), it is clear that the rank condition cannot hold for fewer than n samples. Although, for multiple outputs ($\ell > 1$), it might be possible to have fewer than n samples, in general, we cannot take fewer than n samples. Consideration of more than n samples is superfluous, because Theorem 2.2 on page 21 (Cayley–Hamilton) can be used to express CA^{n+k} for $k \geq 0$ as a linear combination of $C, CA, \ldots, CA^{n-1}$.

To complete the proof, we next show that, if the rank condition does not hold, the LTI system is not observable. If $\mathcal{O}_n$ does not have full column rank, then there exists a nonzero vector $\overline{x}$ such that $\mathcal{O}_n \overline{x} = 0$. This means that the zero-input response is zero. However, the zero-input response is also zero for any vector $\alpha \overline{x}$, with $\alpha \in \mathbb{R}$. Thus, the states $\alpha \overline{x}$ cannot be distinguished on the basis of the zero-input response, and hence the system is not observable. $\qquad\square$

Example 3.12 (Observability) Consider the system (3.23)–(3.24) from Example 3.11 on page 66. Since the output depends solely on the state $x_1(k)$, and since there is no coupling between the states $x_1(k)$ and $x_2(k)$, the state $x_2(k)$ cannot be observed, and hence the system is not observable. The observability matrix is given by

$$
\mathcal{O}_n = \begin{bmatrix} c & 0 \\ ca_1 & 0 \end{bmatrix}.
$$

This matrix does not have full rank.

It is easy to verify that the controllability, reachability, and observability of the system do not change with a similarity transformation of the state. Furthermore, the concepts of observability and reachability are *dual* to one another. This means that, if the pair (A, C) is observable, the pair (A^T, C^T) is reachable and also that, if the pair (A, B) is reachable, the pair (A^T, B^T) is observable. By virtue of this property of duality, we can translate Lemma 3.2 and Theorem 3.7 into their dual counterparts.

Lemma 3.4 (Popov–Belevitch–Hautus observability test)
(Kailath, 1980) The LTI system (3.13)–(3.14) is observable if and only if, for all $\lambda \in \mathbb{C}$ and $x \in \mathbb{C}^n$, $x \neq 0$, such that $Ax = \lambda x$, it holds that $Cx \neq 0$.

Theorem 3.8 (Lyapunov stability for observable systems)
(Kailath, 1980) If the system (3.13)–(3.14) is asymptotically stable and observable, then there exists a unique positive-definite matrix $P \in \mathbb{R}^{n \times n}$ such that $P - A^T P A = C^T C$.

We conclude the discussion on reachability and observability by defining minimality.

Definition 3.19 (Minimality) *The LTI system (3.13)–(3.14) is minimal if it is both reachable and observable.*

The dimension of the state vector $x(k)$ of a minimal LTI system (3.13)–(3.14) is called *the order* of the LTI system.

3.4.4 Input–output descriptions

There are several ways to represent an LTI system. Up to now we have used only the state-space representation (3.13)–(3.14). In this section some other representations are derived.

From (3.13) we can write

$$x(k+1) = qx(k) = Ax(k) + Bu(k),$$

where q is the forward shift operator. Therefore, if the operator (mapping) $(qI - A)$ is boundedly invertible, then

$$x(k) = (qI - A)^{-1} Bu(k). \tag{3.28}$$

If we consider the system (3.13)–(3.14) for $k \geq k_0$ and assume that $u(k)$ is bounded for $k \geq k_0$, the mapping of $Bu(k)$ by the operator $(qI - A)^{-1}$ always exists. Under these conditions, the output satisfies,

$$y(k) = \Big(C(qI - A)^{-1}B + D\Big)u(k) = H(q)u(k).$$

This equation gives an input–output description of the system (3.13)–(3.14). The matrix $H(q)$ is called the *transfer function* of the system. Using the relation between the inverse and the determinant of a matrix, Equation (2.4) on page 20, we can write

$$H(q) = \frac{C \operatorname{adj}(qI - A)B + D \det(qI - A)}{\det(qI - A)}. \tag{3.29}$$

The entries of this matrix are *proper* rational functions of the shift operator q; for a proper rational function the degree of the numerator polynomial is no greater than the degree of the denominator polynomial. The degree of a polynomial in q equals the highest order of q with a nonzero coefficient. Every entry $H_{ij}(q)$ can be written as a quotient of two polynomials in q, that is,

$$H_{ij}(q) = \frac{F_{ij}(q)}{G(q)}, \tag{3.30}$$

where

$$F_{ij}(q) = \sum_{p=0}^{n} f_{ij,p}q^p, \qquad G(q) = \sum_{p=0}^{n} g_p q^p.$$

The roots of the polynomial $G(q)$ that do not cancel out with the roots of F_{ij} are called the *poles* of the transfer function $H_{ij}(q)$. The roots of F_{ij} are called the *zeros*. On comparing Equation (3.30) with (3.29), it follows that the poles of the system equal the eigenvalues of the A matrix if there are no pole–zero cancellations.

From Equation (3.30) we see that every output $y_i(k)$, $i = 1, 2, \ldots, \ell$, can be written as

$$y_i(k) = \sum_{j=1}^{m} \frac{F_{ij}(q)}{G(q)} u_j(k).$$

It is customary to take $g_n = 1$; with this convention we have

$$y_i(k + n) = \sum_{j=1}^{m} \sum_{p=0}^{n} f_{ij,p} u_j(k + p) - \sum_{p=0}^{n-1} g_p y_i(k + p). \tag{3.31}$$

This is called a *difference equation*. For a SISO system it simply becomes

$$y(k+n) = \sum_{p=0}^{n} f_p u(k+p) - \sum_{p=1}^{n-1} g_p y(k+p).$$

Example 3.13 (Difference equation) Consider the LTI system

$$A = \begin{bmatrix} -1.3 & -0.4 \\ 1 & 0 \end{bmatrix}, \qquad B = \begin{bmatrix} 1 \\ 0 \end{bmatrix},$$
$$C = \begin{bmatrix} -2 & 1 \end{bmatrix}, \qquad D = 2.$$

The eigenvalues of the matrix A are -0.8 and -0.5. The inverse of the mapping $(qI - A)$ can be computed with the help of Example 2.5 on page 20,

$$(qI - A)^{-1} = \begin{bmatrix} q + 1.3 & 0.4 \\ -1 & q \end{bmatrix}^{-1} = \frac{1}{q(q+1.3) + 0.4} \begin{bmatrix} q & -0.4 \\ 1 & q+1.3 \end{bmatrix},$$

and we get

$$H(q) = C(qI - A)^{-1}B + D = \frac{-2q+1}{q^2 + 1.3q + 0.4} + 2 = \frac{2q^2 + 0.6q + 1.8}{q^2 + 1.3q + 0.4}.$$

The roots of the denominator are indeed -0.8 and -0.5. With the definition of the shift operator q, the relationship $y(k) = H(q)u(k)$ can be written into the difference equation,

$$y(k+2) = 2u(k+2) + 0.6u(k+1) + 1.8u(k) - 1.3y(k+1) - 0.4y(k).$$

Provided that $(qI - A)$ is boundedly invertible, we can expand Equation (3.28) on page 69 into an infinite series, like this:

$$x(k) = (qI - A)^{-1}Bu(k) = q^{-1}(I - q^{-1}A)^{-1}Bu(k) = \sum_{i=0}^{\infty} A^i Bu(k-i-1).$$
$$\tag{3.32}$$

Therefore,

$$\begin{aligned} y(k) &= \sum_{i=0}^{\infty} CA^i Bu(k-i-1) + Du(k) \\ &\overset{(j=i+1)}{=} \sum_{j=1}^{\infty} CA^{j-1} Bu(k-j) + Du(k) \\ &= \sum_{j=0}^{\infty} h(j)u(k-j), \end{aligned} \tag{3.33}$$

where the matrix signal $h(k)$ is called the *impulse response* of the system. The name stems from the fact that, if we take $u(k)$ equal to an impulse signal at time instant zero, we have $y(k) = h(k)$. If the impulse response $h(k)$ becomes zero at a certain time instant k_0 and remains zero for $k > k_0$, the system is called a *finite-impulse-response* (FIR) system, otherwise it is called an *infinite-impulse-response* (IIR) system. Note that the (zero-state response) output $y(k)$ of the system is in fact the convolution of the impulse response and the input signal, that is, $y(k) = h(k) * u(k)$. The impulse response of the system (3.13)–(3.14) is given by

$$h(k) = \begin{cases} 0, & k = -1, -2, \ldots, \\ D, & k = 0, \\ CA^{k-1}B, & k = 1, 2, \ldots \end{cases} \tag{3.34}$$

The matrices $h(0), h(1), \ldots$, are called the *Markov parameters* of the system. Equation (3.33) clearly shows that the output at time instant k depends only on the values of the input at time instants equal to or smaller than k.

The transfer function $H(q)$ and the impulse response $h(k)$ are closely related to each other. By applying the z-transform to Equation (3.13) on page 57, we get

$$X(z) = (zI - A)^{-1}BU(z),$$

which is very similar to Equation (3.28) on page 69. Now, we can write

$$Y(z) = \Big(C(zI - A)^{-1}B + D \Big) U(z) = H(z)U(z),$$

where $H(z)$ equals the transfer function of the system, but now expressed not in the shift operator q but in the complex variable z. Next we show that, under the assumption that the system is initially at rest, $H(z)$ is the z-transform of the impulse response $h(k)$. By applying the z-transform to the output and using the expression (3.33) on page 71, we get

$$
\begin{aligned}
Y(z) \quad &= \quad \sum_{k=-\infty}^{\infty} y(k)z^{-k} \\
&= \quad \sum_{k=-\infty}^{\infty} \sum_{j=0}^{\infty} h(j)u(k-j)z^{-k} \\
&\overset{(i=k-j)}{=} \sum_{j=0}^{\infty} h(j)z^{-j} \sum_{i=-\infty}^{\infty} u(i)z^{-i}.
\end{aligned}
$$

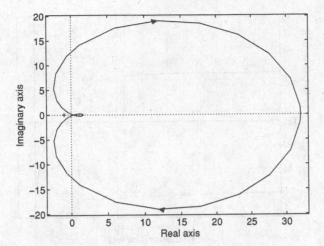

Fig. 3.4. The frequency response of the LTI system described in Example 3.13.

Since $h(k) = 0$ for $k < 0$, we have

$$Y(z) = \sum_{j=-\infty}^{\infty} h(j)z^{-j} \sum_{i=-\infty}^{\infty} u(i)z^{-i} = H(z)U(z).$$

Since the z-transform is related to the DTFT (see Section 3.3.2), we can also write $Y(e^{j\omega T}) = H(e^{j\omega T})U(e^{j\omega T})$. The complex-valued matrix $H(e^{j\omega T})$ equals the DTFT of the impulse-response matrix $h(k)$ of the system; it is called the *frequency-response function* (FRF) of the system. The magnitude of $H(e^{j\omega T})$ is called the *amplitude response* of the system; the phase of $H(e^{j\omega T})$ is called the *phase response* of the system.

Example 3.14 (Frequency-response function) The frequency-response function of the LTI system described in Example 3.13 is plotted in Figure 3.4. The corresponding amplitude and phase response are plotted in Figure 3.5.

We have seen that there are various ways to describe an LTI system and we have seen how they are related. Figure 3.6 summarizes the relations among these different representations. This figure shows how to derive an alternative system representation from a linear state-space description. What we have not considered yet is the reverse: how to derive a state-space representation given any other system representation. This problem is called the *realization* problem.

One way to obtain a state-space realization is by the use of canonical forms. Two popular canonical forms for SISO systems are the *controller*

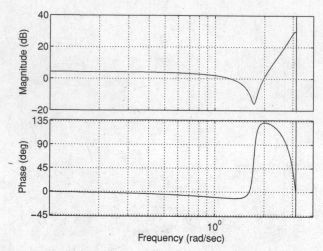

Fig. 3.5. The amplitude and phase response of the LTI system described in Example 3.13.

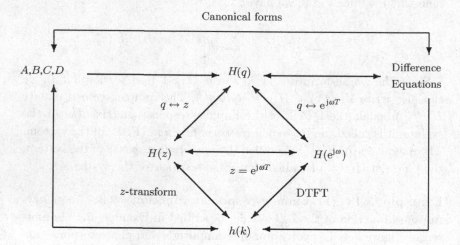

Fig. 3.6. Linear system representations and their relations.

canonical form and the *observer canonical form*. Given the transfer function

$$H(q) = \frac{b_n q^n + b_{n-1} q^{n-1} + \cdots + b_1 q + b_0}{q^n + a_{n-1} q^{n-1} + \cdots + a_1 q + a_0}$$

$$= \frac{c_{n-1} q^{n-1} + \cdots + c_1 q + c_0}{q^n + a_{n-1} q^{n-1} + \cdots + a_1 q + a_0} + d,$$

the controller canonical form is given by

$$A = \begin{bmatrix} 0 & 1 & 0 & \cdots & 0 \\ 0 & 0 & 1 & \cdots & 0 \\ \vdots & \vdots & & \ddots & \vdots \\ 0 & 0 & 0 & \cdots & 1 \\ -a_0 & -a_1 & -a_2 & \cdots & -a_{n-1} \end{bmatrix}, \quad B = \begin{bmatrix} 0 \\ 0 \\ \vdots \\ 0 \\ 1 \end{bmatrix}, \tag{3.35}$$

$$C = \begin{bmatrix} c_0 & c_1 & c_2 & \cdots & c_{n-1} \end{bmatrix}, \qquad D = d,$$

and the observer canonical form by

$$A = \begin{bmatrix} 0 & 0 & \cdots & 0 & -a_0 \\ 1 & 0 & \cdots & 0 & -a_1 \\ 0 & 1 & \cdots & 0 & -a_2 \\ \vdots & \vdots & & \ddots & \vdots \\ 0 & 0 & \cdots & 1 & -a_{n-1} \end{bmatrix}, \quad B = \begin{bmatrix} c_0 \\ c_1 \\ c_2 \\ \vdots \\ c_{n-1} \end{bmatrix},$$

$$C = \begin{bmatrix} 0 & 0 & \cdots & 0 & 1 \end{bmatrix}, \qquad D = d.$$

In Exercise 3.9 on page 84 you are asked to verify this result.

Now we turn to a second way of obtaining state-space models. Below we explain how to obtain a state-space realization (A, B, C, D) from the Markov parameters or impulse response $h(k)$ of the system. The key to this is Lemma 3.5, which is based on work of Ho and Kalman (1966) and Kung (1978).

Lemma 3.5 *Consider the LTI system (3.13)–(3.14) of order n. Define the Hankel matrix $\mathcal{H}_{n+1,n+1}$ constructed from the Markov parameters $h(k)$ given by Equation (3.34) as follows:*

$$\mathcal{H}_{n+1,n+1} = \begin{bmatrix} h(1) & h(2) & \cdots & h(n+1) \\ h(2) & h(3) & \cdots & h(n+2) \\ \vdots & \vdots & \ddots & \vdots \\ h(n+1) & h(n+2) & \cdots & h(2n+1) \end{bmatrix}. \tag{3.36}$$

Provided that the system (3.13)–(3.14) is reachable and observable, the following conditions hold:

(i) *rank*$(\mathcal{H}_{n+1,n+1}) = n$.
(ii) $\mathcal{H}_{n+1,n+1} = \mathcal{O}_{n+1}\mathcal{C}_{n+1}$.

Proof Straightforward substitution of Equation (3.34) on page 72 into the definition of $\mathcal{H}_{n+1,n+1}$ yields

$$
\mathcal{H}_{n+1,n+1} =
\begin{bmatrix}
CB & CAB & \cdots & CA^nB \\
CAB & CA^2B & \cdots & CA^{n+1}B \\
\vdots & \vdots & \ddots & \vdots \\
CA^nB & CA^{n+1}B & \cdots & CA^{2n}B
\end{bmatrix}
$$

$$
=
\begin{bmatrix}
C \\
CA \\
\vdots \\
CA^n
\end{bmatrix}
\begin{bmatrix} B & AB & \cdots & A^nB \end{bmatrix}.
$$

Because of the reachability and observability assumptions, Sylvester's inequality (Lemma 2.1 on page 16) shows that $\text{rank}(\mathcal{H}_{n+1,n+1}) = n$.

□

On the basis of this lemma, the SVD of the matrix $\mathcal{H}_{n+1,n+1}$ can be used to compute the system matrices (A, B, C, D) up to a similarity transformation T, that is, $(T^{-1}AT, T^{-1}B, CT, D) = (A_T, B_T, C_T, D_T)$. Let the SVD of $\mathcal{H}_{n+1,n+1}$ be given by

$$
\mathcal{H}_{n+1,n+1} = \begin{bmatrix} U_n & \overline{U}_n \end{bmatrix} \begin{bmatrix} \Sigma_n & 0 \\ 0 & 0 \end{bmatrix} \begin{bmatrix} V_n^T \\ \overline{V}_n^T \end{bmatrix} = U_n \Sigma_n V_n^T,
$$

with $\Sigma_n \in \mathbb{R}^{n \times n}$ and $\text{rank}(\Sigma_n) = n$, then

$$
U_n = \mathcal{O}_{n+1}T =
\begin{bmatrix}
CT \\
CTT^{-1}AT \\
\vdots \\
CT(T^{-1}AT)^n
\end{bmatrix}
=
\begin{bmatrix}
C_T \\
C_T A_T \\
\vdots \\
C_T A_T^n
\end{bmatrix},
$$

where $T = \mathcal{C}_{n+1}V_n\Sigma_n^{-1}$. Because $\text{rank}(\mathcal{H}_{n+1,n+1}) = n$, the matrix T is invertible. Hence, the matrix C_T equals the first ℓ rows of U_n, that is, $C_T = U_n(1:\ell,:)$. The matrix A_T is obtained by solving the following overdetermined equation:

$$
U_n(1:(n-1)\ell,:)A_T = U_n(\ell+1:n\ell,:).
$$

To determine the matrix B_T, observe that

$$
\Sigma_n V_n^T = T^{-1}\mathcal{C}_{n+1}
$$
$$
= \begin{bmatrix} T^{-1}B & T^{-1}ATT^{-1}B & \cdots & (T^{-1}AT)^nT^{-1}B \end{bmatrix}
$$
$$
= \begin{bmatrix} B_T & A_T B_T & \cdots & A_T^n B_T \end{bmatrix}.
$$

So B_T equals the first m columns of the matrix $\Sigma_n V_n^T$. The matrix $D_T = D$ equals $h(0)$, as indicated by Equation (3.34) on page 72.

In Lemma 3.5 on page 75 it is explicitly assumed that the sequence $h(k)$ is the impulse response of a finite-dimensional LTI system. What if we do not know where the sequence $h(k)$ comes from? In other words, what arbitrary sequences $h(k)$ can be realized by a finite-dimensional LTI system? The following lemma provides an answer to this question.

Lemma 3.6 *Given a sequence $h(k)$, $k = 1, 2, \ldots$, consider the Hankel matrix defined by Equation (3.36). If $\text{rank}(\mathcal{H}_{n+i,n+i}) = n$ for all $i = 0, 1, 2, \ldots$, the sequence $h(k)$ is the impulse response of an LTI system (3.13)–(3.14) of order n.*

Proof We give a constructive proof for SISO systems. A proof for the MIMO case can be found in Rugh (1996). Since $\text{rank}(\mathcal{H}_{n+1,n+1}) = n$, there exists a vector $v \in \mathbb{R}^{n+1}$ with a nonzero last element v_{n+1}, such that $\mathcal{H}_{n+1,n+1}v = 0$. Therefore, we can define

$$a_i = \frac{v_{i+1}}{v_{n+1}}, \quad i = 0, \ldots, n-1,$$

and thus the scalars a_i are uniquely determined by

$$\begin{bmatrix} h(1) & h(2) & \cdots & h(n+1) \\ h(2) & h(3) & \cdots & h(n+2) \\ \vdots & \vdots & \ddots & \vdots \\ h(n+1) & h(n+2) & \cdots & h(2n+1) \end{bmatrix} \begin{bmatrix} a_0 \\ a_1 \\ \vdots \\ a_{n-1} \\ 1 \end{bmatrix} = 0. \quad (3.37)$$

Next we define the scalars b_i as follows:

$$\begin{bmatrix} h(1) & 0 & \cdots & 0 \\ h(2) & h(1) & \cdots & 0 \\ \vdots & \vdots & \ddots & \vdots \\ h(n) & h(n-1) & \cdots & h(1) \end{bmatrix} \begin{bmatrix} 1 \\ a_{n-1} \\ a_{n-2} \\ \vdots \\ a_0 \end{bmatrix} = \begin{bmatrix} b_{n-1} \\ b_{n-2} \\ \vdots \\ b_0 \end{bmatrix}. \quad (3.38)$$

Now we have

$$\sum_{k=1}^{\infty} h(k) q^{-k} = \frac{b_{n-1}q^{n-1} + b_{n-2}q^{n-2} + \cdots + b_1 q + b_0}{q^n + a_{n-1}q^{n-1} + \cdots + a_1 q + a_0} = H(q),$$

which can be shown to be equivalent to (3.37) and (3.38) by equating powers of q. This final equation clearly shows that $h(k)$ is the impulse response of an LTI system with transfer function $H(q)$. This transfer function can be put into the controller canonical form (3.35) on page 75. This system is of order n by the proof of Lemma 3.5. $\quad\square$

Example 3.15 (Realization) Let the sequence $h(k)$ be defined by

$$h(0) = 0,$$
$$h(1) = 1,$$
$$h(k) = h(k-1) + h(k-2).$$

We construct the following Hankel matrices:

$$\mathcal{H}_{2,2} = \begin{bmatrix} 1 & 1 \\ 1 & 2 \end{bmatrix}, \qquad \mathcal{H}_{3,3} = \begin{bmatrix} 1 & 1 & 2 \\ 1 & 2 & 3 \\ 2 & 3 & 5 \end{bmatrix}.$$

It is easy to see that $\mathrm{rank}(\mathcal{H}_{3,3}) = \mathrm{rank}(\mathcal{H}_{2,2}) = 2$. According to Lemma 3.6, $h(k)$ is the impulse response of an LTI system. To determine this system we have to solve Equations (3.37) and (3.38). For this case Equation (3.37) becomes

$$\begin{bmatrix} 1 & 1 & 2 \\ 1 & 2 & 3 \\ 2 & 3 & 5 \end{bmatrix} \begin{bmatrix} a_0 \\ a_1 \\ 1 \end{bmatrix} = 0.$$

The solution is, of course, $a_1 = -1$ and $a_2 = -1$. Equation (3.38) equals

$$\begin{bmatrix} 1 & 0 \\ 1 & 1 \end{bmatrix} \begin{bmatrix} 1 \\ a_1 \end{bmatrix} = \begin{bmatrix} b_1 \\ b_0 \end{bmatrix}.$$

Therefore, $b_1 = 1$ and $b_0 = 0$. We get the following difference equation:

$$y(k+2) = u(k+1) + y(k) + y(k+1).$$

It is easy to verify that, on taking the input $u(k)$ equal to an impulse sequence, the sequence $y(k)$ equals $h(k)$.

3.5 Interaction between systems

In this section we look at the combination of systems. We distinguish three interconnections: parallel, cascade, and feedback. They are shown in Figures 3.7, 3.8, and 3.9, respectively. Let's take a closer look at the system descriptions of the resulting overall system for these three cases.

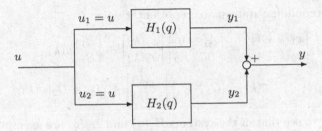

Fig. 3.7. Two systems in parallel connection.

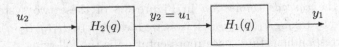

Fig. 3.8. Two systems in cascade connection.

Given two LTI systems

$$x_1(k+1) = A_1x_1(k) + B_1u_1(k),$$
$$y_1(k) = C_1x_1(k) + D_1u_1(k),$$
$$x_2(k+1) = A_2x_2(k) + B_2u_2(k),$$
$$y_2(k) = C_2x_2(k) + D_2u_2(k),$$

with corresponding transfer functions $H_1(q)$ and $H_2(q)$, if these two systems have the same number of outputs and the same number of inputs, we can make a *parallel connection* as in Figure 3.7, by adding the outputs of the two systems and taking the same input for each system, that is, $u(k) = u_1(k) = u_2(k)$. The new output is given by

$$y(k) = y_1(k) + y_2(k) = \Big(H_1(q) + H_2(q)\Big)u(k).$$

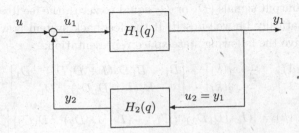

Fig. 3.9. Two systems in feedback connection.

The corresponding state-space representation is

$$\begin{bmatrix} x_1(k+1) \\ x_2(k+1) \end{bmatrix} = \begin{bmatrix} A_1 & 0 \\ 0 & A_2 \end{bmatrix} \begin{bmatrix} x_1(k) \\ x_2(k) \end{bmatrix} + \begin{bmatrix} B_1 \\ B_2 \end{bmatrix} u(k),$$

$$y(k) = \begin{bmatrix} C_1 & C_2 \end{bmatrix} \begin{bmatrix} x_1(k) \\ x_2(k) \end{bmatrix} + \begin{bmatrix} D_1 + D_2 \end{bmatrix} u(k).$$

It is easy to see that, if the systems $H_1(q)$ and $H_2(q)$ are asymptotically stable, so is their parallel connection.

If the number of inputs to $H_1(q)$ equals the number of outputs of $H_2(q)$, the *cascade connection* in Figure 3.8 of these two systems is obtained by taking $u_1(k) = y_2(k)$. Now we have $y_1(k) = H_1(q)H_2(q)u_2(k)$. The corresponding state-space representation is the following:

$$\begin{bmatrix} x_1(k+1) \\ x_2(k+1) \end{bmatrix} = \begin{bmatrix} A_1 & B_1C_2 \\ 0 & A_2 \end{bmatrix} \begin{bmatrix} x_1(k) \\ x_2(k) \end{bmatrix} + \begin{bmatrix} B_1D_2 \\ B_2 \end{bmatrix} u_2(k),$$

$$y_1(k) = \begin{bmatrix} C_1 & D_1C_2 \end{bmatrix} \begin{bmatrix} x_1(k) \\ x_2(k) \end{bmatrix} + D_1D_2u_2(k).$$

From this we see that, if the systems $H_1(q)$ and $H_2(q)$ are asymptotically stable, so is their cascade connection.

If the number of inputs to $H_2(q)$ equals the number of outputs of $H_1(q)$, and the number of inputs to $H_1(q)$ equals the number of outputs of $H_2(q)$, the *feedback connection* in Figure 3.9 of these two systems is obtained by taking $u_2(k) = y_1(k)$ and $u_1(k) = u(k) - y_2(k)$, where $u(k)$ is the input signal to the feedback-connected system. Now we can write

$$x_1(k+1) = A_1x_1(k) - B_1C_2x_2(k) + B_1u(k) - B_1D_2y_1(k),$$
$$x_2(k+1) = A_2x_2(k) + B_2y_1(k),$$
$$y_1(k) = C_1x_1(k) - D_1C_2x_2(k) + D_1u(k) - D_1D_2y_1(k).$$

The feedback system is called *well-posed* if the matrix $(I + D_1D_2)$ is invertible. Note that, if one of the systems has a delay between all of its input and output signals (D_1 or D_2 equal to zero), the feedback system will automatically be well-posed. If the feedback system is well-posed, we can derive the following state-space representation:

$$\begin{bmatrix} x_1(k+1) \\ x_2(k+1) \end{bmatrix} = \overline{A} \begin{bmatrix} x_1(k) \\ x_2(k) \end{bmatrix} + \begin{bmatrix} B_1 - B_1D_2(I + D_1D_2)^{-1}D_1 \\ B_2(I + D_1D_2)^{-1}D_1 \end{bmatrix} u(k),$$

$$y_1(k) = \begin{bmatrix} (I + D_1D_2)^{-1}C_1 & -(I + D_1D_2)^{-1}D_1C_2 \end{bmatrix} \begin{bmatrix} x_1(k) \\ x_2(k) \end{bmatrix}$$

$$+ (I + D_1D_2)^{-1}D_1u(k),$$

where

$$\overline{A} = \begin{bmatrix} A_1 - B_1 D_2 (I + D_1 D_2)^{-1} C_1 & -B_1 C_2 + B_1 D_2 (I + D_1 D_2)^{-1} D_1 C_2 \\ B_2 (I + D_1 D_2)^{-1} C_1 & A_2 - B_2 (I + D_1 D_2)^{-1} D_1 C_2 \end{bmatrix}.$$

A closer look at the matrix $\overline{A}$ reveals that stability properties can change under feedback connections. This is illustrated in the following example.

Example 3.16 (Stability under feedback) This example shows that an unstable system can be stabilized by introducing feedback. Feedback connections are very often used for this purpose. The unstable system

$$x_1(k+1) = 2x_1(k) + 7u_1(k),$$
$$y_1(k) = \frac{1}{4} x_1(k),$$

becomes stable in feedback connection with the system

$$x_2(k+1) = x_2(k) + 5u_2(k),$$
$$y_2(k) = \frac{1}{4} x_2(k).$$

The "A" matrix of the feedback-interconnected system is given by

$$\overline{A} = \begin{bmatrix} 2 & -7/4 \\ 5/4 & 1 \end{bmatrix}.$$

The eigenvalues of this matrix are $1/4$ and $3/4$, hence the feedback system is stable.

Example 3.17 (Instability under feedback) This example shows that, even when two systems are stable, their feedback connection can become unstable. Consider the stable systems

$$x_1(k+1) = 0.5x_1(k) + u_1(k),$$
$$y_1(k) = 3x_1(k),$$
$$x_2(k+1) = 0.1x_2(k) + u_2(k),$$
$$y_2(k) = -0.55x_2(k).$$

The "A" matrix of the feedback-interconnected system is given by

$$\overline{A} = \begin{bmatrix} 0.5 & 0.55 \\ 3 & 0.1 \end{bmatrix}.$$

The eigenvalues of this matrix are -1 and 1.6, hence the feedback system is unstable.

3.6 Summary

In this chapter we have reviewed some theory on discrete-time signals and systems. We showed that a discrete-time signal can be obtained by sampling a continuous-time signal. To measure the "size" of a discrete-time signal, we introduced the signal norms: the ∞-norm, the 2-norm, and the 1-norm. The z-transform was defined as a transform from a discrete-time signal to a complex function defined on the complex z-plane. From this definition the discrete-time Fourier transform (DTFT) was derived. Several properties of both transforms were given. The inverse z-transform was described and it was shown that, without specifying the existence region of the z-transformed signal, the inverse cannot be uniquely determined. To be able to compute the DTFT for sequences that are not absolutely summable, we introduced the Dirac delta function and described some of its most important properties.

After dealing with signals, the focus shifted to discrete-time systems. We introduced a general definition of a state-space system and looked at time invariance and linearity. The remaining part of the chapter dealt with linear time-invariant (LTI) systems. We defined the following properties for these systems: stability, asymptotic stability, bounded-input, bounded-output stability, controllability, and observability. It was shown that a linear system can approximate a nonlinear time-invariant state-space system in the neighborhood of a certain operating point. It was also shown that stability of LTI systems is determined by the eigenvalues of the A matrix and that controllability and observability can be checked from certain rank conditions involving the matrices (A, B) and (A, C), respectively. It was mentioned that the state of an LTI system can be changed with a similarity transform without affecting the input–output behavior, stability, controllability, and observability properties. Then we showed that the state-space representation is not the only description for an LTI system; other descriptions are the transfer function, difference equation, impulse response, and frequency response. We explained how these descriptions are related to each other and we discussed the realization problem. The chapter was concluded by describing parallel, cascade, and feedback connections of two LTI systems. It was shown that a feedback connection of two stable systems can result in an unstable overall system.

Exercises

3.1   Compute the ∞-norm, 2-norm, and 1-norm for the following discrete-time signals:

(a) $y(k) = ks(k) - (k-2)s(k-2)$,
(b) $y(k) = (9 - k^2) + |9 - k^2|$,
(c) $y(k) = 4 + 3\Delta(k) - 7\Delta(k-2)$,
(d) $y(k) = \dfrac{e^k - e^{-k}}{e^k + e^{-k}}$.

3.2   Consider the z-transform (Definition 3.6 on page 48).

(a) Prove the properties of the z-transform given in Table 3.1 on page 49.
(b) Compute the z-transform of the signals listed in Table 3.2 on page 50.

3.3   Consider the DTFT (Definition 3.7 on page 51).

(a) Prove the properties of the DTFT given in Table 3.3 on page 53.
(b) Compute the inverse DTFT of

$$X(e^{j\omega T}) = j\frac{\pi}{T}\delta\left(\omega + \frac{a}{T}\right) - j\frac{\pi}{T}\delta\left(\omega - \frac{a}{T}\right), \quad a \in \mathbb{R}.$$

(c) Compute the inverse DTFT of

$$X(e^{j\omega T}) = \frac{\pi}{T}\delta\left(\omega + \frac{a}{T}\right) + \frac{\pi}{T}\delta\left(\omega - \frac{a}{T}\right), \quad a \in \mathbb{R}.$$

(d) Compute the DTFT of

$$x(k) = \begin{cases} 1, & |k| \leq \ell, \\ 0, & |k| > \ell, \end{cases} \quad \ell \in \mathbb{Z}, \ \ell > 0.$$

(e) Compute the DTFT of the so-called Hamming window

$$w_M(k) = \begin{cases} 0.54 - 0.46\cos(\pi/Mk), & -M \leq k \leq M, \\ 0, & \text{otherwise.} \end{cases}$$

3.4   Prove Theorem 3.3 on page 52.

3.5   Determine the values of α for which the following system is asymptotically stable:

$$x(k+1) = \begin{bmatrix} \alpha - 1 & 0 \\ \alpha & 1/2 \end{bmatrix} x(k) + \begin{bmatrix} \alpha \\ 4 \end{bmatrix} u(k),$$

$$y(k) = \begin{bmatrix} 1 & 2 \end{bmatrix} x(k).$$

3.6   Determine the values of α for which the following system is controllable and for which the system is observable:

$$x(k+1) = \begin{bmatrix} 1 & -\alpha \\ \alpha & 1 \end{bmatrix} x(k) + \begin{bmatrix} \alpha \\ 1 \end{bmatrix} u(k),$$

$$y(k) = \begin{bmatrix} 1 & 0 \end{bmatrix} x(k).$$

3.7   Given the LTI state-space system

$$x(k+1) = Ax(k) + Bu(k), \qquad (E3.1)$$
$$y(k) = Cx(k) + Du(k), \qquad (E3.2)$$

(a) let $U(z)$ and $Y(z)$ denote the z-transform of $u(k)$ and that of $y(k)$, respectively. Show that the transfer function $H(z)$ in the relationship $Y(z) = H(z)U(z)$ is $H(z) = D + C(zI - A)^{-1}B$.

(b) Express the Markov parameters $h(i)$ in the expansion,

$$H(z) = \sum_{i=0}^{\infty} h(i)z^{-i}.$$

in terms of the system matrices A, B, C, and D.

3.8   Consider the solution P of the Lyapunov equation in Theorem 3.5.

(a) Show that P satisfies $(I - A \otimes A)\mathrm{vec}(P) = \mathrm{vec}(Q)$.

(b) Show that, if the matrix A is upper triangular with entries $A(i,j) = 0$ for $i > j$, the matrix P can be determined from

$$p_{n-i}\Big(I - A(n-i, n-i)A^{\mathrm{T}}\Big)$$

$$= \Big(q_{n-i} + \sum_{j=n-i+1}^{n} A(n-i, j)p_j\Big),$$

where p_i and q_i denote the ith row of the matrix P and that of the matrix Q, respectively.

3.9   Given a SISO state-space system,

(a) derive the controllability matrix and the transfer function if the system is given in controller canonical form.

(b) Derive the observability matrix and the transfer function if the system is given in observer canonical form.

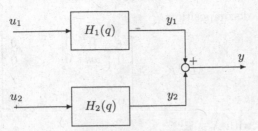

Fig. 3.10. Connection of two systems for Exercise 3.11.

3.10   Given two LTI systems that are both observable and controllable,

   (a) show that the parallel connection of these two systems is again controllable.

   (b) What can you say about the observability of the parallel connection?

3.11   Consider the interconnection of two LTI systems, with transfer functions $H_1(q)$ and $H_2(q)$ given in Figure 3.10. Let (A_1, B_1, C_1, D_1) be a state-space realization of $H_1(q)$ and let (A_2, B_2, C_2, D_2) be a state-space realization of $H_2(q)$. Derive a state-space realization of the interconnection in terms of the system matrices (A_1, B_1, C_1, D_1) and (A_2, B_2, C_2, D_2).

3.12   The dynamic behavior of a mechanical system can be described by the following differential equation:

$$M\frac{d^2x(t)}{dt^2} + Kx(t) = F(t), \qquad (E3.3)$$

with $x(t) \in \mathbb{R}^n$ the displacements of n nodes in the system, $F(t) \in \mathbb{R}^n$ the driving forces, $M \in \mathbb{R}^{n\times n}$ the mass matrix, and $K \in \mathbb{R}^{n\times n}$ the stiffness matrix. The mass and stiffness matrices are positive-definite. Consider the generalized eigenvalue decomposition given by

$$KX = MX\Lambda,$$

where the columns of the matrix $X \in \mathbb{C}^{n\times n}$ are the generalized eigenvectors, and the diagonal entries of the diagonal matrix $\Lambda \in \mathbb{C}^{n\times n}$ are the generalized eigenvalues.

   (a) Show that the eigenvalues of the matrix $M^{-1}K$ are positive and therefore this matrix has the following eigenvalue decomposition:

$$M^{-1}K = X \begin{bmatrix} \omega_1^2 & 0 & 0 & \cdots & 0 \\ 0 & \omega_2^2 & 0 & \cdots & 0 \\ \vdots & \vdots & \vdots & \ddots & \vdots \\ 0 & 0 & 0 & \cdots & \omega_n^2 \end{bmatrix} X^{-1},$$

with $X \in \mathbb{R}^{n \times n}$.

(b) Using the result of part (a), define a coordinate change for the displacements and the forces as

$$\xi(t) = X^{-1}x(t), \qquad u(t) = X^{-1}M^{-1}F(t),$$

and define the n decoupled second-order systems

$$\frac{d^2\xi_i(t)}{dt^2} + \omega_i^2 \xi_i(t) = u_i(t),$$

where $\xi_i(t)$ is the ith entry of $\xi(t)$, and $u_i(t)$ is the ith entry of $u(t)$; then show that the solution $x(t)$ to Equation (E3.3) is given by $x(t) = X\xi(t)$.

4

Random variables and signals

After studying this chapter you will be able to

- define random variables and signals;
- describe a random variable by the cumulative distribution function and by the probability density function;
- compute the expected value, mean, variance, standard deviation, correlation, and covariance of a random variable;
- define a Gaussian random signal;
- define independent and identically distributed (IID) signals;
- describe the concepts of stationarity, wide-sense stationarity, and ergodicity;
- compute the power spectrum and the cross-spectrum;
- relate the input and output spectra of an LTI system;
- describe the stochastic properties of linear least-squares estimates and weighted linear least-squares estimates;
- solve the stochastic linear least-squares problem; and
- describe the concepts of unbiased, minimum-variance, and maximum-likelihood estimates.

4.1 Introduction

In Chapter 3 the response of an LTI system to various deterministic signals, such as a step input, was considered. A characteristic of a deterministic signal or sequence is that it can be reproduced exactly. On the other hand, a random signal, or a sequence of random variables, cannot be exactly reproduced. The randomness or unpredictability of the value of a certain variable in a modeling context arises generally from the limitations of the modeler in predicting a measured value by applying the "laws of Nature." These limitations can be a consequence of the limits of scientific knowledge or of the desire of the modeler to work with models of low complexity. Measurements, in particular, introduce an unpredictable part because of their finite accuracy.

There are excellent textbooks that cover a formal treatment of random signals and the filtering of such signals by deterministic systems, such as Leon-Gracia (1994) and Grimmett and Stirzaker (1983). In this chapter a brief review is made of the necessary statistical concepts to understand the signal-analysis problems treated in later chapters.

The chapter is organized as follows. In Section 4.2 we review elementary concepts from probability theory that are used to characterize a random variable. Only continuous-valued random variables are considered. In Section 4.3 the concept and properties of random signals are discussed. The study of random signals in the frequency domain through power spectra is the topic of Section 4.4. Section 4.5 concludes the chapter with an analysis of the properties of linear least-squares estimates in a stochastic setting. Throughout this chapter the adjectives "random" and "stochastic" will both be used to indicate non-determinism.

4.2 Description of a random variable

The deterministic property is an ideal mathematical concept, since in real-life situations signals and the behavior of systems are often not predictable exactly. An example of an unpredictable signal is the acceleration measured on the wheel axis of a compact car. Figure 4.1 displays three sequences of the recorded acceleration during a particular time interval when a car is driving at constant speed on different test tracks. The nondeterministic nature of these time records stems from the fact that there is no prescribed formula to generate such a time record synthetically for the same or a different road surface. A consequence of this nondeterministic nature is that the recording of the acceleration will be different when it is measured for a different period in time with the same sensor mounted at the same location, while the car is driving at the same speed over the same road segment. Artificial generation of the acceleration signals like the ones in Figure 4.1 may be of interest in a road simulator that simulates a car driving over a particular road segment for an arbitrary length of time. Since these signals are nondeterministic,

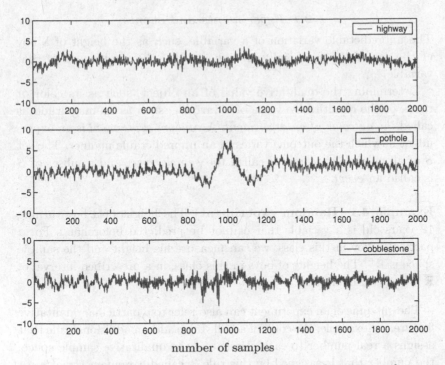

Fig. 4.1. Real-life recordings of measurements of the accelerations on the rear wheel axis of a compact-size car driving on three different road surfaces. From top to bottom: highway, road with a pothole, and cobblestone road.

generating exactly the same signals is not possible. However, we might not be interested in the exact reproduction of the recorded acceleration sequence. For example, in evaluating the durability properties of a new prototype vehicle using the road simulator, we need only a time sequence that has "similar features" to the original acceleration signals. An example of such a feature is the sample mean of all the 2000 samples of each time record in Figure 4.1. Let the acceleration sequence at the top of Figure 4.1 be denoted by $x(k)$, with $k = 0, 1, 2, \ldots$ The sample mean $\widehat{m}_x$ is then defined as

$$\widehat{m}_x = \frac{1}{2000} \sum_{k=0}^{1999} x(k). \tag{4.1}$$

For that purpose it is of interest first to determine features from time records acquired in real life and then develop tools that can generate signals that possess the same features. Such tools are built upon notions and instruments introduced in *statistics*.

4.2.1 Experiments and events

The unpredictable variation of a variable, such as the height of a boy at the age of 12, is generally an indication of the randomness of that variable.

Determining the qualitative value of an object, such as its color or taste, or the quantitative value of a variable, such as its magnitude, is called *the outcome of an experiment*. A *random experiment* is an experiment in which the outcome varies in an unpredictable manner. The set S of all possible outcomes is called the *sample space* and a subset of S is called an *event*.

Example 4.1 (Random experiment) The height of a boy turning 12 years old is a variable that cannot be predicted beforehand. For a particular boy in this class, we can measure his height and the sample space is $\mathbb{R}^+$. The heights of boys in this class, in a prescribed interval of $\mathbb{R}^+$, constitute an event.

The outcome of an experiment can also refer to a particular qualitative feature, for example, the color of a ball. If we add an additional rule that assigns a real number to each element of the qualitative sample space, the number that is assigned by this rule is called a *random variable*.

Example 4.2 (Random variable) This year's students who take the exam of this course can pass (P) or fail (F). We design an experiment in which we arbitrarily (randomly) select three students from the group of students who take the exam this year. The sample space that corresponds to this experiment is

$$S = \{\text{PPP}, \text{PPF}, \text{PFP}, \text{PFF}, \text{FPP}, \text{FPF}, \text{FFP}, \text{FFF}\}.$$

The number of passes in each set of three students is a random variable. It assigns a number to each outcome $s \in S$. This number is a random variable that can be described as "the number of students who pass the exam, out of a group of three arbitrarily selected students who take the course this year."

4.2.2 The probability model

Suppose that we throw three dice at a time and do this n times. Let $N_i(n)$ for $i = 1, 2, \ldots, 6$ be the number of times the outcome is a die with i spots. Then the relative frequency that the outcome is i is defined

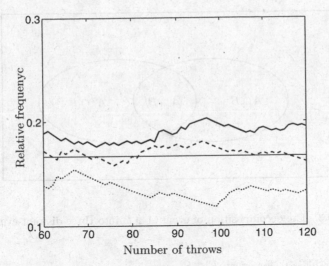

Fig. 4.2. The relative frequency of the number of eyes when throwing three similar dice; for one (solid line), three (dashed line), and five (dotted line) spots.

as

$$f_i(n) = \frac{N_i(n)}{n}. \tag{4.2}$$

The limit

$$\lim_{n \to \infty} f_i(n) = p_i,$$

if it exists, it is called the probability of the outcome i.

Example 4.3 (Relative frequencies) The above experiment of throwing three dice is done 120 times by a child's fair hand. The relative frequency defined in (4.2) is plotted in Figure 4.2 for respectively one, three, and five spots, that is $i = 1, 3$, and 5. It is clear that, for $i = 3$, the relative frequency approaches $1/6 \approx 0.167$; however, for the other values of i this is not the case. This may be a sign that either one of the dice used is "not perfect" or the child's hand is not so fair after all.

Formally, let s be the outcome of a random experiment and let X be the corresponding random variable with sample spaces S and S_X, respectively. A probability law for this random variable is a rule that assigns to each event E a positive number $\Pr[E]$, called the *probability of E*, that satisfies the following axioms

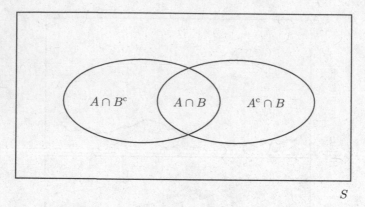

S

Fig. 4.3. The decomposition of event $A \cup B$ into three disjoint events.

(i) $\Pr[E] \geq 0$, for every $E \in S_X$.
(ii) $\Pr[S_X] = 1$, for the certain event S_X.
(iii) For any two mutually exclusive events E_1 and E_2,

$$\Pr[E_1 \cup E_2] = \Pr[E_1] + \Pr[E_2].$$

These axioms allow us to derive the probability of an event from the already-defined probabilities of other events. This is illustrated in the following example.

Example 4.4 (Derivation of probabilities) If the probabilities of events A, B, and their intersection $A \cap B$ are defined, then we can find the probability of $A \cup B$ as

$$\Pr[A \cup B] = \Pr[A] + \Pr[B] - \Pr[A \cap B].$$

To see this, we decompose $A \cup B$ into three disjoint sets, as displayed in Figure 4.3. In this figure each set represents an event. Let A^c denote the complement of A, that is S without A: $A \cup A^c = S$, and let B^c denote the complement of B. We then have

$$\Pr[A \cup B] = \Pr[A \cap B^c] + \Pr[B \cap A^c] + \Pr[A \cap B],$$
$$\Pr[A] = \Pr[A \cap B^c] + \Pr[A \cap B],$$
$$\Pr[B] = \Pr[A^c \cap B] + \Pr[A \cap B].$$

From this set of relations we can easily find the desired probability of $\Pr[A \cup B]$.

The probability of a random variable is an important means to characterize its behavior. The empirical way to determine probabilities via relative frequencies is a cumbersome approach, as illustrated in Example 4.3. A more systematic approach based on counting methods can be used, as is illustrated next.

Example 4.5 (Derivation of probabilities based on counting methods) An urn contains four balls numbered 1 to 4. We select two balls in succession without putting the selected balls back into the urn. We are interested in the probability of selecting a pair of balls for which the number of the first selected ball is smaller than or equal to that of the second.

The total number of distinct ordered pairs is $4 \cdot 3 = 12$. From these, only six ordered pairs have their first ball with a number smaller than that of the second one; thus the probability of the event is $6/12 = 1/2$.

In the above example, the probability would change if the selection of the second ball were preceded by putting the first selected ball back into the urn. So the probability of an event B may be conditioned on that of another event A that has already happened. This is denoted by $\Pr[B|A]$. According to Bayes' rule, we can express this probability as

$$\Pr[B|A] = \frac{\Pr[A \cap B]}{\Pr[A]}. \tag{4.3}$$

If the two events are *independent*, then we know that the probability of B is not affected by whether we know that event A has happened or not. In that case, we have $\Pr[B|A] = \Pr[B]$ and, according to Bayes' rule,

$$\Pr[A \cap B] = \Pr[A]\Pr[B].$$

Instead of assigning probabilities by counting methods or deriving them from basic axioms via notions from set theory, a formal way to assign probabilities is via the cumulative distribution function. In this chapter we will consider such functions only for random variables that take continuous values. Similar concepts exist for discrete random variables (Leon-Gracia, 1994).

Definition 4.1 (Cumulative distribution function) *The cumulative distribution function (CDF) $F_X(\alpha)$ of a random variable X yields the probability of the event $\{X \leq \alpha\}$, which is denoted by*

$$F_X(\alpha) = \Pr[X \leq \alpha], \quad for -\infty < \alpha < \infty.$$

The axioms of probability imply that the CDF has the following properties (Leon-Gracia, 1994).

(i) $0 \leq F_X(\alpha) \leq 1$.
(ii) $\lim_{\alpha \to \infty} F_X(\alpha) = 1$.
(iii) $\lim_{\alpha \to -\infty} F_X(\alpha) = 0$.
(iv) $F_X(\alpha)$ is a nondecreasing function of α:

$$F_X(\alpha) \leq F_X(\beta) \quad \text{for } \alpha < \beta.$$

(v) The probability of the event $\{\alpha < X \leq \beta\}$ is given by

$$\Pr[\alpha < X \leq \beta] = F_X(\beta) - F_X(\alpha).$$

(vi) The probability of the event $\{X > \alpha\}$ is

$$\Pr[X > \alpha] = 1 - F_X(\alpha).$$

Exercise 4.1 on page 122 requests a proof of the above properties.

The cumulative distribution function is a piecewise-continuous function that may contain jumps.

Another, more frequently used, characterization of a random variable
is the *probability density function* (PDF).

Definition 4.2 (Probability density function) *The probability density function (PDF) $f_X(\alpha)$ of a random variable X, if it exists, is equal
to the derivative of the cumulative distribution function $F_X(\alpha)$, which is
denoted by*

$$f_X(\alpha) = \frac{\mathrm{d}F_X(\alpha)}{\mathrm{d}\alpha}.$$

The CDF can be obtained by integrating the PDF:

$$F_X(\alpha) = \int_{-\infty}^{\alpha} f_X(\beta)\mathrm{d}\beta.$$

The PDF has the property $f_X(\alpha) \geq 0$ and

$$\int_{-\infty}^{\infty} f_X(\alpha)\mathrm{d}\alpha = 1.$$

We can derive the probability of the event $\{a < X \leq b\}$ by using

$$\Pr[a < X \leq b] = \int_{a}^{b} f_X(\alpha)\mathrm{d}\alpha.$$

4.2.3 Linear functions of a random variable

Consider the definition of a random variable Y in terms of another random variable X as

$$Y = aX + b,$$

where $a \in \mathbb{R}$ is a positive constant and $b \in \mathbb{R}$. Let X have a CDF, denoted by $F_X(\alpha)$, and a PDF, denoted by $f_X(\alpha)$. We are going to determine the CDF and PDF of the random variable Y.

The event $\{Y \leq \beta\}$ is equivalent to the event $\{aX + b \leq \beta\}$. Since $a > 0$, the event $\{aX + b \leq \beta\}$ can also be written as $\{X \leq (\beta - b)/a\}$ and thus

$$F_Y(\beta) = \Pr\left[X \leq \frac{\beta - b}{a}\right] = F_X\left(\frac{\beta - b}{a}\right).$$

Using the chain rule for differentiation, the PDF of the random variable Y is equal to

$$f_Y(\beta) = \frac{1}{a} f_X\left(\frac{\beta - b}{a}\right).$$

4.2.4 The expected value of a random variable

The CDF and PDF fully specify the behavior of a random variable in the sense that they determine the probabilities of events corresponding to that random variable. Since these functions cannot be determined experimentally in a trivial way, in many engineering problems the specification of the behavior of a random variable is restricted to its expected value or to the expected value of a function of this random variable.

Definition 4.3 (Expected value) *The expected value of a random variable X is given by*

$$E[X] = \int_{-\infty}^{\infty} \alpha f_X(\alpha) d\alpha.$$

The expected value is often called the *mean* of a random variable or the *first-order moment*. Higher-order moments of a random variable can also be obtained.

Definition 4.4 *The nth-order moment of a random variable X is given by*

$$E[X^n] = \int_{-\infty}^{\infty} \alpha^n f_X(\alpha) d\alpha.$$

A useful quantity related to the second-order moment of a random variable is the *variance*.

Definition 4.5 (Variance) *The variance of a random variable X is given by*

$$\text{var}[X] = E\Big[(X - E[X])^2\Big].$$

Sometimes the *standard deviation* is used, which equals the square root of the variance:

$$\text{std}[X] = \text{var}[X]^{1/2}.$$

The expression for the variance can be simplified as follows:

$$\begin{aligned}
\text{var}[X] &= E\Big[X^2 - 2E[X]X + E[X]^2\Big] \\
&= E[X^2] - 2E[X]E[X] + E[X]^2 \\
&= E[X^2] - E[X]^2.
\end{aligned}$$

This shows that, for a zero-mean random variable ($E[X] = 0$), the variance equals its second-order moment $E[X^2]$.

4.2.5 Gaussian random variables

Many natural phenomena involve a random variable X that is the consequence of a large number of events that have occurred on a minuscule level. An example of such a phenomenon is measurement noise due to the thermal movement of electrons. When the random variable X is the sum of a large number of random variables, then, under very general conditions, the law of large numbers (Grimmett and Stirzaker, 1983) implies that the probability density function of X approaches that of a Gaussian random variable.

Definition 4.6 *A Gaussian random variable X is a random variable that has the following probability density function:*

$$f_X(\alpha) = \frac{1}{\sqrt{2\pi}\sigma} \exp\left(-\frac{(\alpha - m)^2}{2\sigma^2}\right), \quad -\infty < \alpha < \infty,$$

where $m \in \mathbb{R}$ and $\sigma \in \mathbb{R}^+$.

Gaussian random variables are sometimes also called *normal random variables*. A graph of the PDF is given in Figure 4.4.

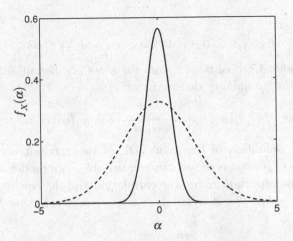

Fig. 4.4. The probability density function $f_X(\alpha)$ of a Gaussian random random variable for $m = 0$, $\sigma = 0.5$ (solid line) and for $m = 0$, $\sigma = 1.5$ (dashed line).

The PDF of a Gaussian random variable is completely specified by the two constants m and σ. These constants can be obtained as

$$E[X] = m,$$
$$\mathrm{var}[X] = \sigma^2.$$

You are asked to prove this result in Exercise 4.2 on page 122. Since the PDF of a Gaussian random variable is fully specified by m and σ, the following specific notation is introduced to indicate a Gaussian random variable X with mean m and variance σ^2:

$$X \sim (m, \sigma^2). \tag{4.4}$$

4.2.6 Multiple random variables

It often occurs in practice that, in a single engineering problem, several random variables are measured at the same time. This may be an indication that these random variables are related. The probability of events that involve the joint behavior of multiple random variables is described by the joint cumulative distribution function or the joint probability density function.

Definition 4.7 (Joint cumulative distribution function) *The joint cumulative distribution function of two random variables X_1 and X_2 is* *defined as*

$$F_{X_1,X_2}(\alpha_1,\alpha_2) = \Pr[X_1 \le \alpha_1 \text{ and } X_2 \le \alpha_2].$$

When the joint CDF of two random variables is differentiable, then we can define the probability density function as

$$f_{X_1,X_2}(\alpha_1,\alpha_2) = \frac{\partial^2}{\partial\alpha_1\,\partial\alpha_2}F_{X_1,X_2}(\alpha_1,\alpha_2).$$

With the definition of the joint PDF of two random variables, the expectation of functions of two random variables can be defined as well. Two relevant expectations are the correlation and the covariance of two random variables. The *correlation* of two random variables X_1 and X_2 is

$$R_{X_1,X_2} = E[X_1X_2] = \int_{-\infty}^{\infty}\int_{-\infty}^{\infty}\alpha_1\alpha_2 f_{X_1,X_2}(\alpha_1,\alpha_2)\mathrm{d}\alpha_1\,\mathrm{d}\alpha_2.$$

Let $m_{X_1} = E[X_1]$ and $m_{X_2} = E[X_2]$ denote the means of the random variables X_1 and X_2, respectively. Then the *covariance* of the two random variables X_1 and X_2 is

$$\begin{aligned}C_{X_1,X_2} &= E[(X_1 - m_{X_1})(X_2 - m_{X_2})] \\ &= R_{X_1,X_2} - m_{X_1}m_{X_2}.\end{aligned}$$

On the basis of the above definitions for two random variables, we can define the important notions of independent, uncorrelated, and orthogonal random variables.

Definition 4.8 *Two random variables X_1 and X_2 are independent if*

$$f_{X_1,X_2}(\alpha_1,\alpha_2) = f_{X_1}(\alpha_1)f_{X_2}(\alpha_2),$$

where the marginal PDFs are given by

$$f_{X_1}(\alpha_1) = \int_{-\infty}^{\infty}f_{X_1,X_2}(\alpha_1,\alpha_2)\mathrm{d}\alpha_2,$$

$$f_{X_2}(\alpha_2) = \int_{-\infty}^{\infty}f_{X_1,X_2}(\alpha_1,\alpha_2)\mathrm{d}\alpha_1.$$

Definition 4.9 *Two random variables X_1 and X_2 are uncorrelated if*

$$E[X_1X_2] = E[X_1]E[X_2].$$

This definition can also be written as

$$R_{X_1,X_2} = m_{X_1}m_{X_2}.$$

Therefore, when X_1 and X_2 are uncorrelated, their covariance equals zero. Note that their correlation R_{X_1,X_2} can still be nonzero. Exercise 4.3 on page 122 requests you to show that the variance of the sum of two uncorrelated random variables equals the sum of the variances of the individual random variables.

Definition 4.10 *Two random variables X_1 and X_2 are orthogonal if*

$$E[X_1 X_2] = 0.$$

Zero-mean random variables are orthogonal when they are uncorrelated. However, orthogonal random variables are not necessarily uncorrelated.

The presentation for the case of two random variables can be extended to the vector case. Let X be a vector with entries X_i for $i = 1, 2, \ldots, n$ that jointly have a Gaussian distribution with mean equal to:

$$m_X = \begin{bmatrix} E[X_1] \\ \vdots \\ E[X_n] \end{bmatrix}$$

and covariance matrix C_X equal to

$$C_X = \begin{bmatrix} C_{X_1,X_1} & C_{X_1,X_2} & \cdots & C_{X_1,X_n} \\ C_{X_2,X_1} & C_{X_2,X_2} & & C_{X_2,X_n} \\ \vdots & \vdots & \ddots & \vdots \\ C_{X_n,X_1} & C_{X_n,X_2} & \cdots & C_{X_n,X_n} \end{bmatrix},$$

then the joint probability density function is given by

$$f_X(\alpha) = f_{X_1,X_2,\ldots,X_n}(\alpha_1, \alpha_2, \ldots, \alpha_n)$$
$$= \frac{1}{(2\pi)^{n/2} \det(C_X)^{1/2}} \exp\left(-\frac{1}{2}(\alpha - m_X)^{\mathrm{T}} C_X^{-1}(\alpha - m_X)\right), \quad (4.5)$$

where α is a vector with entries α_i, $i = 1, 2, \ldots, n$. A linear transformation of a Gaussian random vector preserves the Gaussianity (Leon-Gracia, 1994). Let A be an invertible matrix in $\mathbb{R}^{n \times n}$ and let the random vectors X and Y, with entries X_i and Y_i for $i = 1, 2, \ldots, n$, be related by

$$Y = AX.$$

Then, if the entries of X are jointly Gaussian-distributed random variables, the entries of the vector Y are again jointly Gaussian random variables.

4.3 Random signals

A random signal or a stochastic process arises on measuring a random variable at particular time instances, such as the acceleration on the wheel axis of a car. Such discrete-time records were displayed in Figure 4.1 on page 89. In the example of Figure 4.1, we record a different sequence each time (each run) we drive the same car under equal circumstances (that is, with the same driver, over the same road segment, at the same speed, during a time interval of equal length, etc.). These records are called *realizations* of that stochastic process. The collection of realizations of a random signal is called the *ensemble* of discrete-time signals.

Let the time sequence of the acceleration on the wheel axis during the ξ_jth run be denoted by

$$\{x(k, \xi_j)\}_{k=0}^{N-1}.$$

Then the kth sample $\{x(k, \xi_j)\}$ of each run is a random variable, denoted briefly by $x(k)$, that can be characterized by its cumulative distribution function

$$F_{x(k)}(\alpha) = \Pr[x(k) \leq \alpha],$$

and, assuming that this function is continuous, its probability density function equals

$$f_{x(k)}(\alpha) = \frac{\partial}{\partial \alpha} F_{x(k)}(\alpha, k).$$

For two different time instants k_1 and k_2, we can characterize the two random variables $x(k_1)$ and $x(k_2)$ by their joint CDF or PDF.

For a fixed value j the sequence $\{x(k, \xi_j)\}$ is called a realization of a random signal. The family of time sequences $\{x(k, \xi)\}$ is called a random signal or a stochastic process.

4.3.1 Expectations of random signals

Each entry of the discrete-time vector random signal $\{x(k, \xi)\}$, with $x(k, \xi) \in \mathbb{R}^n$ for a fixed k, is a random variable. When we indicate this sequence for brevity by the time sequence $x(k)$, the mean is also a time sequence and is given by

$$m_x(k) = E[x(k)].$$

On the basis of the joint probability density function of the two random variables $x(k)$ and $x(\ell)$, the *auto-covariance* (matrix) function is defined as

$$C_x(k,\ell) = E\left[\left(x(k) - m_x(k)\right)\left(x(\ell) - m_x(\ell)\right)^{\mathrm{T}}\right].$$

Note that $C_x(k,k) = \mathrm{var}[x(k)]$. The *auto-correlation* function of $x(k)$ is defined as

$$R_x(k,\ell) = E\left[x(k)x(\ell)^{\mathrm{T}}\right].$$

Considering two random signals $x(k)$ and $y(k)$, the *cross-covariance* function is defined as

$$C_{xy}(k,\ell) = E\left[(x(k) - m_x(k))(y(\ell) - m_y(\ell))^{\mathrm{T}}\right].$$

The *cross-correlation* function is defined as

$$R_{xy}(k.\ell) = E[x(k)y(\ell)^{\mathrm{T}}].$$

Following Definitions 4.9 and 4.10, the random signals $x(k)$ and $y(k)$ are uncorrelated if

$$C_{xy}(k,\ell) = 0, \quad \text{for all } k, \ell,$$

and orthogonal if

$$R_{xy}(k,\ell) = 0, \quad \text{for all } k, \ell.$$

4.3.2 Important classes of random signals

4.3.2.1 Gaussian random signals

Definition 4.11 *A discrete-time random signal $x(k)$ is a Gaussian random signal if every collection of a finite number of samples of this random signal is jointly Gaussian.*

Let $C_x(k,\ell)$ denote the auto-covariance function of the random signal $x(k)$. Then, according to Definition 4.11, the probability density function of the samples $x(k)$, $k = 0, 1, 2, \ldots, N-1$ of a Gaussian random signal is given by

$$f_{x(0),x(1),\ldots,x(N-1)}(\alpha_0, \alpha_1, \ldots, \alpha_{N-1})$$
$$= \frac{1}{(2\pi)^{N/2}\det(C_x)^{1/2}} \exp\left(-\frac{1}{2}(\alpha - m_x)C_x^{-1}(\alpha - m_x)^{\mathrm{T}}\right),$$

with

$$m_x = \begin{bmatrix} E[x(0)] \\ E[x(1)] \\ \vdots \\ E[x(N-1)] \end{bmatrix}$$

and

$$C_x = \begin{bmatrix} C_x(0,0) & C_x(0,1) & \cdots & C_x(0,N-1) \\ C_x(1,0) & C_x(1,1) & & C_x(1,N-1) \\ \vdots & \vdots & \ddots & \vdots \\ C_x(N-1,0) & C_x(N-1,1) & \cdots & C_x(N-1,N-1) \end{bmatrix}.$$

4.3.2.2 IID random signals

The abbreviation IID stands for *independent, identically distributed*. An IID random signal $x(k)$ is a sequence of independent random variables in which each random variable has the same probability density function. Thus, the joint probability distribution function for any number of finite samples of the IID random signal satisfies

$$f_{x(0),x(1),\ldots,x(N-1)}(\alpha_0, \alpha_1, \ldots, \alpha_{N-1})$$
$$= f_{x(0)}(\alpha_0) f_{x(1)}(\alpha_1) \ldots f_{x(N-1)}(\alpha_{N-1}),$$

and

$$f_{x(0)}(\alpha) = f_{x(1)}(\alpha) = \cdots = f_{x(N-1)}(\alpha).$$

4.3.3 Stationary random signals

If we compare the signals in Figure 4.1 on page 89, we observe that the randomness of the first one looks rather constant over time, whereas this is not the case for the second one. This leads to the postulation that random signals similar to the one in the top graph of Figure 4.1 have probabilistic properties that are time-independent. Such signals belong to the class of stationary random signals, which are defined formally as follows.

Definition 4.12 *A discrete-time random signal $x(k)$ is stationary if the joint cumulative distribution function of any finite number of samples does not depend on the placement of the time origin, that is,*

$$F_{x(k_0),x(k_1),\ldots,x(k_{N-1})}(\alpha_0, \alpha_1, \ldots, \alpha_{N-1}) = F_{x(k_0+\tau),x(k_1+\tau),\ldots,x(k_{N-1}+\tau)}, \tag{4.6}$$

for all time shifts $\tau \in \mathbb{Z}$.

If the above definition holds only for $N = 1$, then the random signal $x(k)$ is called *first-order stationary*. In that case, we have that the mean of $x(k)$ is constant. In a similar way, if Equation (4.6) holds for $N = 2$, the random signal $x(k)$ is *second-order stationary*. Let $R_x(k, \ell)$ denote the auto-covariance function of such a second-order stationary random signal $x(k)$, then this function satisfies

$$R_x(k, \ell) = R_x(k + \tau, \ell + \tau). \qquad (4.7)$$

Therefore, the auto-correlation function depends only on the difference between k and ℓ. This difference is called the *lag* and Equation (4.7) is denoted compactly by

$$R_x(k, \ell) = R_x(k - \ell).$$

When we are interested only in the mean and the correlation function of a random signal, we can consider a more restricted form of stationarity, namely *wide-sense stationarity (WSS)*.

Definition 4.13 (Wide-sense stationarity) *A random signal $x(k)$ is wide-sense stationary (WSS) if the following three conditions are satisfied.*

(i) *Its mean is constant:* $m_x(k) = E[x(k)] = m_x$.

(ii) *Its auto-correlation function $R_x(k, \ell)$ depends only on the lag $k - \ell$.*

(iii) *Its variance is finite:* $\mathrm{var}[x(k)] = E\left[(x(k) - m_x)^2\right] < \infty$.

In the case of random vector signals, the above notions can easily be extended. In that case the random signals that form the entries of the vector are said to be jointly (wide-sense) stationary.

A specific WSS random signal is the white-noise signal. The auto-covariance function of a white-noise random signal $x(k) \in \mathbb{R}$ satisfies

$$C_x(k_1, k_2) = \sigma_x^2 \Delta(k_1 - k_2), \quad \text{for all } k_1, k_2,$$

where $\sigma_x^2 = \mathrm{var}[x(k)]$.

The definition of the auto-correlation function of a WSS random signal leads to a number of basic properties. These are summarized next.

Lemma 4.1 *The auto-correlation function $R_x(\tau)$ of a WSS random signal $x(k)$ is symmetric in its argument τ, that is,*

$$R_x(\tau) = R_x(-\tau).$$

Proof The corollary follows directly from the definition of the auto-correlation function (4.7),

$$R_x(\tau) = E[x(k)x(k - \tau)] = E[x(k - \tau)x(k)] = R_x(-\tau).$$

$\square$

Corollary 4.1 *The auto-correlation function $R_x(\tau)$ of a WSS random signal $x(k)$ satisfies, for $\tau = 0$,*

$$R_x(0) = E[x(k)x(k)] \geq 0.$$

Lemma 4.2 *The maximum of the auto-correlation function $R_x(\tau)$ of a WSS random signal $x(k)$ occurs at $\tau = 0$,*

$$R_x(0) \geq R_x(k), \quad \text{for all } k.$$

Proof Note that

$$E\left[\left(x(k-\tau) - x(k)\right)^2\right] \geq 0.$$

This is expanded as

$$2R_x(0) - 2\gamma R_x(\tau) \geq 0,$$

as desired.   □

4.3.4 Ergodicity and time averages of random signals

The strong law of large numbers (Grimmett and Stirzaker, 1983) is an important law in statistics that gives rise to the property of *ergodicity* of a random signal. It states that, for a stationary IID random signal $x(k)$ with mean $E[x(k)] = m_x$, the *time average* (see also Equation (4.1) on page 89) converges with probability unity to the mean value m_x, provided that the number of observations N goes to infinity. This is denoted by

$$\Pr\left[\lim_{N\to\infty} \frac{1}{N} \sum_{k=0}^{N-1} x(k) = m_x\right] = 1.$$

The strong law of large numbers offers an empirical tool with which to derive an estimate of the mean of a random signal, that in practice can be observed only via (a single) realization.

In general, an ergodic theorem states under what conditions statistical quantities characterizing a stationary random signal, such as its covariance function, can be derived with probability unity from a single realization of that random signal. Since such conditions are often difficult to verify in applications, the approach taken is simply to assume under the ergodic argument that time averages can be used to compute (with probability unity) the expectation of interest.

Let $\{x(k)\}_{k=0}^{N-1}$ and $\{y(k)\}_{k=0}^{N-1}$ be two realizations of the stationary random signals $x(k)$ and $y(k)$, respectively. Then, under the ergodicity argument, we obtain relationships of the following kind:

$$\Pr\left[\lim_{N\to\infty}\frac{1}{N}\sum_{k=0}^{N-1}x(k)=E[x(k)]\right]=1,$$

$$\Pr\left[\lim_{N\to\infty}\frac{1}{N}\sum_{k=0}^{N-1}y(k)=E[y(k)]\right]=1.$$

If $E[x(k)]$ and $E[y(k)]$ are denoted by m_x and m_y, respectively, then

$$\Pr\left[\lim_{N\to\infty}\frac{1}{N}\sum_{k=0}^{N-1}\Big(x(k)-m_x\Big)\Big(x(k-\tau)-m_x\Big)=C_x(\tau)\right]=1,$$

$$\Pr\left[\lim_{N\to\infty}\frac{1}{N}\sum_{k=0}^{N-1}\Big(x(k)-m_x\Big)\Big(y(k-\tau)-m_y\Big)=C_{xy}(\tau)\right]=1.$$

4.4 Power spectra

In engineering, the frequency content of a discrete-time signal is computed via the Fourier transform of the signal. However, since a random signal is not a single time sequence, that is, not a single realization, the Fourier transform of a random signal would remain a random signal. To get a deterministic notion of the frequency content for a random signal, the power-spectral density function, or the *power spectrum*, is used. The discrete-time Fourier transform introduced in Section 3.3.2 enables us to determine the power spectrum of a signal. The spectrum of a signal can be thought of as the distribution of the signal's energy over the whole frequency band. Signal spectra are defined for WSS time sequences.

Definition 4.14 (Signal spectra) *Let $x(k)$ and $y(k)$ be two zero-mean WSS sequences with sampling time T. The (power) spectrum of $x(k)$ is*

$$\Phi^x(\omega)=\sum_{\tau=-\infty}^{\infty}R_x(\tau)e^{-j\omega\tau T}, \tag{4.8}$$

and the cross-spectrum between $x(k)$ and $y(k)$ is

$$\Phi^{xy}(\omega)=\sum_{\tau=-\infty}^{\infty}R_{xy}(\tau)e^{-j\omega\tau T}. \tag{4.9}$$

The inverse DTFT applied to (4.8) yields

$$R_x(\tau) = \frac{T}{2\pi} \int_{-\pi/T}^{\pi/T} \Phi^x(\omega) e^{j\omega\tau T} \, d\omega.$$

The power spectrum of a WSS random signal has a number of interesting properties. The symmetry of the auto-correlation function $R_x(\tau)$ suggests symmetry of the power spectrum.

Property 4.1 (Real-valued) *Let a WSS random signal $x(k) \in \mathbb{R}$ with real-valued auto-correlation function $R_x(\tau)$ be given. Then its power spectrum $\Phi^x(\omega)$, when it exists, is real-valued and symmetric with respect to ω, that is*

$$\Phi^x(-\omega) = \Phi^x(\omega).$$

For $\tau = 0$, we obtain a stochastic variant of Parseval's identity (compare the following with Equation (3.5) on page 52).

Property 4.2 (Parseval) *Let a WSS random signal $x(k) \in \mathbb{R}$ with sampling time T and power spectrum $\Phi^x(\omega)$ be given. Then*

$$E[x(k)^2] = \frac{T}{2\pi} \int_{-\pi/T}^{\pi/T} \Phi^x(\omega) d\omega. \tag{4.10}$$

This property shows that the total energy of the signal $x(k)$ given by $E[x(k)^2]$ is distributed over the frequency band $-\pi/T \leq \omega \leq \pi/T$. Therefore, the spectrum $\Phi^x(\omega)$ can be viewed as the distribution of the signal's energy over the whole frequency band, as stated already at the beginning of this section.

In the identification of dynamic LTI systems from sampled input and output sequences using the DTFT (or DFT, see Section 6.2), it is important to relate the frequency-response function (FRF) of the system to the signal spectra of the input and output sequences. Some important relationships are summarized in the following lemma.

Lemma 4.3 (Filtering WSS random signals) *Let $u(k)$ be WSS and the input to the BIBO-stable LTI system with transfer function*

$$G(q) = \sum_{k=0}^{\infty} g(k) q^{-k},$$

such that $y(k) = G(q)u(k)$. Then

   (i) $y(k)$ *is WSS,*
   (ii) $\Phi^{yu}(\omega) = G(e^{j\omega T})\Phi^u(\omega)$, *and*
   (iii) $\Phi^y(\omega) = |G(e^{j\omega T})|^2 \Phi^u(\omega)$.

Proof From the input–output relationship and linearity of the expectation operator it follows that

$$E[y(k)] = \sum_{\ell=0}^{\infty} g(\ell)E[u(k-\ell)] = 0.$$

The WSS property follows from a derivation of the auto-correlation function $R_y(\tau) = E[y(k)y(k-\tau)]$. This is done in two steps. First we evaluate $R_{yu}(\tau) = E[y(k)u(k-\tau)]$, then we evaluate $R_y(\tau)$:

$$R_{yu}(\tau) = \sum_{p=0}^{\infty} g(p)E[u(k-p)u(k-\tau)]$$

$$= \sum_{p=0}^{\infty} g(p)R_u(\tau - p)$$

$$= g(\tau) * R_u(\tau).$$

Note that $R_{yu}(\tau)$ depends only on τ, not on k. Since the DTFT of a convolution of two time sequences equals the product of their individual DTFTs, we have proven point (ii). Evaluation of $R_y(\tau)$ yields

$$R_y(\tau) = \sum_{p=0}^{\infty} g(p)E[y(k-\tau)u(k-p)]$$

$$= \sum_{p=0}^{\infty} g(p)R_{yu}(p-\tau)$$

$$= g(-\tau) * R_{yu}(\tau)$$

$$= g(-\tau) * g(\tau) * R_u(\tau).$$

This proves point (iii), if again the convolution property of the DTFT is used. Since the system is BIBO-stable, a bounded $R_u(\tau)$ implies a bounded $R_y(\tau)$ and therefore $y(k)$ is WSS.   □

An alternative proof in which the system $G(q)$ is given by a multi-variable state-space model is requested in Exercise 4.10 on page 124. An important application of the result of Lemma 4.3 is the generation of a random signal with a certain desired spectrum. According to this lemma, the spectrum of a random signal can be changed by applying a linear filter. If a white-noise signal is filtered by a linear filter, the spectrum of the filtered signal can be controlled by changing the FRF of the filter.

Property 4.3 (Nonnegativity) *Let a WSS random signal $x(k) \in \mathbb{R}$ with sampling time T and power spectrum $\Phi^x(\omega)$ be given, then*

$$\Phi^x(\omega) \geq 0.$$

Proof To prove this property, we consider the sequence $y(k)$ obtained by filtering the sequence $x(k)$ by an ideal bandpass filter $H(q)$ given by

$$H(e^{j\omega T}) = \begin{cases} 1, & \omega_1 < \omega < \omega_2, \\ 0, & \text{otherwise.} \end{cases}$$

We can use Lemma 4.3, which can be shown to hold also for ideal filters,

$$H(q) = \sum_{-\infty}^{\infty} h(k)q^{-k}$$

(see for example Papoulis (1991), Chapter 10), and Property 4.2 to derive

$$E[y(k)^2] = \frac{T}{2\pi} \int_{\omega_1}^{\omega_2} \Phi^x(\omega) \mathrm{d}\omega \geq 0.$$

This must hold for any choice of ω_1 and ω_2, with $\omega_1 < \omega_2$. The only possibility is that $\Phi^x(\omega)$ must be nonnegative in every interval.   □

The definitions given above apply only to random signals. To be able to analyze deterministic time sequences in a similar matter, we can replace the statistical expectation values in these definitions by their sample averages, when these sample averages exist. For example, in Equation (4.8) on page 105, $R_x(\tau) = E[x(k)x(k - \tau)]$ is then replaced by

$$\lim_{N \to \infty} \frac{1}{N} \sum_{k=0}^{N-1} x(k)x(k - \tau).$$

4.5 Properties of least-squares estimates

In Sections 2.6 and 2.7, the solutions to various least-squares problems were derived. In these problems the goal is to find the best solution with respect to a quadratic cost function. However, in Chapter 2 nothing was said about the accuracy of the solutions derived. For that it is necessary to provide further constraints on the quantities involved in the problem specifications. On the basis of notions of random variables, we derive in this section the covariance matrix of the least-squares solution, and show how this concept can be used to give additional conditions under which the least-squares estimate is an optimal estimate.

4.5.1 *The linear least-squares problem*

Recall from Section 2.6 the linear least-squares problem (2.8) on page 29:

$$\min_x \epsilon^T \epsilon \quad \text{subject to } y = Fx + \epsilon, \tag{4.11}$$

with $F \in \mathbb{R}^{m \times n}$ $(m \geq n)$, $y \in \mathbb{R}^m$, and $x \in \mathbb{R}^n$. Provided that F has full column rank, the solution to this problem is given by

$$\hat{x} = (F^T F)^{-1} F^T y.$$

Assume that the vector ϵ is a random signal with mean zero and covariance matrix $C_\epsilon = I_m$, denoted by

$$\epsilon \sim (0, I_m).$$

Assume also that x and F are deterministic. Clearly, the vector y is also a random signal. Therefore, the solution $\hat{x}$ which is a linear combination of the entries of the vector y is also a random signal. The statistical properties of the solution $\hat{x}$ are given in the next lemma.

Lemma 4.4 (Statistical properties of least-squares estimates)
Given $y = Fx + \epsilon$ with $x \in \mathbb{R}^n$ an unknown deterministic vector, $\epsilon \in \mathbb{R}^m$ a random variable with the statistical properties

$$\epsilon \sim (0, I_m),$$

and the matrix $F \in \mathbb{R}^{m \times n}$ deterministic and of full column rank, the vector

$$\hat{x} = (F^T F)^{-1} F^T y$$

has mean

$$E[\hat{x}] = x$$

and covariance matrix

$$E\left[\left(\hat{x} - x \right) \left(\hat{x} - x \right)^T \right] = (F^T F)^{-1}. \tag{4.12}$$

Proof We can express

$$\begin{aligned}
\hat{x} &= (F^T F)^{-1} F^T y \\
&= (F^T F)^{-1} F^T (Fx + \epsilon) \\
&= x + (F^T F)^{-1} F^T \epsilon.
\end{aligned}$$

The mean of $\hat{x}$ follows from the fact that $E[\epsilon] = 0$:

$$E[\hat{x}] = E[x] + (F^T F)^{-1} F^T E[\epsilon] = x.$$

The covariance matrix of $\widehat{x}$ becomes

$$
\begin{aligned}
E\left[\left(\widehat{x}-x\right)\left(\widehat{x}-x\right)^{\mathrm{T}}\right] &= E\left[\left(x+(F^{\mathrm{T}}F)^{-1}F^{\mathrm{T}}\epsilon - x\right)\right.\\
&\qquad\qquad \left. \times\left(x+(F^{\mathrm{T}}F)^{-1}F^{\mathrm{T}}\epsilon - x\right)^{\mathrm{T}}\right]\\
&= (F^{\mathrm{T}}F)^{-1}F^{\mathrm{T}}\,E[\epsilon\epsilon^{\mathrm{T}}]\,F(F^{\mathrm{T}}F)^{-1}\\
&= (F^{\mathrm{T}}F)^{-1},
\end{aligned}
$$

where the last equation is obtained using the fact that $E[\epsilon\epsilon^{\mathrm{T}}] = I_m$. $\square$

The solution $\widehat{x}$ to the linear least-squares problem is called a *linear estimator*, because it is linear in the data vector y. A linear estimator is of the form $\widetilde{x} = My$ with $M \in \mathbb{R}^{n\times m}$. Among all possible linear estimators, the linear least-squares estimator $\widehat{x} = (F^{\mathrm{T}}F)^{-1}F^{\mathrm{T}}y$ has some special statistical properties. First, it is an *unbiased estimator*.

Definition 4.15 (Unbiased linear estimator) *A linear estimator* $\widetilde{x} = My$, *with* $M \in \mathbb{R}^{n\times m}$, *is called an unbiased linear estimator if*

$$
E[\widetilde{x}] = x.
$$

The linear least-squares estimator $\widehat{x}$ in Lemma 4.4 is an unbiased linear estimator. For an unbiased estimator, the mean of the estimate equals the real value of the variable to be estimated. Hence, there is no systematic error.

Lemma 4.4 gives an expression for the covariance matrix of $\widehat{x}$. This is an important property of the estimator, because it provides us with a measure for the variance between different experiments. This can be interpreted as a measure of the uncertainty of the estimator. The smaller the variance, the smaller the uncertainty. In this respect, we could be interested in the unbiased linear estimator with the smallest possible variance.

Definition 4.16 (Minimum-variance unbiased linear estimator) *A linear estimator* $\widetilde{x}_1 = M_1 y$ *with* $M_1 \in \mathbb{R}^{n\times m}$ *is called the minimum-variance unbiased linear estimator of the least-squares problem (4.11) if it is an unbiased linear estimator and if its covariance matrix satisfies*

$$
E\left[\left(\widetilde{x}_1-x\right)\left(\widetilde{x}_1-x\right)^{\mathrm{T}}\right] \leq E\left[\left(\widetilde{x}_2-x\right)\left(\widetilde{x}_2-x\right)^{\mathrm{T}}\right]
$$

for any unbiased linear estimate $\widetilde{x}_2 = M_2 y$ *of (4.11).*

The Gauss–Markov theorem given below states that the linear least-squares estimator $\widehat{x}$ is the minimum-variance unbiased linear estimator; in other words, among all possible unbiased linear estimators, the linear least-squares estimator $\widehat{x}$ has the smallest variance.

Theorem 4.1 (Minimum-variance estimation property) *(Kailath et al., 2000). For the least-squares problem (4.11) with $x \in \mathbb{R}^n$ an unknown deterministic vector, $\epsilon \in \mathbb{R}^m$ a random variable satisfying $\epsilon \sim (0, I_m)$, and the matrix $F \in \mathbb{R}^{m \times n}$ deterministic and of full column rank, the linear least-squares estimate*

$$\widehat{x} = (F^{\mathrm{T}}F)^{-1}F^{\mathrm{T}}y$$

is the minimum-variance unbiased linear estimator.

Proof Lemma 4.4 shows that the linear least-squares estimate $\widehat{x}$ is unbiased and has covariance matrix

$$E\left[(\widehat{x} - x)(\widehat{x} - x)^{\mathrm{T}}\right] = (F^{\mathrm{T}}F)^{-1}.$$

Any other linear estimator $\widetilde{x} = My$ can be written as

$$\widetilde{x} = My = MFx + M\epsilon.$$

Such a linear estimator is unbiased if $MF = I_n$, since

$$E[\widetilde{x} - x] = E[MFx + M\epsilon - x] = (MF - I_n)x.$$

The covariance matrix of an unbiased linear estimator equals

$$E\left[(\widetilde{x} - x)(\widetilde{x} - x)^{\mathrm{T}}\right] = ME[\epsilon\epsilon^{\mathrm{T}}]M^{\mathrm{T}} = MM^{\mathrm{T}}.$$

The relation with the covariance matrix of the linear least squares estimate is given below:

$$\begin{aligned} E\left[(\widehat{x} - x)(\widehat{x} - x)^{\mathrm{T}}\right] &= (F^{\mathrm{T}}F)^{-1} \\ &= MF(F^{\mathrm{T}}F)^{-1}F^{\mathrm{T}}M^{\mathrm{T}} \\ &\leq MM^{\mathrm{T}} = E\left[(\widetilde{x} - x)(\widetilde{x} - x)^{\mathrm{T}}\right]. \end{aligned}$$

The inequality stems from the fact that $F(F^{\mathrm{T}}F)^{-1}F^{\mathrm{T}} = \Pi_F$ is an orthogonal projection matrix. From the properties of Π_F discussed in Section 2.6.1, we know that

$$||z||_2 \geq ||\Pi_F z||_2.$$

Taking $z = M^T\eta$, this is equivalent to

$$\eta^T M(I - \Pi_F)M^T\eta \geq 0,$$

for all η, or alternatively $MM^T \geq MF(F^TF)^{-1}F^TM^T$.   □

4.5.2 The weighted linear least-squares problem

Recall from Section 2.7 the weighted least-squares problem (2.17) on page 35 as

$$\min_x \epsilon^T\epsilon \quad \text{subject to } y = Fx + L\epsilon, \tag{4.13}$$

with $F \in \mathbb{R}^{m \times n}$ ($m \geq n$) of full rank, $y \in \mathbb{R}^m$, $x \in \mathbb{R}^n$, and $L \in \mathbb{R}^{m \times m}$ a nonsingular matrix defining the weighting matrix W as $(LL^T)^{-1}$.

Assume that the vector ϵ is a random signal with mean zero and covariance matrix $C_\epsilon = I_m$. The vector $\mu = L\epsilon$ is a zero-mean random vector with covariance matrix $C_\mu = LL^T$. Hence, the matrix L can be used to incorporate additive disturbances, with a general covariance matrix, to the term Fx in (4.11) on page 109. The problem (4.13) can be converted into

$$\min_x \epsilon^T\epsilon \quad \text{subject to } L^{-1}y = L^{-1}Fx + \epsilon. \tag{4.14}$$

As shown in Section 2.7, the solution to the weighted least-squares problem (4.13) is given by

$$\hat{x} = (F^TWF)^{-1}F^TWy, \tag{4.15}$$

where $W = (LL^T)^{-1}$. By realizing that (4.14) is a problem of the form (4.11) on page 109, it follows from Theorem 4.1 that $\hat{x}$ given by (4.15) is the minimum-variance unbiased linear estimator for the problem (4.13). This is summarized in the following theorem.

Theorem 4.2 (Minimum-variance estimation property)   *For the weighted least-squares problem (4.13) with $x \in \mathbb{R}^n$ an unknown deterministic vector, $\epsilon \in \mathbb{R}^m$ a random variable satisfying $\epsilon \sim (0, I_m)$, the matrix $F \in \mathbb{R}^{m \times n}$ deterministic and of full column rank, and L a nonsingular matrix, the linear least-squares estimate $\hat{x}$,*

$$\hat{x} = (F^TWF)^{-1}F^TWy,$$

with $W = (LL^T)^{-1}$, is the minimum-variance unbiased linear estimator.

In Exercise 4.7 on page 123 you are asked to prove this result.

4.5.3 The stochastic linear least-squares problem

In the previous discussion of least-squares problems the vector x was assumed to be *deterministic*. In the stochastic least-squares problem x is a random vector and prior information on its mean and covariance matrix is available. The stochastic least-squares problem is stated as follows.

We are given the mean $\bar{x}$ and the covariance matrix $P \geq 0$ of the unknown random vector x, and the observations y which are related to x as

$$y = Fx + L\epsilon, \tag{4.16}$$

with $\epsilon \sim (0, I_m)$ and $E[(x - \bar{x})\epsilon^{\mathrm{T}}] = 0$. The matrices $F \in \mathbb{R}^{m \times n}$ and $L \in \mathbb{R}^{m \times m}$ are deterministic, with L assumed to have full (column) rank. The problem is to determine a linear estimate $\tilde{x}$,

$$\tilde{x} = \begin{bmatrix} M & N \end{bmatrix} \begin{bmatrix} y \\ \bar{x} \end{bmatrix}, \tag{4.17}$$

such that $E[(x - \tilde{x})(x - \tilde{x})^{\mathrm{T}}]$ is minimized and $E[\tilde{x}] = \bar{x}$.

The estimate (4.17) is a linear estimate since it transforms the given data y and $\bar{x}$ linearly. The fact that the estimate $\tilde{x}$ minimizes $E[(x - \tilde{x})(x - \tilde{x})^{\mathrm{T}}]$ and has the property $E[\tilde{x}] = \bar{x}$ makes it a minimum-variance unbiased estimate. The solution to the stochastic least-squares problem is given in the following theorem.

Theorem 4.3 (Solution to the stochastic linear least-squares problem) *The minimum-variance unbiased estimate $\hat{x}$ that solves the stochastic linear least-squares problem is given by*

$$\hat{x} = PF^{\mathrm{T}}(FPF^{\mathrm{T}} + W^{-1})^{-1}y + \left(I_n - PF^{\mathrm{T}}(FPF^{\mathrm{T}} + W^{-1})^{-1}F\right)\bar{x}, \tag{4.18}$$

where the weight matrix W is defined as $(LL^{\mathrm{T}})^{-1}$. The covariance matrix of this estimate equals

$$E[(x - \hat{x})(x - \hat{x})^{\mathrm{T}}] = P - PF^{\mathrm{T}}(FPF^{\mathrm{T}} + W^{-1})^{-1}FP. \tag{4.19}$$

If the covariance matrix P is positive-definite, the minimum-variance unbiased estimate and its covariance matrix can be written as

$$\widehat{x} = \left(P^{-1} + F^{\mathrm{T}}WF\right)^{-1}F^{\mathrm{T}}Wy + \left(I_n - \left(P^{-1} + F^{\mathrm{T}}WF\right)^{-1}F^{\mathrm{T}}WF\right)\overline{x},$$

$$\tag{4.20}$$

$$E[(x - \widehat{x})(x - \widehat{x})^{\mathrm{T}}] = \left(P^{-1} + F^{\mathrm{T}}WF\right)^{-1}. \tag{4.21}$$

Proof Since $E[\widetilde{x}] = MF\overline{x} + N\overline{x}$, the property that $E[\widetilde{x}] = \overline{x}$ holds, provided that $MF + N = I_n$. Therefore,

$$x - \widetilde{x} = (I_n - MF)(x - \overline{x}) - ML\epsilon$$

and

$$E[(x - \widetilde{x})(x - \widetilde{x})^{\mathrm{T}}] = (I_n - MF)P(I_n - MF)^{\mathrm{T}} + MW^{-1}M^{\mathrm{T}}$$

$$= \begin{bmatrix} I_n & -M \end{bmatrix} \underbrace{\begin{bmatrix} P & PF^{\mathrm{T}} \\ FP & FPF^{\mathrm{T}} + W^{-1} \end{bmatrix}}_{Q} \begin{bmatrix} I_n \\ -M^{\mathrm{T}} \end{bmatrix}.$$

$$\tag{4.22}$$

Since the weighting matrix W^{-1} is positive-definite, we conclude that $FPF^{\mathrm{T}} + W^{-1}$ is also positive-definite. Similarly to in the application of the "completion-of-squares" argument in Section 2.6, the Schur complement (Lemma 2.3 on page 19) can be used to factorize the underbraced matrix Q as

$$Q = \begin{bmatrix} I_n & PF^{\mathrm{T}}(FPF^{\mathrm{T}} + W^{-1})^{-1} \\ 0 & I_m \end{bmatrix}$$

$$\times \begin{bmatrix} P - PF^{\mathrm{T}}(FPF^{\mathrm{T}} + W^{-1})^{-1}FP & 0 \\ 0 & FPF^{\mathrm{T}} + W^{-1} \end{bmatrix}$$

$$\times \begin{bmatrix} I_n & 0 \\ (FPF^{\mathrm{T}} + W^{-1})^{-1}FP & I_m \end{bmatrix}.$$

Substituting this factorization into Equation (4.22) yields

$$E[(x - \widetilde{x})(x - \widetilde{x})^{\mathrm{T}}] = P - PF^{\mathrm{T}}(FPF^{\mathrm{T}} + W^{-1})^{-1}FP$$

$$+ \left(PF^{\mathrm{T}}(FPF^{\mathrm{T}} + W^{-1})^{-1} - M\right)(FPF^{\mathrm{T}} + W^{-1})$$

$$\times \left((FPF^{\mathrm{T}} + W^{-1})^{-1}FP - M^{\mathrm{T}}\right).$$

By application of the "completion-of-squares" argument, the matrix M that minimizes the covariance matrix $E[(x - \tilde{x})(x - \tilde{x})^{\mathrm{T}}]$ is given by

$$M = PF^{\mathrm{T}}(FPF^{\mathrm{T}} + W^{-1})^{-1}.$$

This defines the estimate $\hat{x}$ as in Equation (4.18). The corresponding minimal-covariance matrix is given by Equation (4.19).

When the prior covariance matrix P is positive-definite, we can apply the matrix-inversion lemma (Lemma 2.2 on page 19) to $(FPF^{\mathrm{T}} + W^{-1})^{-1}$, to obtain

$$
\begin{aligned}
M &= PF^{\mathrm{T}}\left(W - WF(P^{-1} + F^{\mathrm{T}}WF)^{-1}F^{\mathrm{T}}W\right) \\
&= P\left(I_n - F^{\mathrm{T}}WF(P^{-1} + F^{\mathrm{T}}WF)^{-1}\right)F^{\mathrm{T}}W \\
&= P\left((P^{-1} + F^{\mathrm{T}}WF) - F^{\mathrm{T}}WF\right)(P^{-1} + F^{\mathrm{T}}WF)^{-1}F^{\mathrm{T}}W \\
&= (P^{-1} + F^{\mathrm{T}}WF)^{-1}F^{\mathrm{T}}W.
\end{aligned}
$$

This defines the estimate $\hat{x}$ as in Equation (4.20). Again using $P > 0$, the matrix-inversion lemma shows that the corresponding minimal-covariance matrix given by Equation (4.19) can be written as in Equation (4.21). $\qquad\square$

The solution of the stochastic linear least-squares problem can also be obtained by solving a deterministic weighted least-squares problem. The formulation of such a problem and a square-root solution are given in the next section.

4.5.4 A square-root solution to the stochastic linear least-squares problem

The information needed to solve the stochastic least-squares problem consists of the data equation (4.16) and the mean $\bar{x}$ and covariance information P of the random vector x. To represent the statistical information on the random variable x in an equation format, we introduce an auxiliary random variable ξ with mean zero and covariance matrix I_n, that is $\xi \sim (0, I_n)$. Since we assumed the covariance matrix P to be semi-positive-definite (see Section 2.4), we can compute its square root $P^{1/2}$ (see Section 2.5) such that

$$P = P^{1/2}P^{\mathrm{T}/2}.$$

This square root can be chosen to be *upper-* or *lower-*triangular (Golub and Van Loan, 1996). With ξ, $\overline{x}$, and $P^{1/2}$ so defined, the random variable x can be modeled through the following matrix equation:

$$x = \overline{x} - P^{1/2}\xi \quad \text{with } \xi \sim (0, I_n). \tag{4.23}$$

This equation is called a *generalized covariance representation* (Duncan and Horn, 1972; Paige, 1985). It is easy to verify that this representation results in a mean $\overline{x}$ for x; the covariance matrix of x follows from the calculations:

$$E\left[\left(x - \overline{x}\right)\left(x - \overline{x}\right)^{\mathrm{T}}\right] = P^{1/2}E\left[\xi\xi^{\mathrm{T}}\right]P^{\mathrm{T}/2}$$

$$= P^{1/2}P^{\mathrm{T}/2}$$

$$= P.$$

On the basis of the data equation (4.16) and the representation of the information on the random vector x in (4.23), we state the following weighted least squares problem:

$$\min_{x} \nu^{\mathrm{T}}\nu \quad \text{subject to } \begin{bmatrix} \overline{x} \\ y \end{bmatrix} = \begin{bmatrix} I_n \\ F \end{bmatrix} x + \begin{bmatrix} -P^{1/2} & 0 \\ 0 & L \end{bmatrix} \nu, \text{ with } \nu \sim (0, I_n).$$

$$\tag{4.24}$$

The next theorem shows that the (square-root) solution to this problem is the solution to the stochastic least-squares problem stated in Section 4.5.3.

Theorem 4.4 (Square-root solution) *Consider the following RQ factorization,*

$$\begin{bmatrix} -FP^{1/2} & -L \\ -P^{1/2} & 0 \end{bmatrix} T_{\mathrm{r}} = \begin{bmatrix} R & 0 \\ G & S \end{bmatrix} \tag{4.25}$$

with T_{r} orthogonal and the right-hand side lower-triangular. The solution to the stochastic linear least-squares problem given in Equation (4.18) equals

$$\widehat{x} = GR^{-1}y + \left(I_n - GR^{-1}F\right)\overline{x}. \tag{4.26}$$

The covariance matrix of this estimate is

$$E[(x - \widehat{x})(x - \widehat{x})^{\mathrm{T}}] = SS^{\mathrm{T}}, \tag{4.27}$$

and is equal to the one given in Equation (4.19).

If the covariance matrix P is positive-definite, then the estimate $\widehat{x}$ is the unique solution to the weighted least-squares problem (4.24) with the weight matrix W defined as $(LL^{\mathrm{T}})^{-1}$.

Proof First, we seek a solution of the weighted least-squares problem
(4.24), then we show that this solution equals the solution (4.18) of
the stochastic least-squares problem. Next, we show that its covariance
matrix equals (4.19). Following the strategy for solving weighted least-
squares problems in Section 2.7, we first apply a left transformation T_ℓ,
given by

$$T_\ell = \begin{bmatrix} F & -I_m \\ I_n & 0 \end{bmatrix},$$

to the following set of equations from (4.24):

$$\begin{bmatrix} \bar{x} \\ y \end{bmatrix} = \begin{bmatrix} I_n \\ F \end{bmatrix} x + \begin{bmatrix} -P^{1/2} & 0 \\ 0 & L \end{bmatrix} \nu.$$

This set of equations hence becomes

$$\begin{bmatrix} F\bar{x} - y \\ \bar{x} \end{bmatrix} = \begin{bmatrix} 0 \\ I_n \end{bmatrix} x + \begin{bmatrix} -FP^{1/2} & -L \\ -P^{1/2} & 0 \end{bmatrix} \nu.$$

Second, we use the orthogonal transformation T_r of Equation (4.25) to
transform the set of equations into

$$\begin{bmatrix} F\bar{x} - y \\ \bar{x} \end{bmatrix} = \begin{bmatrix} 0 \\ I_n \end{bmatrix} x + \begin{bmatrix} R & 0 \\ G & S \end{bmatrix} \begin{bmatrix} \kappa \\ \delta \end{bmatrix} \quad \text{with } T_r^T \nu = \begin{bmatrix} \kappa \\ \delta \end{bmatrix}. \qquad (4.28)$$

Multiplying each side of the equality in (4.25) by its transpose yields

$$\begin{bmatrix} FPF^T + LL^T & FP \\ PF^T & P \end{bmatrix} = \begin{bmatrix} RR^T & RG^T \\ GR^T & GG^T + SS^T \end{bmatrix}.$$

As a result, we find the following relationship between the given matrices
F, P, and L and the obtained matrices R, G, and S:

$$(FPF^T + LL^T) = (FPF^T + W^{-1}) = RR^T, \qquad (4.29)$$

$$PF^T = GR^T, \qquad (4.30)$$

$$P - GG^T = SS^T. \qquad (4.31)$$

The first equation shows that the matrix R is full rank since the weighting matrix W was assumed to be full rank. Equation (4.31) can, with
the help of the matrix-inversion lemma (Lemma 2.2 on page 19), be used
to show that, when in addition P is full rank, the matrix S is full rank.

On the basis of the invertibility of the matrix R, we can explicitly
write κ in Equation (4.28) as

$$\kappa = R^{-1}(F\bar{x} - y).$$

This reduces the second block row of Equation (4.28) to

$$\bar{x} = x + GR^{-1}(F\bar{x} - y) + S\delta.$$

Reordering terms yields

$$x = \underbrace{(I_n - GR^{-1}F)\bar{x} + GR^{-1}y} + S\delta. \tag{4.32}$$

Now we show that the underbraced term is the solution of the weighted least-squares problem (4.24), provided that the matrix S is invertible. We also show that the covariance matrix of this estimate is SS^T. The transformed set of equations (4.28) and the invertibility of the matrices R and S can be used to express the cost function in Equation (4.24) as

$$\begin{aligned}
\min_x \nu^T\nu &= \min_x \kappa^T\kappa + \delta^T\delta \\
&= \min_x (F\bar{x} - y)^T R^{-T} R^{-1}(F\bar{x} - y) \\
&\quad + (x - (I_n - GR^{-1}F)\bar{x} - GR^{-1}y)^T S^{-T} \\
&\quad \times S^{-1}(x - (I_n - GR^{-1}F)\bar{x} - GR^{-1}y).
\end{aligned}$$

An application of the "completion-of-squares" argument to this cost function shows that the solution of (4.24) is given by $\hat{x}$ in Equation (4.26). With the property $\delta \sim (0, I)$ and Equation (4.32), the covariance matrix of this estimate is obtained and it equals Equation (4.27).

We are now left to prove that the solution $\hat{x}$ obtained above and its covariance matrix SS^T equal Equations (4.18) and (4.19), respectively. Using Equations (4.29) and (4.30), and the fact that R is invertible, we can write GR^{-1} as $PF^T(FPF^T + W^{-1})^{-1}$ and therefore the solution $\hat{x}$ can be written as

$$\hat{x} = \left(I - PF^T(FPF^T + W^{-1})^{-1}F\right)\bar{x} + PF^T(FPF^T + W^{-1})^{-1}y.$$

Again using Equations (4.29) and (4.30), we can write the product GG^T as $PF^T(FPF^T + W^{-1})^{-1}FP$. Substituting this expression into Equation (4.31) shows that the covariance matrix SS^T equals the one given in Equation (4.19). $\square$

The proof of the previous theorem shows that the calculation of the solution $\hat{x}$ and its covariance matrix *does not* require the prior covariance matrix P to be full rank. The latter property is required only in order to show that $\hat{x}$ is the solution of the weighted least-squares problem (4.24). When P (or its square root S) is not full rank, the weighted least-squares problem (4.24) no longer has a unique solution.

Table 4.1. *The estimated mean value of the relative error*
$\|\widehat{x} - x\|_2 / \|x\|_2$ *with the minimum-variance unbiased estimate* $\widehat{x}$
numerically computed in three different ways

	Equation used		
	(4.18)	(4.20)	(4.26)
$\|\widehat{x} - x\|_2 / \|x\|_2$	0.0483	0.1163	0.0002

The algorithm presented in Theorem 4.4 is called a *square-root algorithm*, since it estimates the vector x using the square root of the covariance matrix and it does not need the square of the weighting matrix L. Thus it works directly with the original data. Although, when P is invertible, the three approaches to compute the estimate of x as given by Equations (4.18), (4.20), and (4.26) presented here are *analytically equivalent*, they can differ significantly in numerical calculations. This is illustrated in the next example.

Example 4.6 (Numerical solution of a stoschastic least-squares problem) Consider the stochastic least-squares problem for the following data:

$$\overline{x} = 0, \qquad P = 10^7 \cdot I_3, \qquad L = 10^{-6} \cdot I_3, \qquad x = \begin{bmatrix} 1 \\ -1 \\ 0.1 \end{bmatrix}.$$

The matrices P and L indicate that the prior information is inaccurate and that the relationship between x and y (the measurement) is very accurate. The matrix F is generated randomly by the Matlab (Math-Works, 2000b) command

$$F = \text{gallery}(\text{'randsvd'},3);$$

For 100 random generations of the matrix F, the estimate of x is computed using equations (4.18), (4.20), and (4.26). The means of the relative error

$$\frac{\|\widehat{x} - x\|_2}{\|x\|_2}$$

are listed in Table 4.1 for the three estimates.

The results show that, for these (extremely) simple data, the calculations via the square root are a factor of 25 more accurate than the calculations via the covariance matrices. When using the covariance matrices, the calculation of the estimate $\widehat{x}$ via (4.18) is preferred over the one given by (4.20). The fact that the matrix F is ill-conditioned shows that Equation (4.20), which requires the matrix P to be invertible, performs worse.

The previous example shows that the preferred computational method for the minimum-variance unbiased estimate of the stochastic linear least-squares problem is the square-root algorithm of Theorem 4.4 on page 116. The square-root algorithm will also be used in addressing the Kalman-filter problem in Chapter 5.

4.5.5 *Maximum-likelihood interpretation of the weighted linear least-squares problem*

Consider the problem of determining x from the measurements y given by

$$y = Fx + \mu, \tag{4.33}$$

with μ a zero-mean random vector with jointly Gaussian-distributed entries and covariance matrix C_μ. According to Section 4.3.2, the probability density function of μ is given by

$$f_\mu(\alpha_1, \alpha_2, \ldots, \alpha_m) = \frac{1}{(2\pi)^{m/2} \det(C_\mu)^{1/2}} \exp\left(-\frac{1}{2}\alpha^{\mathrm{T}} C_\mu^{-1}\alpha\right),$$

with $\alpha = [\alpha_1, \alpha_2, \ldots, \alpha_m]^{\mathrm{T}}$. If we combine this probability density function with the signal model (4.33), we get the *likelihood function*

$$l(y|Fx) = \frac{1}{(2\pi)^{m/2} \det(C_\mu)^{1/2}} \exp\left(-\frac{1}{2}(y - Fx)^{\mathrm{T}} C_\mu^{-1}(y - Fx)\right).$$

It expresses the likelihood of the measurement y as a function of the parameters Fx. On the basis of this likelihood function, the idea is to take as an estimate for x the value of x that makes the observation y most likely. Therefore, to determine an estimate of x, the likelihood function is maximized with respect to x. This particular estimate is called the *maximum-likelihood estimate*. Often, for ease of computation, the logarithm of the likelihood function is maximized. The maximum-likelihood estimate $\widehat{x}_{\mathrm{ML}}$ is obtained as follows:

$$\widehat{x}_{\mathrm{ML}} = \arg\max_x \ln l(y|Fx)$$

$$= \arg\min_x \left(\frac{1}{2}(y - Fx)^{\mathrm{T}} C_\mu^{-1}(y - Fx) + \ln(2\pi)^{m/2} \det(C_\mu)^{1/2}\right).$$

The solution $\widehat{x}_{\mathrm{ML}}$ is independent of the value of $\ln(2\pi)^{m/2} \det(C_\mu)^{1/2}$, therefore the same $\widehat{x}_{\mathrm{ML}}$ results from

$$\widehat{x}_{\mathrm{ML}} = \arg\min_x (y - Fx)^{\mathrm{T}} C_\mu^{-1}(y - Fx), \tag{4.34}$$

where the scaling by $1/2$ has also been dropped and maximization of $-(y - Fx)^{\mathrm{T}} C_\mu^{-1}(y - Fx)$ has been replaced by minimization of $(y - Fx)^{\mathrm{T}} C_\mu^{-1}(y - Fx)$.

Comparing Equation (4.34) with (4.13) shows that $\widehat{x}_{\mathrm{ML}}$ is the solution to a weighted linear least-squares problem of the form (4.13) with $L = C_\mu^{1/2}$.

4.6 Summary

This chapter started off with a review of some basic concepts of probability theory. The definition of a random variable was given, and it was explained how the cumulative distribution function and the probability density function describe a random variable. Other concepts brought to light were the expected value, mean, variance, standard deviation, covariance, and correlation. Relations between multiple random variables can be described by the joint cumulative distribution and the joint probability density function. Two random variables are independent if their joint probability density function can be factored. Two random variables are uncorrelated if the expected value of their product equals the product of their expected values; they are orthogonal if the expected value of their product equals zero.

A random signal or random process is a sequence of random variables. Some quantities that are often used to characterize random signals are the auto-covariance, auto-correlation, cross-covariance, and cross-correlation. Important classes of random signals are the Gaussian signals and the IID signals. A random signal is called stationary if the joint cumulative distribution function of any finite number of samples does not depend on the placement of the time origin. A random signal is called wide-sense stationary (WSS) if its mean does not depend on time and its auto-correlation function depends only on the time lag. For WSS sequences we defined the power spectrum and the cross-spectrum and we gave the relations between input and output spectra for a linear time-invariant system.

Next we turned our attention to stochastic properties of linear least-squares estimates. Under certain conditions the solution to the linear least-squares problem is an unbiased linear estimator and it has the minimum-variance estimation property. A similar result was presented for the weighted least-squares problem. We concluded the chapter by discussing the stochastic linear least-squares problem and the maximum-likelihood interpretation of the weighted least-squares problem.

Exercises

4.1   Prove the six properties of the cumulative distribution function listed on page 94.

4.2   Let X be a Gaussian random variable with PDF given by

$$f_x(\alpha) = \frac{1}{\sqrt{2\pi}\sigma}e^{-(\alpha-m)^2/(2\sigma^2)},$$

with $-\infty < \alpha < \infty$. Prove that

$$E[X] = m,$$
$$\text{var}[X] = \sigma^2.$$

4.3   Let $x(1)$ and $x(2)$ be two uncorrelated random variables with mean values m_1 and m_2, respectively. Show that

$$E\left[\left(x(1) + x(2) - (m_1 + m_2)\right)^2\right]$$
$$= E[(x(1) - m_1)^2] + E[(x(2) - m_2)^2].$$

4.4   Let the sequence $y(k)$ be generated by filtering the zero-mean WSS random signal $w(k)$ by the filter with transfer function

$$H(z) = \frac{1}{z - a}, \quad \text{with } a < 1.$$

Further assume that $w(k)$ is exponentially correlated, with auto-correlation function

$$R_w(\tau) = c^{|\tau|}, \quad \text{with } c < 1,$$

and take a sampling time $T = 1$.

    (a) Determine the power spectrum $\Phi^w(\omega)$ of $w(k)$.
    (b) Determine the power spectrum $\Phi^y(\omega)$ of $y(k)$.

4.5   Let the output of an LTI system be given by

$$y(k) = G(q)u(k) + v(k),$$

with $v(k)$ an unknown disturbance signal. The system is operated in closed-loop mode with the input determined by the feedback connection

$$u(k) = r(k) - C(q)y(k),$$

with a controller $C(q)$ and $r(k)$ an external reference signal.

    (a) Express $y(k)$ in terms of the signals $r(k)$ and $v(k)$.

    (b) Show that for $\Phi^v(\omega) \neq 0$ and $C(e^{j\omega T})G(e^{j\omega T}) \neq -1$ we have

$$\frac{\Phi^{yu}(\omega)}{\Phi^u(\omega)} \neq G(e^{j\omega T}).$$

    (c) Determine the FRFs $G(e^{j\omega T})$ and $C(e^{j\omega T})$ from the power spectra and cross-spectra of the signals $r(k)$, $u(k)$, and $y(k)$.

4.6   Let the sequence $y(k)$ be generated by filtering the WSS zero-mean white-noise sequence $w(k)$ with $E[w(k)^2] = 1$ by the filter with transfer function

$$H(z) = \frac{1}{z - 0.9}.$$

Take a sampling time $T = 1$.

    (a) Determine the auto-correlation function $R_y(\tau)$ analytically.

    (b) Generate N samples of the sequence $y(k)$ and use these batches subsequently to generate the estimates $\widehat{R}_y^N(\tau)$ by

$$\widehat{R}_y^N(\tau) = \frac{1}{N - \tau} \sum_{i=\tau+1}^{N} y(i)y(i - \tau),$$

for $N = 1000$ and $10\,000$. Compare these sample estimates with their analytical equivalent.

4.7   Prove Theorem 4.2 on page 112.

4.8   Under the assumption that the matrices $P^{1/2}$ and L in weighted least-squares problem (4.24) on page 116 are invertible, show that this weighted least-squares problem is equivalent to the optimization problem

$$\min_x (\overline{x} - x)^T P^{-1}(\overline{x} - x) + (y - Fx)^T W(y - Fx), \qquad \text{(E4.1)}$$

with $W = (LL^T)^{-1}$.

4.9   Let the cost function $J(K)$ be given as

$$J(K) = E\left[(x - Ky)(x - Ky)^T\right],$$

with $x \in \mathbb{R}$ and $y \in \mathbb{R}^N$ random vectors having the following covariance matrices:

$$E[xx^T] = R_x,$$
$$E[yy^T] = R_y,$$
$$E[xy^T] = R_{xy}.$$

(a) Determine the optimal matrix k_0 such that

$$K_0 = \arg\min_K J(K).$$

(b) Determine the value of $J(K_0)$.

(c) The random vector x has zero mean and a variance equal to 1. The vector y contains N noisy measurements of x such that

$$y = \begin{bmatrix} y(0) \\ y(1) \\ \vdots \\ y(N-1) \end{bmatrix},$$

with

$$y(k) = x + v(k), \quad k = 0, 1, \ldots, N - 1,$$

where $v(k)$ is a zero-mean white-noise sequence that is independent of x and has a variance equal to σ_v^2. Determine for $N = 5$ the matrices R_{xy} and R_y.

(d) Under the same conditions as in part (c), determine the optimal estimate of x as $\widehat{x} = K_0 y$, with K_0 minimizing $J(K)$ as defined in part (a).

4.10   Consider the signal $y(k)$ generated by filtering a stochastic signal $w(k)$ by the LTI system given by

$$x(k+1) = Ax(k) + Bw(k), \quad x(0) = 0, \qquad \text{(E4.2)}$$
$$y(k) = Cx(k) + Dw(k), \qquad \text{(E4.3)}$$

with $x(k) \in \mathbb{R}^n$ and $y(k) \in \mathbb{R}^\ell$. The input $w(k)$ to this filter has an auto-correlation given by

$$R_w(\tau) = E[w(k)w(k-\tau)] \qquad \text{(E4.4)}$$

and a power spectrum given by

$$\phi^w(\omega) = \sum_{\tau=-\infty}^{\infty} R_w(\tau)e^{-j\omega\tau}. \qquad \text{(E4.5)}$$

Assuming that the system (E4.2)–(E4.3) is asymptotically stable, show that the following expressions hold:

(a) $R_{yw}(\tau) = \sum_{i=0}^{\infty} CA^i BR_w(\tau - i - 1) + DR_w(\tau),$

(b) $R_y(\tau) = \sum_{i=0}^{\infty} \sum_{j=0}^{\infty} CA^i BR_w(\tau + j - i) B^{\mathrm{T}}(A^j)^{\mathrm{T}} C^{\mathrm{T}}$

$\qquad + DR_w(\tau) D^{\mathrm{T}}$

$\qquad + \sum_{i=0}^{\infty} CA^i BR_w(\tau - i - 1) D^{\mathrm{T}}$

$\qquad + \sum_{j=0}^{\infty} DR_w(\tau - j - 1) B^{\mathrm{T}}(A^j)^{\mathrm{T}} C^{\mathrm{T}},$

(c) $\phi^{yw}(\omega) = \left[D + C(e^{j\omega} - A)^{-1} B \right] \phi^w(\omega)$

(d) $\phi^y(\omega) = \left[D + C(e^{j\omega} - A)^{-1} B \right] \phi^w(\omega)$

$\qquad \times \left[D^{\mathrm{T}} + B^{\mathrm{T}}(e^{-j\omega} - A^{\mathrm{T}})^{-1} C^{\mathrm{T}} \right].$

5

Kalman filtering

After studying this chapter you will be able to

- use an observer to estimate the state vector of a linear time-invariant system;
- use a Kalman filter to estimate the state vector of a linear system using knowledge of the system matrices, the system input and output measurements, and the covariance matrices of the disturbances in these measurements;
- describe the difference among the predicted, filtered, and smoothed state estimates;
- formulate the Kalman-filter problem as a stochastic and a weighted least-squares problem;
- solve the stochastic least-squares problem by application of the completion-of-squares argument;
- solve the weighted least-squares problem in a recursive manner using elementary insights of linear algebra and the mean and covariance of a stochastic process;
- derive the square-root covariance filter (SRCF) as the recursive solution to the Kalman-filter problem;
- verify the optimality of the Kalman filter via the white-noise property of the innovation process; and
- use the Kalman-filter theory to estimate unknown inputs of a linear dynamical system in the presence of noise perturbations on the model (process noise) and the observations (measurement noise).

5.1 Introduction

Imagine that you are measuring a scalar quantity $x(k)$, say a temperature. Your sensor measuring this quantity produces $y(k)$. Since the measurement is not perfect, some (stochastic) measurement errors are introduced. If we let $v(k)$ be a zero-mean white-noise sequence with variance R, then a plausible model for the observed data is

$$y(k) = x(k) + v(k). \qquad (5.1)$$

A relevant question is that of whether an algorithm can be devised that processes the observations $y(k)$ (measured during a certain time interval) to produce an estimate of $x(k)$ that has a smaller variance than R. Going even further, we could then ask whether it is possible to tune the algorithm such that it minimizes the variance of the error in the estimate of $x(k)$.

The answers to these questions very much depend on how the temperature $x(k)$ *changes* with time. In other words, all depends on the dynamics of the underlying process. When the temperature is "almost" constant, we could represent its dynamics as

$$x(k+1) = x(k) + w(k), \qquad (5.2)$$

where $w(k)$ is a zero-mean white-noise sequence with variance Q, and $w(k)$ is possibly correlated with $v(k)$.

Using Equations (5.1) and (5.2), the problem of finding a minimum-error variance estimate of the quantity $x(k)$ (the state) is a special case of the well-known *Kalman-filter problem*.

The Kalman filter is a computational scheme to reconstruct the state of a given state-space model in a statistically optimal manner, which is generally expressed as the minimum variance of the state-reconstruction error conditioned on the acquired measurements. The filter may be derived in a framework based on conditional probability theory. The theoretical foundations used from statistics are quite involved and complex. Alternative routes have been explored in order to provide simpler derivations and implementations of the same "optimal-statistical-state observer." An overview of these different theoretical approaches would be very interesting, but would lead us too far. To name only a few, there are, for example, the innovations approach (Kailath, 1968), scattering theory (Friedlander *et al.*, 1976), and the work of Duncan and Horn (1972). Duncan and Horn (1972) formulated a relation between the minimum-error variance-estimation problem for general time-varying discrete-time systems and *weighted least squares*. Paige (1985) used this approach in combination with a generalization of the covariance representation used in the Kalman filter to derive the *square-root* recursive Kalman filter. This filter is known to possess better computational properties than those of the classical covariance formulation (Verhaegen and Van Dooren, 1986). In this chapter, the approach of Paige will be followed in deriving the square-root Kalman-filter recursions.

The Kalman filter is a member of the class of filters used to reconstruct missing information, such as part of the state vector, from measured quantities in a state-space model. This class of filters is known as *observers*. In Section 5.2, we treat as an introduction to Kalman filters the special class of linear observers used to reconstruct the state of a linear time-invariant state-space model. In Section 5.3 the Kalman-filter problem is introduced, and it is formulated as a stochastic least-squares problem. Its solution based on the completion-of-squares argument is presented in Section 5.4. The solution, known as the conventional Kalman filter, gives rise to the definition of the so-called (one-step) ahead predicted and filtered state estimate. The square-root variants of the conventional Kalman-filter solution are discussed in Section 5.5. The derivation of the numerically preferred square-root variants is performed as in Section 4.5.4, via the analysis of a (deterministic) weighted least-squares problem. This recursive solution both of the stochastic and of the weighted least-squares problem formulation is initially presented in a so-called *measurement-update* step, followed by a *time-update* step. Further interesting insights into Kalman filtering are developed in this section. These include the combined measurement-update and time-update formulation of the well-known *square-root covariance filter* (SRCF) implementation, and the innovation representation. The derivation of fixed-interval smoothing solutions to the state-estimation problem is discussed in the framework of weighted least-squares problems in Section 5.6. Section 5.7 presents the Kalman filter for linear time-invariant systems, and the conditions for which the Kalman-filter recursions become stationary are given. Section 5.8 applies Kalman filtering to estimate unknown inputs of linear dynamic systems.

5.2 The asymptotic observer

An observer is a filter that approximates the state vector of a dynamical system from measurements of the input and output sequences. It requires a model of the system under consideration. The first contribution regarding observers for LTI systems in a state-space framework was made by Luenberger (1964). He considered the approximation of the state-vector sequence $x(k)$ for $k \geq 0$, with $x(0)$ unknown, of the following LTI state-space model:

$$x(k+1) = Ax(k) + Bu(k), \qquad (5.3)$$

$$y(k) = Cx(k) + Du(k), \qquad (5.4)$$

where $x(k) \in \mathbb{R}^n$ is the state vector, $u(k) \in \mathbb{R}^m$ the input vector, and $y(k) \in \mathbb{R}^\ell$ the output vector.

We can approximate the state in Equation (5.3) by driving the equation

$$\widehat{x}(k+1) = A\widehat{x}(k) + Bu(k),$$

with the same input sequence as Equation (5.3). If the initial state $\widehat{x}(0)$ equals $x(0)$, then the state sequences $x(k)$ and $\widehat{x}(k)$ will be equal. If the initial conditions differ, $x(k)$ and $\widehat{x}(k)$ will become equal after some time, provided that the A matrix is asymptotically stable. To see this, note that the difference between the state $\widehat{x}(k)$ and the real state given by

$$x_e(k) = \widehat{x}(k) - x(k)$$

satisfies $x_e(k+1) = Ax_e(k)$. Although the states become equal after some time, we have no control over the rate at which $\widehat{x}(k)$ approaches $x(k)$.

Instead of using only the input $u(k)$ to reconstruct the state, we can also use the output. The estimate of the state can be improved by introducing a correction based on the difference between the measured output $y(k)$ and the estimated output $\widehat{y}(k) = C\widehat{x}(k) + Du(k)$, as follows:

$$\widehat{x}(k+1) = A\widehat{x}(k) + Bu(k) + K\Big(y(k) - C\widehat{x}(k) - Du(k)\Big), \qquad (5.5)$$

where K is a gain matrix. The system represented by this equation is often called an *observer*. We have to chose the matrix K in an appropriate way. The difference $x_e(k)$ between the estimated state $\widehat{x}(k)$ and the real state $x(k)$ satisfies

$$x_e(k+1) = (A - KC)x_e(k),$$

and thus, if K is chosen such that $A - KC$ is asymptotically stable, the difference between the real state $x(k)$ and the estimated state $\widehat{x}(k)$ goes to zero for $k \to \infty$:

$$\lim_{k \to \infty} x_e(k) = \lim_{k \to \infty} \Big(\widehat{x}(k) - x(k)\Big) = 0.$$

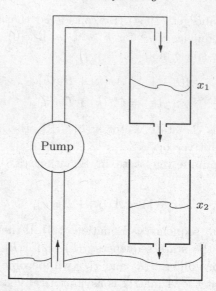

Fig. 5.1. A schematic view of the double-tank process of Example 5.1.

The choice of the matrix K also determines the rate at which $\widehat{x}(k)$ goes to zero.

The observer (5.5) is called an *asymptotic observer*. Regarding the choice of the matrix K in the asymptotic observer, we have the following important result.

Lemma 5.1 (Observability) *(Kailath, 1980) Given matrices $A \in \mathbb{R}^{n \times n}$ and $C \in \mathbb{R}^{\ell \times n}$, if the pair (A, C) is observable, then there exists a matrix $K \in \mathbb{R}^{n \times \ell}$ such that $A - KC$ is asymptotically stable.*

This lemma illustrates the importance of the observability condition on the pair (A, C) in designing an observer. In Exercise 5.4 on page 174 you are requested to prove this lemma for a special case. We conclude this section with an illustrative example of the use of the observer.

Example 5.1 (Observer for a double-tank process) Consider the double-tank process depicted in Figure 5.1 (Eker and Malmborg, 1999). This process is characterized by the two states x_1 and x_2, which are the height of the water in the upper tank and that in the lower tank, respectively. The input signal u is the voltage to the pump and the output signal $y = x_2$ is the level in the lower tank. A nonlinear continuous-time state-space description derived from Bernoulli's energy equation of the process is

$$\frac{d}{dt}\begin{bmatrix} x_1(t) \\ x_2(t) \end{bmatrix} = \begin{bmatrix} -\alpha_1\sqrt{x_1(t)} + \beta u(t) \\ \alpha_1\sqrt{x_1(t)} - \alpha_2\sqrt{x_2(t)} \end{bmatrix},$$

where α_1 and α_2 are the areas of the output flows of both tanks. The equilibrium point $(\overline{x}_1, \overline{x}_2)$ of this system, for a constant input $\overline{u}$, is defined by $dx_1(t)/dt = 0$ and $dx_2(t)/dt = 0$, yielding

$$\begin{bmatrix} \sqrt{\overline{x}_1} \\ \sqrt{\overline{x}_2} \end{bmatrix} = \begin{bmatrix} \dfrac{\beta}{\alpha_1}\overline{u} \\ \dfrac{\beta}{\alpha_2}\overline{u} \end{bmatrix}.$$

In the following we take $\alpha_1 = 1$, $\alpha_2 = 1$, and $\beta = 1$. Linearizing the state-space model around its equilibrium point with $\overline{u} = 1$ (see Section 3.4.1) yields the following linear state-space model:

$$\frac{d}{dt}\begin{bmatrix} x_1(t) \\ x_2(t) \end{bmatrix} = \begin{bmatrix} -\dfrac{1}{2\sqrt{\overline{x}_1}} & 0 \\ \dfrac{1}{2\sqrt{\overline{x}_1}} & -\dfrac{1}{2\sqrt{\overline{x}_2}} \end{bmatrix}\begin{bmatrix} x_1(t) \\ x_2(t) \end{bmatrix} + \begin{bmatrix} 1 \\ 0 \end{bmatrix}u(t),$$

$$\begin{bmatrix} \overline{x}_1 \\ \overline{x}_2 \end{bmatrix} = \begin{bmatrix} 1 \\ 1 \end{bmatrix}$$

where the same symbols are used to represent the individual variation of the state and input quantities around their equilibrium values. The linear continuous-time state-space model then reads

$$\frac{d}{dt}\begin{bmatrix} x_1(t) \\ x_2(t) \end{bmatrix} = \begin{bmatrix} -1/2 & 0 \\ 1/2 & -1/2 \end{bmatrix}\begin{bmatrix} x_1(t) \\ x_2(t) \end{bmatrix} + \begin{bmatrix} 1 \\ 0 \end{bmatrix}u(t),$$

$$y(t) = \begin{bmatrix} 0 & 1 \end{bmatrix}\begin{bmatrix} x_1(t) \\ x_2(t) \end{bmatrix}.$$

This continuous-time description is discretized using a zeroth-order hold assumption on the input for a sampling period equal to 0.1 s (Åström and Wittenmark, 1984). This yields the following discrete-time state-space model:

$$x(k+1) = \begin{bmatrix} 0.9512 & 0 \\ 0.0476 & 0.9512 \end{bmatrix}x(k) + \begin{bmatrix} 0.0975 \\ 0.0024 \end{bmatrix}u(k),$$

$$y(k) = \begin{bmatrix} 0 & 1 \end{bmatrix}x(k).$$

The input $u(k)$ is a periodic block-input sequence with a period of 20 s, as shown in Figure 5.2.

The output of the discrete-time model of the double tank is simulated with this input sequence and with the initial conditions taken equal

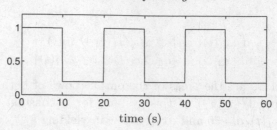

Fig. 5.2. The periodic block-input sequence for the double-tank process of Example 5.1.

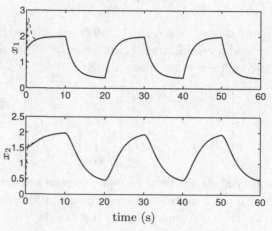

Fig. 5.3. True (solid line) and reconstructed (broken line) state-vector sequences of the double-tank process of Example 5.1 obtained with an asymptotic observer.

to

$$x(0) = \begin{bmatrix} 1.5 \\ 1.5 \end{bmatrix}.$$

The gain matrix K in an observer of the form (5.5) is designed (using the Matlab Control Toolbox (MathWorks, 2000c) command `place`) such that the poles of the matrix $A - KC$ are equal to 0.7 and 0.8. The true and reconstructed state-vector sequences are plotted in Figure 5.3. We clearly observe that after only a few seconds the state vector of the observer (5.5) becomes equal to the true state vector. Changing the eigenvalues of the matrix $A - KC$ will influence the speed at which the difference between the observer and the true states becomes zero. Decreasing the magnitude of the eigenvalues makes the interval wherein the state of the observer and the true system differ *smaller*. Therefore, when there is no noise in the measurements and the model of the system is perfectly known, the example seems to suggest making the magnitudes of the eigenvalues *zero*. This is then called a dead-beat observer. However, when the measurements contain noise, as will almost always be the case in practice, the selection of the eigenvalues is not trivial. An answer regarding how to locate the eigenvalues "optimally" is provided by the Kalman filter, which is discussed next.

5.3 The Kalman-filter problem

Like the observer treated in the previous section, the Kalman filter is a filter that approximates the state vector of a dynamical system from measurements of the input and output sequences. The main difference from the observer is that the Kalman filter takes noise disturbances into account. Consider the LTI model (5.3)–(5.4) of the previous section, but now corrupted by two noise sequences $w(k)$ and $v(k)$:

$$x(k+1) = Ax(k) + Bu(k) + w(k), \tag{5.6}$$
$$y(k) = Cx(k) + Du(k) + v(k). \tag{5.7}$$

The vector $w(k) \in \mathbb{R}^n$ is called the *process noise* and $v(k) \in \mathbb{R}^\ell$ is called the *measurement noise*.

If we now use the observer (5.5) to reconstruct the state of the system (5.6)–(5.7), the difference $x_e(k) = \widehat{x}(k) - x(k)$ between the estimated state $\widehat{x}(k)$ and the real state $x(k)$ satisfies

$$x_e(k+1) = (A - KC)x_e(k) - w(k) + Kv(k).$$

Even if $A - KC$ were asymptotically stable, $x_e(k)$ would not go to zero for $k \to \infty$, because of the presence of $w(k)$ and $v(k)$. Now, the goal is to make $x_e(k)$ "small," because then the state estimate will be close to the real state. Since $x_e(k)$ is a random signal, we can try to make the mean of $x_e(k)$ equal to zero: $E[x_e(k)] = E[\widehat{x}(k) - x(k)] = 0$. In other words, we may look for an *unbiased* state estimate $\widehat{x}(k)$ (compare this with Definition 4.15 on page 110). However, this does not mean that $x_e(k)$ will be "small"; it can still vary wildly around its mean zero. Therefore, in addition, we want to make the state-error covariance matrix $E[x_e(k)x_e(k)^{\mathrm{T}}]$ as small as possible; that is, we are looking for a minimum-error variance estimate (compare this with Definition 4.16 on page 110).

In the following we will discuss the Kalman filter for a linear time-varying system of the form

$$x(k+1) = A(k)x(k) + B(k)u(k) + w(k), \qquad (5.8)$$
$$y(k) = C(k)x(k) + v(k). \qquad (5.9)$$

Note that the output Equation (5.9) does not contain a direct feed-through term $D(k)u(k)$. This is not a serious limitation, because this term can easily be incorporated into the derivations that follow. In Exercise 5.1 on page 172 the reader is requested to adapt the Kalman-filter derivations of this chapter when the feed-through term $D(k)u(k)$ is added to the output equation (5.9).

We are now ready to state the Kalman-filter problem, which will be treated in Sections 5.4 and 5.5.

We are given the signal-generation model (5.8) and (5.9) with the process noise $w(k)$ and measurement noise $v(k)$ assumed to be zero-mean white-noise sequences with joint covariance matrix

$$E\left[\begin{bmatrix} v(k) \\ w(k) \end{bmatrix} \begin{bmatrix} v(j)^{\mathrm{T}} & w(j)^{\mathrm{T}} \end{bmatrix}\right] = \begin{bmatrix} R(k) & S(k)^{\mathrm{T}} \\ S(k) & Q(k) \end{bmatrix} \Delta(k-j) \geq 0, \quad (5.10)$$

with $R(k) > 0$ and where $\Delta(k)$ is the unit pulse (see Section 3.2). At time instant $k-1$, we have an estimate of $x(k)$, which is denoted by $\widehat{x}(k|k-1)$ with properties

$$E[x(k)] = E[\widehat{x}(k|k-1)], \qquad (5.11)$$
$$E\left[\Big(x(k) - \widehat{x}(k|k-1)\Big)\Big(x(k) - \widehat{x}(k|k-1)\Big)^{\mathrm{T}}\right] = P(k|k-1) \geq 0. \qquad (5.12)$$

This estimate is uncorrelated with the noise $w(k)$ and $v(k)$. The problem is to determine a linear estimate of $x(k)$ and $x(k+1)$ based on the given data $u(k)$, $y(k)$, and $\widehat{x}(k|k-1)$, which have the following form:

$$\begin{bmatrix} \widehat{x}(k|k) \\ \widehat{x}(k+1|k) \end{bmatrix} = M \begin{bmatrix} y(k) \\ -B(k)u(k) \\ \widehat{x}(k|k-1) \end{bmatrix}, \qquad (5.13)$$

with $M \in \mathbb{R}^{2n \times (\ell + 2n)}$, such that both estimates are minimum-variance unbiased estimates; that is, estimates with the properties

$$E[\widehat{x}(k|k)] = E[x(k)], \qquad E[\widehat{x}(k+1|k)] = E[x(k+1)], \qquad (5.14)$$

and the expressions below are minimal:

$$E\left[\Big(x(k) - \widehat{x}(k|k)\Big)\Big(x(k) - \widehat{x}(k|k)\Big)^{\mathrm{T}}\right],$$

$$E\left[\Big(x(k+1) - \widehat{x}(k+1|k)\Big)\Big(x(k+1) - \widehat{x}(k+1|k)\Big)^{\mathrm{T}}\right]. \qquad (5.15)$$

5.4 The Kalman filter and stochastic least squares

The signal-generating model in Equations (5.8) and (5.9) can be denoted by the set of equations

$$\begin{bmatrix} y(k) \\ -B(k)u(k) \end{bmatrix} = \begin{bmatrix} C(k) & 0 \\ A(k) & -I_n \end{bmatrix} \begin{bmatrix} x(k) \\ x(k+1) \end{bmatrix} + L(k)\epsilon(k), \quad \epsilon(k) \sim (0, I_{\ell+n}),$$

$$(5.16)$$

with $L(k)$ a lower-triangular square root of the joint covariance matrix

$$\begin{bmatrix} R(k) & S(k)^{\mathrm{T}} \\ S(k) & Q(k) \end{bmatrix} = L(k)L(k)^{\mathrm{T}}$$

and $\epsilon(k)$ an auxiliary variable representing the noise sequences. We observe that this set of equations has the same form as the stochastic least-squares problem formulation (4.16) on page 113, which was analyzed in Section 4.5.3.

On the basis of this insight, it would be tempting to treat the Kalman-filter problem as a stochastic least-squares problem. However, there are two key differences. First, in formulating a stochastic least-squares problem based on the set of Equations (5.16) we need prior estimates of the mean and covariance matrix of the vector

$$\begin{bmatrix} x(k) \\ x(k+1) \end{bmatrix}.$$

In the Kalman-filter problem formulation it is assumed that only prior information is given about the state $x(k)$. Second, in the Kalman-filter problem we seek to minimize the individual covariance matrices of the state estimate of $x(k)$ and $x(k+1)$, rather than their joint covariance matrix

$$E\left[\begin{bmatrix} x(k) - \widehat{x}(k|k) \\ x(k+1) - \widehat{x}(k+1|k) \end{bmatrix}\begin{bmatrix} x(k) - \widehat{x}(k|k) \\ x(k+1) - \widehat{x}(k+1|k) \end{bmatrix}^{\mathrm{T}}\right].$$

Minimizing the joint covariance matrix would be the objective of the stochastic least-squares problem.

In conclusion, the Kalman-filter problem cannot be solved by a straight-forward application of the solution to the stochastic least-squares problem. A specific variant needs to be developed, as is demonstrated in the following theorem.

Theorem 5.1 (Conventional Kalman filter) *Let the conditions stipulated in the Kalman-filter problem hold, then the minimum-variance unbiased estimate for $x(k)$ is given by*

$$\widehat{x}(k|k) = P(k|k-1)C(k)^{\mathrm{T}}\left(C(k)P(k|k-1)C(k)^{\mathrm{T}} + R(k)\right)^{-1}y(k)$$
$$+ \left(I_n - P(k|k-1)C(k)^{\mathrm{T}}\left(C(k)P(k|k-1)C(k)^{\mathrm{T}}\right.\right.$$
$$\left.\left. + R(k)\right)^{-1}C(k)\right)\widehat{x}(k|k-1), \tag{5.17}$$

with covariance matrix

$$E\left[\left(x(k) - \widehat{x}(k|k)\right)\left(x(k) - \widehat{x}(k|k)\right)^{\mathrm{T}}\right]$$
$$= P(k|k-1) - P(k|k-1)C(k)^{\mathrm{T}}\left(C(k)P(k|k-1)C(k)^{\mathrm{T}} + R(k)\right)^{-1}$$
$$\times C(k)P(k|k-1). \tag{5.18}$$

The minimum-variance unbiased estimate for $x(k+1)$ is given by

$$\widehat{x}(k+1|k) = \left(A(k)P(k|k-1)C(k)^{\mathrm{T}} + S(k)\right)$$
$$\times \left(C(k)P(k|k-1)C(k)^{\mathrm{T}} + R(k)\right)^{-1}y(k)$$
$$+ B(k)u(k) + \left(A(k) - \left(A(k)P(k|k-1)C(k)^{\mathrm{T}} + S(k)\right)\right.$$
$$\left.\times \left(C(k)P(k|k-1)C(k)^{\mathrm{T}} + R(k)\right)^{-1}C(k)\right)\widehat{x}(k|k-1), \tag{5.19}$$

with covariance matrix

$$E\left[\left(x(k+1) - \widehat{x}(k+1|k)\right)\left(x(k+1) - \widehat{x}(k+1|k)\right)^{\mathrm{T}}\right]$$

$$= A(k)P(k|k-1)A(k)^{\mathrm{T}} + Q(k) - \left(A(k)P(k|k-1)C(k)^{\mathrm{T}} + S(k)\right)$$

$$\times \left(C(k)P(k|k-1)C(k)^{\mathrm{T}} + R(k)\right)^{-1}$$

$$\times \left(C(k)P(k|k-1)A(k)^{\mathrm{T}} + S(k)^{\mathrm{T}}\right). \tag{5.20}$$

Proof We first seek an explicit expression for the covariance matrices that we want to minimize in terms of the matrix M in Equation (5.13). Next, we perform the minimization of these covariance matrices with respect to M.

We partition the matrix M in Equation (5.13) as follows:

$$\begin{bmatrix} \widehat{x}(k|k) \\ \widehat{x}(k+1|k) \end{bmatrix} = \begin{bmatrix} M_{11} & M_{12} \\ M_{21} & M_{22} \end{bmatrix} \begin{bmatrix} y(k) \\ -B(k)u(k) \end{bmatrix} + \begin{bmatrix} M_{13} \\ M_{23} \end{bmatrix} \widehat{x}(k|k-1),$$

with $M_{11}, M_{21} \in \mathbb{R}^{n \times \ell}$, and $M_{12}, M_{22}, M_{13}, M_{23} \in \mathbb{R}^{n \times n}$. On substituting the data equation (5.16) into the linear estimate (5.13), we obtain

$$\begin{bmatrix} \widehat{x}(k|k) \\ \widehat{x}(k+1|k) \end{bmatrix} = \begin{bmatrix} M_{11}C(k) + M_{12}A(k) & -M_{12} \\ M_{21}C(k) + M_{22}A(k) & -M_{22} \end{bmatrix} \begin{bmatrix} x(k) \\ x(k+1) \end{bmatrix}$$

$$+ \begin{bmatrix} M_{11} & M_{12} \\ M_{21} & M_{22} \end{bmatrix} L(k)\epsilon(k) + \begin{bmatrix} M_{13} \\ M_{23} \end{bmatrix} \widehat{x}(k|k-1).$$

$$\tag{5.21}$$

By taking the mean, we obtain

$$E[\widehat{x}(k|k)] = \left(M_{11}C(k) + M_{12}A(k)\right)E[x(k)] - M_{12}E[x(k+1)]$$

$$+ M_{13}E[\widehat{x}(k|k-1)],$$

$$E[\widehat{x}(k+1|k)] = \left(M_{21}C(k) + M_{22}A(k)\right)E[x(k)] - M_{22}E[x(k+1)]$$

$$+ M_{23}E[\widehat{x}(k|k-1)].$$

Using Equation (5.11) on page 134, we see that both estimates satisfy the unbiasedness condition (5.14), provided that

$$\begin{aligned} M_{13} &= I_n - M_{11}C(k), & M_{12} &= 0, \\ M_{23} &= A(k) - M_{21}C(k), & M_{22} &= -I_n. \end{aligned}$$

Using these expressions in the linear estimate (5.21) yields

$$\begin{bmatrix} \widehat{x}(k|k) \\ \widehat{x}(k+1|k) \end{bmatrix} = \begin{bmatrix} M_{11}C(k) & 0 \\ M_{21}C(k) - A(k) & I_n \end{bmatrix} \begin{bmatrix} x(k) \\ x(k+1) \end{bmatrix}$$

$$+ \begin{bmatrix} I_n - M_{11}C(k) \\ A(k) - M_{21}C(k) \end{bmatrix} \widehat{x}(k|k-1)$$

$$+ \begin{bmatrix} M_{11} & 0 \\ M_{21} & -I_n \end{bmatrix} L(k)\epsilon(k),$$

or, equivalently,

$$\begin{bmatrix} x(k) - \widehat{x}(k|k) \\ x(k+1) - \widehat{x}(k+1|k) \end{bmatrix} = \begin{bmatrix} I_n - M_{11}C(k) \\ A(k) - M_{21}C(k) \end{bmatrix} \Big(x(k) - \widehat{x}(k|k-1) \Big)$$

$$- \begin{bmatrix} M_{11} & 0 \\ M_{21} & -I_n \end{bmatrix} L(k)\epsilon(k).$$

Using the property that $\epsilon(k)$ and $(x(k) - \widehat{x}(k|k-1))$ are uncorrelated, we find the following expression for the joint covariance matrix:

$$E\left[\begin{bmatrix} x(k) - \widehat{x}(k|k) \\ x(k+1) - \widehat{x}(k+1|k) \end{bmatrix} \begin{bmatrix} x(k) - \widehat{x}(k|k) \\ x(k+1) - \widehat{x}(k+1|k) \end{bmatrix}^{\mathrm{T}} \right]$$

$$= \begin{bmatrix} I_n - M_{11}C(k) \\ A(k) - M_{21}C(k) \end{bmatrix} P(k|k-1) \begin{bmatrix} I_n - M_{11}C(k) \\ A(k) - M_{21}C(k) \end{bmatrix}^{\mathrm{T}}$$

$$+ \begin{bmatrix} M_{11} & 0 \\ M_{21} & -I_n \end{bmatrix} \begin{bmatrix} R(k) & S(k)^{\mathrm{T}} \\ S(k) & Q(k) \end{bmatrix} \begin{bmatrix} M_{11}^{\mathrm{T}} & M_{21}^{\mathrm{T}} \\ 0 & -I_n \end{bmatrix}.$$

As a result, the covariance matrices that we want to minimize can be denoted by

$$E\left[\Big(x(k) - \widehat{x}(k|k) \Big) \Big(x(k) - \widehat{x}(k|k) \Big)^{\mathrm{T}} \right]$$

$$= P(k|k-1) - M_{11}C(k)P(k|k-1) - P(k|k-1)C(k)^{\mathrm{T}} M_{11}^{\mathrm{T}}$$

$$+ M_{11}\Big(C(k)P(k|k-1)C(k)^{\mathrm{T}} + R(k) \Big) M_{11}^{\mathrm{T}} \qquad (5.22)$$

and

$$E\left[(x(k+1) - \widehat{x}(k+1|k))(x(k+1) - \widehat{x}(k+1|k))^{\mathrm{T}} \right]$$

$$= A(k)P(k|k-1)A(k)^{\mathrm{T}} + Q(k)$$

$$- M_{21}\Big(C(k)P(k|k-1)A(k)^{\mathrm{T}} + S(k)^{\mathrm{T}} \Big)$$

$$- \Big(A(k)P(k|k-1)C(k)^{\mathrm{T}} + S(k) \Big) M_{21}^{\mathrm{T}}$$

$$+ M_{21}\Big(C(k)P(k|k-1)C(k)^{\mathrm{T}} + R(k) \Big) M_{21}^{\mathrm{T}}. \qquad (5.23)$$

Now we perform the minimization step by application of the completion-of-squares argument to both expressions for the covariance matrices. This yields the following expressions for M_{11} and M_{21} (see the proof of Theorem 4.3 on page 113):

$$M_{11} = P(k|k-1)C(k)^{\mathrm{T}} \left(C(k)P(k|k-1)C(k)^{\mathrm{T}} + R(k) \right)^{-1},$$

$$M_{21} = \left(A(k)P(k|k-1)C(k)^{\mathrm{T}} + S(k) \right)$$
$$\times \left(C(k)P(k|k-1)C(k)^{\mathrm{T}} + R(k) \right)^{-1}.$$

Using (5.22), and subsequently substituting the resulting expression for M into the linear estimate equations (5.13), yields the estimate for $\widehat{x}(k|k)$ as given by (5.17) and that for $\widehat{x}(k+1|k)$ as given by (5.19). Finally, if we substitute the above expressions for M_{11} and M_{21} into the expressions (5.22) and (5.23), we obtain the optimal covariance matrices (5.18) and (5.20). □

The solution given by Theorem 5.1 is recursive in nature. The state update $\widehat{x}(k+1|k)$ and its covariance matrix in (5.20) can be used as prior estimates for the next time instant $k + 1$. Upon replacing k by $k + 1$ in the theorem, we can continue estimating $\widehat{x}(k+1|k+1)$ and $\widehat{x}(k+2|k+1)$ and so on.

The estimates $\widehat{x}(k + 1|k)$, $\widehat{x}(k + 2|k + 1), \ldots$ obtained in this way are called the one-step-ahead predicted state estimates. This is because the predicted state for the particular time instants $k + 1, k + 2, \ldots$ makes use of (measurement) data up to time instant $k, k + 1, \ldots$, respectively. This explains the notation used in the argument of $x(\cdot|\cdot)$. In general the updating process for computing these estimates and their corresponding covariance matrices is called the *time update*.

The estimates $\widehat{x}(k|k)$, $\widehat{x}(k + 1|k + 1), \ldots$ obtained by successive application of Theorem 5.1 are called the filtered state estimates. The updating of the state at a particular time instant k is done at exactly the same moment as that at which the input–output measurements are collected. Therefore, this updating procedure is described as the *measurement update*.

The state-estimate update equations and their covariance matrices constitute the conventional Kalman filter, as originally derived by Kalman (1960). To summarize these update equations, we denote the state-error covariance matrix $E[(x(k)-\widehat{x}(k|k))(x(k)-\widehat{x}(k|k))^{\mathrm{T}}]$ of the filtered state estimate by $P(k|k)$. Here the notation used for the argument corresponds to that used for the state estimate. Similarly, the state-error covariance matrix $E[(x(k+1) - \widehat{x}(k+1|k))(x(k+1) - \widehat{x}(k+1|k))^{\mathrm{T}}]$ of the one-step-ahead predicted state estimate is denoted by $P(k+1|k)$. The recursive method to compute the minimum-error variance estimate of the state vector of a linear system using the Kalman filter is summarized below.

Summary of the conventional Kalman filter (filtered state)
Given the prior information $\widehat{x}(k|k-1)$ and $P(k|k-1)$, the update equation for the state-error covariance matrix $P(k|k)$ is given by a so-called *Riccati*-type difference equation:

$$P(k|k) = P(k|k-1) - P(k|k-1)C(k)^{\mathrm{T}}$$
$$\times \left(R(k) + C(k)P(k|k-1)C(k)^{\mathrm{T}}\right)^{-1} C(k)P(k|k-1).$$

$$(5.24)$$

The calculation of the filtered state estimate is usually expressed in terms of the Kalman gain, denoted by $K'(k)$:

$$K'(k) = P(k|k-1)C(k)^{\mathrm{T}} \left(R(k) + C(k)P(k|k-1)C(k)^{\mathrm{T}}\right)^{-1}.$$

The filtered state estimate is given by

$$\widehat{x}(k|k) = \widehat{x}(k|k-1) + K'(k)\left(y(k) - C(k)\widehat{x}(k|k-1)\right),$$

and the covariance matrix is given by

$$E\left[\left(x(k) - \widehat{x}(k|k)\right)\left((x(k) - \widehat{x}(k|k))^{\mathrm{T}}\right)\right] = P(k|k).$$

Summary of the conventional Kalman filter (one-step-ahead predicted state)
Given the prior information $\widehat{x}(k|k-1)$ and $P(k|k-1)$, the update equation for the state-error covariance matrix $P(k+1|k)$ is given by the *Riccati* difference equation:

$$P(k+1|k) = A(k)P(k|k-1)A(k)^{\mathrm{T}} + Q(k)$$
$$-\left(S(k) + A(k)P(k|k-1)C(k)^{\mathrm{T}}\right)$$
$$\times \left(R(k) + C(k)P(k|k-1)C(k)^{\mathrm{T}}\right)^{-1}$$
$$\times \left(S(k) + A(k)P(k|k-1)C(k)^{\mathrm{T}}\right)^{\mathrm{T}}. \qquad (5.25)$$

The Kalman gain, denoted by $K(k)$, is defined as

$$K(k) = \Big(S(k) + A(k)P(k|k-1)C(k)^{\mathrm{T}}\Big)$$
$$\times \Big(R(k) + C(k)P(k|k-1)C(k)^{\mathrm{T}}\Big)^{-1}.$$

The state-update equation is given by

$$\widehat{x}(k+1|k)$$
$$= A(k)\widehat{x}(k|k-1) + B(k)u(k) + K(k)\Big(y(k) - C(k)\widehat{x}(k|k-1)\Big),$$

and the covariance matrix is given by

$$E\Big[\big(x(k+1) - \widehat{x}(k+1|k)\big)\big((x(k+1) - \widehat{x}(k+1|k)\big)^{\mathrm{T}}\Big] = P(k+1|k).$$

5.5 The Kalman filter and weighted least squares

In Section 4.5.4 it was shown that an analytically equivalent, though numerically superior, way to solve (stochastic) least-squares problems is by making use of square-root algorithms. In this section we derive such square-root algorithms for updating the filtered and one-step-ahead predicted state estimates.

5.5.1 A weighted least-squares problem formulation

The square-root solution to the Kalman-filter problem is derived in relation to a dedicated weighted least-squares problem that differs (slightly) from the generic one treated on page 115 in Section 4.5.4. The statement of the weighted least-squares problem begins with the formulation of the set of constraint equations that represents a perturbed linear transformation of the unknown states. For that purpose we explicitly list the square root $L(k)$, defined in (5.16), to represent the process and measurement noise as

$$\begin{bmatrix} v(k) \\ w(k) \end{bmatrix} = \underbrace{\begin{bmatrix} R(k)^{1/2} & 0 \\ X(k) & Q_x(k)^{1/2} \end{bmatrix}}_{L(k)} \begin{bmatrix} \widetilde{v}(k) \\ \widetilde{w}(k) \end{bmatrix} \quad \text{with} \quad \begin{bmatrix} \widetilde{v}(k) \\ \widetilde{w}(k) \end{bmatrix} \sim (0, I_{\ell+n}). \tag{5.26}$$

The matrices $X(k)$ and $Q_x(k)$ satisfy

$$X(k) = S(k)R(k)^{-\mathrm{T}/2}, \tag{5.27}$$
$$Q_x(k) = Q(k) - S(k)R(k)^{-1}S(k)^{\mathrm{T}}. \tag{5.28}$$

You are asked to show that Equation (5.26) is equivalent to (5.10) on page 134 in Exercise 5.2 on page 173.

This expression for $L(k)$ is inserted into the data equation (5.16) on page 135. This particular expression of the data equation is then combined with the prior statistical information on $x(k)$, also presented in the data-equation format, as was done in Equation (4.23) on page 116:

$$x(k) = \widehat{x}(k|k-1) - P(k|k-1)^{1/2}\widetilde{x}(k).$$

As a result, we obtain the following set of constraint equations on the unknowns $x(k)$ and $x(k+1)$:

$$
\begin{bmatrix} \widehat{x}(k|k-1) \\ y(k) \\ -B(k)u(k) \end{bmatrix} = \begin{bmatrix} I_n & 0 \\ C(k) & 0 \\ A(k) & -I_n \end{bmatrix} \begin{bmatrix} x(k) \\ x(k+1) \end{bmatrix}
$$

$$
+ \begin{bmatrix} P(k|k-1)^{1/2} & 0 & 0 \\ 0 & R(k)^{1/2} & 0 \\ 0 & X(k) & Q_x(k)^{1/2} \end{bmatrix} \begin{bmatrix} \widetilde{x}(k) \\ \widetilde{v}(k) \\ \widetilde{w}(k) \end{bmatrix}.
$$

$$(5.29)$$

Let this set of equations be denoted compactly by

$$\overline{y}(k) = \overline{F}(k)\overline{x}(k) + \overline{L}(k)\overline{\mu}(k). \tag{5.30}$$

The weighted least-squares problem for the derivation of the square-root Kalman-filter algorithms is denoted by

$$\min_{\overline{x}(k)} \overline{\mu}(k)^{\mathrm{T}}\overline{\mu}(k) \quad \text{subject to } \overline{y}(k) = \overline{F}(k)\overline{x}(k) + \overline{L}(k)\overline{\mu}(k). \tag{5.31}$$

The goal of the analysis of the weighted least-squares problem is the derivation of square-root solutions for the filtered and one-step-ahead predicted state estimates. Therefore, we will address the numerical transformations involved in solving (5.30) in two consecutive parts. We start with the derivation of the square-root algorithm for computing the filtered state estimate in Section 5.5.2. The derivation for the computation of the one-step-ahead predication is presented in Section 5.5.3.

5.5.2 The measurement update

Following the strategy for solving weighted least-squares problems in Section 2.7 on page 35, we select a left transformation denoted by T_ℓ^{m},

$$
T_\ell^{\mathrm{m}} = \begin{bmatrix} C(k) & -I_\ell & 0 \\ I_n & 0 & 0 \\ 0 & 0 & I_n \end{bmatrix}, \tag{5.32}
$$

to transform the set of constraint equations (5.30) into

$$T_\ell^m \overline{y}(k) = T_\ell^m \overline{F}(k)\overline{x}(k) + T_\ell^m \overline{L}(k)\overline{\mu}(k).$$

The superscript m in the transformation matrix T_ℓ^m refers to the fact that this transformation is used in the derivation of the measurement update. The resulting transformed set of constraint equations is

$$\begin{bmatrix} C(k)\widehat{x}(k|k-1) - y(k) \\ \widehat{x}(k|k-1) \\ -B(k)u(k) \end{bmatrix} = \begin{bmatrix} 0 & 0 \\ I_n & 0 \\ A(k) & -I_n \end{bmatrix} \begin{bmatrix} x(k) \\ x(k+1) \end{bmatrix}$$

$$+ \begin{bmatrix} C(k)P(k|k-1)^{1/2} & -R(k)^{1/2} & 0 \\ P(k|k-1)^{1/2} & 0 & 0 \\ 0 & X(k) & Q_x(k)^{1/2} \end{bmatrix}$$

$$\times \begin{bmatrix} \widetilde{x}(k) \\ \widetilde{v}(k) \\ \widetilde{w}(k) \end{bmatrix}. \qquad (5.33)$$

The difference $y(k) - C(k)\widehat{x}(k|k-1)$ in the first row in (5.33) defines the kth sample of the so-called *innovation sequence* and is labeled by the symbol $e(k)$. This sequence plays a key role in the operation and performance of the state reconstruction. An illustration of this role is given later on in Example 5.2 on page 153.

Next, we apply an orthogonal transformation $\overline{T}_r^m \in \mathbb{R}^{(n+\ell)\times(n+\ell)}$, such that the right-hand side of

$$\begin{bmatrix} C(k)P(k|k-1)^{1/2} & -R(k)^{1/2} \\ P(k|k-1)^{1/2} & 0 \end{bmatrix} \overline{T}_r^m = \begin{bmatrix} R^e(k)^{1/2} & 0 \\ G'(k) & P(k|k)^{1/2} \end{bmatrix} \qquad (5.34)$$

is lower-triangular. This transformation can be computed using a RQ factorization. It yields the matrix $R^e(k)^{1/2} \in \mathbb{R}^{\ell\times\ell}$, which equals the square root of the covariance matrix of the innovation sequence $e(k)$.

If we take the transformation T_r^m equal to

$$T_r^m = \begin{bmatrix} \overline{T}_r^m & 0 \\ 0 & I_n \end{bmatrix},$$

the above compact notation of the set of constraint equations can be further transformed into

$$T_\ell^m \overline{y}(k) = T_\ell^m \overline{F}(k)\overline{x}(k) + T_\ell^m \overline{L}(k) T_r^m (T_r^m)^{\mathrm{T}} \overline{\mu}(k). \qquad (5.35)$$

In addition to the transformation of matrices as listed in (5.34), the orthogonal transformation $\overline{T}_r^m$ leads to the following modification of the zero-mean white-noise vector

$$\begin{bmatrix} \widetilde{x}(k) \\ \widetilde{v}(k) \end{bmatrix}$$

and part of its weight matrix in (5.33):

$$\begin{bmatrix} 0 & X(k) \end{bmatrix}\overline{T}_{\mathrm{r}}^{\mathrm{m}} = \begin{bmatrix} X_1(k) & X_2(k) \end{bmatrix}, \qquad (T_{\mathrm{r}}^{\mathrm{m}})^{\mathrm{T}}\begin{bmatrix} \widetilde{x}(k) \\ \widetilde{v}(k) \end{bmatrix} = \begin{bmatrix} \nu(k) \\ \widetilde{x}'(k) \end{bmatrix}.$$

Hence, the transformed set of constraint equations (5.35) can be written as

$$\begin{bmatrix} C(k)\widehat{x}(k|k-1) - y(k) \\ \widehat{x}(k|k-1) \\ -B(k)u(k) \end{bmatrix} = \begin{bmatrix} 0 & 0 \\ I_n & 0 \\ A(k) & -I_n \end{bmatrix}\begin{bmatrix} x(k) \\ x(k+1) \end{bmatrix}$$

$$+ \begin{bmatrix} R^{\mathrm{e}}(k)^{1/2} & 0 & 0 \\ G'(k) & P'(k|k)^{1/2} & 0 \\ X_1(k) & X_2(k) & Q_x(k)^{1/2} \end{bmatrix}\begin{bmatrix} \nu(k) \\ \widetilde{x}'(k) \\ \widetilde{w}(k) \end{bmatrix}.$$

$$(5.36)$$

Since it was assumed that the matrix $R(k)$ is positive-definite, the matrix $R^{\mathrm{e}}(k)^{1/2}$ is invertible (this follows from (5.39) below). Therefore, the first block row of the set of equations (5.36) defines the *whitened* version of the innovation sequence at time instant k explicitly as

$$\nu(k) = R^{\mathrm{e}}(k)^{-1/2}\Big(C(k)\widehat{x}(k|k-1) - y(k)\Big). \qquad (5.37)$$

The second block row of (5.36) can be written as

$$\widehat{x}(k|k-1) - G'(k)\nu(k) = x(k) + P'(k|k)^{1/2}\widetilde{x}'(k). \qquad (5.38)$$

The next theorem shows that this equation delivers both the filtered state estimate $\widehat{x}(k|k)$ and a square root of its covariance matrix.

Theorem 5.2 (Square-root measurement update) *Let the conditions stipulated in the Kalman-filter problem hold, and let the transformations T_ℓ^{m} and $T_{\mathrm{r}}^{\mathrm{m}}$ transform the set of constraint equations (5.29) into (5.36), then the minimum-variance unbiased filtered state estimate $\widehat{x}(k|k)$ and its covariance matrix are embedded in the transformed set of equations (5.36) as*

$$\widehat{x}(k|k) = \widehat{x}(k|k-1) + G'(k)R^{\mathrm{e}}(k)^{-1/2}\Big(y(k) - C(k)\widehat{x}(k|k-1)\Big),$$

$$E\Big[\big(x(k) - \widehat{x}(k|k)\big)\big(x(k) - \widehat{x}(k|k)\big)^{\mathrm{T}}\Big] = P'(k|k)^{1/2}P'(k|k)^{\mathrm{T}/2}$$

$$= P(k|k).$$

Proof We first show the equivalence between the minimum-covariance matrix $P(k|k)$, derived in Theorem 5.1, and $P'(k|k)$. In the transformation of Equation (5.29) into (5.36), we focus on the following relation:

$$\begin{bmatrix} C(k)P(k|k-1)^{1/2} & -R(k)^{1/2} & 0 \\ P(k|k-1)^{1/2} & 0 & 0 \\ 0 & X(k) & Q_x(k)^{1/2} \end{bmatrix} T_r^m$$

$$= \begin{bmatrix} R^e(k)^{1/2} & 0 & 0 \\ G'(k) & P'(k|k)^{1/2} & 0 \\ X_1(k) & X_2(k) & Q_x(k)^{1/2} \end{bmatrix}.$$

By exploiting the orthogonality of the matrix T_r^m and multiplying both sides on the right by their transposes, we obtain the following expressions:

$$R^e(k) = R(k) + C(k)P(k|k-1)C(k)^{\mathrm{T}}, \tag{5.39}$$

$$G'(k) = P(k|k-1)C(k)^{\mathrm{T}}R^e(k)^{-\mathrm{T}/2}, \tag{5.40}$$

$$P'(k|k) = P(k|k-1) - G'(k)G'(k)^{\mathrm{T}}, \tag{5.41}$$

$$X_1(k) = -X(k)R(k)^{\mathrm{T}/2}R^e(k)^{-\mathrm{T}/2}, \tag{5.42}$$

$$X_2(k)P'(k|k)^{\mathrm{T}/2} = -X_1(k)G'(k)^{\mathrm{T}}, \tag{5.43}$$

$$Q(k) = X(k)X(k)^{\mathrm{T}} + Q_x(k)$$
$$= X_1(k)X_1(k)^{\mathrm{T}} + X_2(k)X_2(k)^{\mathrm{T}} + Q_x(k). \tag{5.44}$$

On substituting the expression for $G'(k)$ of Equation (5.40) into Equation (5.41) and making use of (5.39), the right-hand side of (5.41) equals the covariance matrix $E[(x(k) - \hat{x}(k|k))(x(k) - \hat{x}(k|k))^{\mathrm{T}}]$ in Equation (5.18) of Theorem 5.1 on page 136. The foregoing minimal-covariance matrix was denoted by $P(k|k)$. Therefore we conclude that

$$P'(k|k) = P(k|k).$$

A combination of Equations (5.40) and (5.39) allows us to express the product $G'(k)R^e(k)^{-1/2}$ as

$$G'(k)R^e(k)^{-1/2} = P(k|k-1)C(k)^{\mathrm{T}}\left(R(k) + C(k)P(k|k-1)C(k)^{\mathrm{T}}\right)^{-1}.$$

Using the definition of the white-noise vector $\nu(k)$ in (5.37) in the left-hand side of (5.38), and comparing this result with the expression for

$\hat{x}(k|k)$ in (5.17) on page 136, we conclude that

$$\hat{x}(k|k-1) - G'(k)\nu(k) = \hat{x}(k|k-1) + P(k|k-1)C(k)^{\mathrm{T}}$$
$$\times \Big(R(k) + C(k)P(k|k-1)C(k)^{\mathrm{T}} \Big)^{-1}$$
$$\times \Big(y(k) - C(k)\hat{x}(k|k-1) \Big)$$
$$= \hat{x}(k|k),$$

and the proof is completed. □

The proof of Theorem 5.2 shows that the matrix $P'(k|k)^{1/2}$ can indeed be taken as a square root of the matrix $P(k|k)$. Therefore, when continuing with the transformed set of equations (5.36) we will henceforth replace $P'(k|k)^{1/2}$ by $P(k|k)^{1/2}$. This results in the following starting point of the analysis of the time update in Section 5.5.3:

$$\begin{bmatrix} C(k)\hat{x}(k|k-1) - y(k) \\ \hat{x}(k|k-1) \\ -B(k)u(k) \end{bmatrix} = \begin{bmatrix} 0 & 0 \\ I_n & 0 \\ A(k) & -I_n \end{bmatrix} \begin{bmatrix} x(k) \\ x(k+1) \end{bmatrix}$$
$$+ \begin{bmatrix} R^{\mathrm{e}}(k)^{1/2} & 0 & 0 \\ G'(k) & P(k|k)^{1/2} & 0 \\ X_1(k) & X_2(k) & Q_x(k)^{1/2} \end{bmatrix} \begin{bmatrix} \nu(k) \\ \tilde{x}'(k) \\ \tilde{w}(k) \end{bmatrix}.$$
$$(5.45)$$

The proof further shows that the steps taken in solving the weighted least-squares problem (5.31) result in an update of the prior information on $x(k)$ in the following data-equation format:

$$\hat{x}(k|k-1) = x(k) + P(k|k-1)^{1/2}\tilde{x}(k), \quad \tilde{x}(k) \sim (0, I_n),$$
$$\hat{x}(k|k) = x(k) + P(k|k)^{1/2}\tilde{x}'(k), \quad \tilde{x}'(k) \sim (0, I_n). \quad (5.46)$$

5.5.3 The time update

Starting from the set of equations (5.45), we now seek allowable transformations of the form T_ℓ and T_r as explained in Section 2.7 on page 35, such that we obtain an explicit expression for $x(k+1)$ as

$$\hat{x}(k+1|k) = x(k+1) + P(k+1|k)^{1/2}\tilde{x}(k+1).$$

A first step in achieving this goal is to select a left transformation T_ℓ^{t},

$$T_\ell^{\mathrm{t}} = \begin{bmatrix} I_\ell & 0 & 0 \\ 0 & I_n & 0 \\ 0 & -A(k) & I_n \end{bmatrix}, \quad (5.47)$$

where the superscript t refers to the fact that this transformation is used in the derivation of the time update. Multiplying Equation (5.45) on the left by this transformation matrix yields

$$
\begin{bmatrix} C(k)\widehat{x}(k|k-1) - y(k) \\ \widehat{x}(k|k-1) \\ -B(k)u(k) - A(k)\widehat{x}(k|k-1) \end{bmatrix}
$$
$$
= \begin{bmatrix} 0 & 0 \\ I_n & 0 \\ 0 & -I_n \end{bmatrix} \begin{bmatrix} x(k) \\ x(k+1) \end{bmatrix}
$$
$$
+ \begin{bmatrix} R^e(k)^{1/2} & 0 & 0 \\ G'(k) & P(k|k)^{1/2} & 0 \\ -A(k)G'(k) + X_1(k) & -A(k)P(k|k)^{1/2} + X_2(k) & Q_x(k)^{1/2} \end{bmatrix}
$$
$$
\times \begin{bmatrix} \nu(k) \\ \widetilde{x}'(k) \\ \widetilde{w}(k) \end{bmatrix}.
$$

If we need only an estimate of $x(k+1)$, the second block row of this equation can be discarded. To bring the third block row into conformity with the generalized covariance representation presented in Section 4.5.4, we apply a second (orthogonal) transformation $\overline{T}_r^t$, such that the right-hand side of

$$
\begin{bmatrix} -A(k)P(k|k)^{1/2} + X_2(k) & Q_x(k)^{1/2} \end{bmatrix} \overline{T}_r^t = \begin{bmatrix} -P'(k+1|k)^{1/2} & 0 \end{bmatrix}
$$
$$(5.48)$$

is lower-triangular and $\overline{T}_r^t \in \mathbb{R}^{2n \times 2n}$ and $P(k+1|k)^{1/2} \in \mathbb{R}^{n \times n}$. This compression can again be computed with the RQ factorization. Then, if we take a transformation of the type T_r^t equal to

$$
T_r^t = \begin{bmatrix} I_\ell & 0 \\ 0 & \overline{T}_r^t \end{bmatrix},
$$

the original set of constraint equations (5.30) is finally transformed into

$$
T_\ell^t T_\ell^m \overline{y}(k) = T_\ell^t T_\ell^m \overline{F}(k)\overline{x}(k) + T_\ell^t T_\ell^m \overline{L}(k) T_r^m T_r^t \widetilde{\mu}(k),
$$

with $\widetilde{\mu}(k) = (T_r^t)^T (T_r^m)^T \overline{\mu}(k)$. By virtue of the orthogonality of both T_r^t and T_r^m, we have also that $\widetilde{\mu}(k) \sim (0, I)$. If we discard the second block row of this transformed set of constraint equations, we obtain

$$
\begin{bmatrix} C(k)\widehat{x}(k|k-1) - y(k) \\ B(k)u(k) + A(k)\widehat{x}(k|k-1) \end{bmatrix}
$$

$$
= \begin{bmatrix} 0 \\ I_n \end{bmatrix} x(k+1) + \begin{bmatrix} R^e(k)^{1/2} & 0 & 0 \\ A(k)G'(k) - X_1(k) & P'(k+1|k)^{1/2} & 0 \end{bmatrix} \begin{bmatrix} \nu(k) \\ \widetilde{x}(k+1) \\ \widetilde{w}'(k) \end{bmatrix}.
$$

$$(5.49)$$

The bottom block row of this set of equations is a generalized covariance representation of the random variable $x(k+1)$ and it can be written as

$$
A(k)\widehat{x}(k|k-1) + B(k)u(k) - \Big(A(k)G'(k) - X_1(k)\Big)\nu(k)
$$
$$
= x(k+1) + P'(k+1|k)^{1/2}\widetilde{x}(k+1).
$$

In the next theorem, it is shown that this expression delivers the minimum-variance estimate of the one-step-ahead predicted state vector and (a square root of) its covariance matrix.

Theorem 5.3 (Square-root time update) *Let the conditions stipulated in the Kalman-filter problem hold, and let the transformations T_ℓ^t and T_r^t transform the first and last block row of the set of constraint equations (5.45) into (5.49), then the minimum-variance unbiased one-step-ahead predicted state estimate $\widehat{x}(k+1|k)$ and its covariance matrix are embedded in the transformed set of equations (5.49) as*

$$
\widehat{x}(k+1|k) = A(k)\widehat{x}(k|k-1) + B(k)u(k)
$$
$$
+ \Big(A(k)G'(k) - X_1(k)\Big)R^e(k)^{-1/2}
$$
$$
\times \Big(y(k) - C(k)\widehat{x}(k|k-1)\Big), \qquad (5.50)
$$

$$
E\Big[\Big(x(k) - \widehat{x}(k|k-1)\Big)\Big(x(k) - \widehat{x}(k|k-1)\Big)^{\mathrm{T}}\Big]
$$
$$
= P'(k+1|k)^{1/2}P'(k+1|k)^{\mathrm{T}/2}
$$
$$
= P(k+1|k). \qquad (5.51)
$$

Proof The proof starts, as in the proof of Theorem 5.2, with establishing first the equivalence between the minimal-covariance matrix $P(k+1|k)$ and $P'(k+1|k)$. This result is then used to derive the expression for $\widehat{x}(k+1|k)$.

Multiplying the two sides of the equality in (5.48) results in the following expression for $P'(k+1|k)$:

$$
\begin{aligned}
P'(k+1|k) &= \left(-A(k)P(k|k)^{1/2} + X_2(k)\right)\left(-A(k)P(k|k)^{1/2} + X_2(k)\right)^{\mathrm{T}} \\
&\quad + Q_x(k) \\
&= A(k)P(k|k)A(k)^{\mathrm{T}} - X_2(k)P(k|k)^{\mathrm{T}/2}A(k)^{\mathrm{T}} \\
&\quad - A(k)P(k|k)^{1/2}X_2(k)^{\mathrm{T}} + X_2(k)X_2(k)^{\mathrm{T}} + Q_x(k).
\end{aligned}
$$

Using the expression for $P(k|k)$ $(= P'(k|k))$ in Equation (5.41) and the expression for $X_2(k)R(\check{k}|k)^{\mathrm{T}/2}$ in (5.43), $P'(k+1|k)$ can be written as

$$
\begin{aligned}
P'(k+1|k) &= A(k)\left(P(k|k-1) - G'(k)G'(k)^{\mathrm{T}}\right)A(k)^{\mathrm{T}} \\
&\quad + X_1(k)G'(k)A(k)^{\mathrm{T}} + A(k)G'(k)^{\mathrm{T}}X_1(k)^{\mathrm{T}} \\
&\quad + X_2(k)X_2(k)^{\mathrm{T}} + Q_x(k) \\
&= A(k)P(k|k-1)A(k)^{\mathrm{T}} \hspace{4cm} (5.52) \\
&\quad - \left(A(k)G'(k) - X_1(k)\right)\left(A(k)G'(k) - X_1(k)\right)^{\mathrm{T}} \\
&\quad + X_1(k)X_1(k)^{\mathrm{T}} + X_2(k)X_2(k)^{\mathrm{T}} + Q_x(k). \hspace{1.5cm} (5.53)
\end{aligned}
$$

Finally, using the expressions for $G'(k), X_1(k)$, and $Q(k)$ in Equations (5.40), (5.42), and (5.44) results in

$$
\begin{aligned}
P'(k+1|k) &= A(k)P(k|k-1)A(k)^{\mathrm{T}} + Q(k) \\
&\quad - \left(A(k)P(k|k-1)C(k)^{\mathrm{T}} + S(k)\right) \\
&\quad \times \left(C(k)P(k|k-1)C(k)^{\mathrm{T}} + R(k)\right)^{-1} \\
&\quad \times \left(C(k)P(k|k-1)A(K)^{\mathrm{T}} + S(k)^{\mathrm{T}})\right) \\
&= P(k+1|k). \hspace{5cm} (5.54)
\end{aligned}
$$

The last equality results from Equation (5.20) in Theorem 5.1.

The proof that the left-hand side of (5.50) is indeed the one-step-ahead predicted minimum-variance state estimate follows when use is made of the relationship

$$
\left(A(k)G'(k) - X_1(k)\right) = \left(A(k)P(k|k-1)C(k)^{\mathrm{T}} + S(k)\right)R^{\mathrm{e}}(k)^{-\mathrm{T}/2},
$$

which stems from the transition from Equation (5.53) to (5.54). In conclusion, the gain matrix

$$
\left(A(k)G'(k) - X_1(k)\right)R^{\mathrm{e}}(k)^{-1/2}
$$

multiplying the innovation sequence in Equation (5.50) equals

$$\Big(A(k)G'(k)-X_1(k)\Big)R^e(k)^{-1/2} = \Big(A(k)P(k|k-1)C(k)^{\mathrm{T}}+S(k)\Big)R^e(k)^{-1}$$

and the proof of the theorem is completed.   □

The proof of Theorem 5.3 shows that the matrix $P'(k+1|k)^{1/2}$ can indeed be taken as a square root of the matrix $P(k+1|k)$.

The proof further shows that the steps taken in solving the weighted least-squares problem (5.31) result in an update of the prior information on $x(k)$ in the following data-equation format:

$$\widehat{x}(k|k) = x(k) + P(k|k)^{1/2}\widetilde{x}'(k), \quad \widetilde{x}'(k) \sim (0, I_n),$$
$$\widehat{x}(k+1|k) = x(k+1) + P(k+1|k)^{1/2}\widetilde{x}(k+1), \quad \widetilde{x}(k+1) \sim (0, I_n).$$

$$(5.55)$$

5.5.4 The combined measurement–time update

In order to simplify the implementation, which may be crucial in some real-time applications, the above series of transformations can be combined into a single orthogonal transformation on a specific matrix array. This combined measurement and time update leads to the so-called *square-root covariance filter* (SRCF), a numerically reliable recursive implementation of the Kalman filter. To derive this combined update, we apply the product of T_ℓ^t and T_ℓ^m defined in Equation (5.47) on page 146 and Equation (5.32) on page 142 as follows to the original set of constraint equations (5.30) on page 142, that is,

$$T_\ell^t T_\ell^m \overline{y}(k) = T_\ell^t T_\ell^m \overline{F}(k)\overline{x}(k) + T_\ell^t T_\ell^m \overline{L}(k)\overline{\mu}(k).$$

The set of constraint equations transformed in this manner becomes

$$\begin{bmatrix} C(k)\widehat{x}(k|k-1) - y(k) \\ \widehat{x}(k|k-1) \\ -B(k)u(k) - A(k)\widehat{x}(k|k-1) \end{bmatrix}$$

$$= \begin{bmatrix} 0 & 0 \\ I_n & 0 \\ 0 & -I_n \end{bmatrix}\begin{bmatrix} x(k) \\ x(k+1) \end{bmatrix}$$

$$+ \begin{bmatrix} C(k)P(k|k-1)^{1/2} & -R(k)^{1/2} & 0 \\ P(k|k-1)^{1/2} & 0 & 0 \\ -A(k)P(k|k-1)^{1/2} & X(k) & Q_x(k)^{1/2} \end{bmatrix}\begin{bmatrix} \widetilde{x}(k) \\ \widetilde{v}(k) \\ \widetilde{w}(k) \end{bmatrix}.$$

The second block row can be discarded if just an estimate of $x(k+1)$ is needed. Multiplying the last row by -1 on both sides yields

$$\begin{bmatrix} C(k)\hat{x}(k|k-1) - y(k) \\ B(k)u(k) + A(k)\hat{x}(k|k-1) \end{bmatrix}$$

$$= \begin{bmatrix} 0 \\ I_n \end{bmatrix} x(k+1)$$

$$+ \begin{bmatrix} C(k)P(k|k-1)^{1/2} & -R(k)^{1/2} & 0 \\ A(k)P(k|k-1)^{1/2} & -X(k) & -Q_x(k)^{1/2} \end{bmatrix} \begin{bmatrix} \tilde{x}(k) \\ \tilde{v}(k) \\ \tilde{w}(k) \end{bmatrix}.$$

The orthogonal transformation

$$\begin{bmatrix} I_n & 0 & 0 \\ 0 & -I_\ell & 0 \\ 0 & 0 & -I_n \end{bmatrix} T_r,$$

applied in the second step of the time update and the second step of the measurement update, can now be combined as follows:

$$\begin{bmatrix} C(k)P(k|k-1)^{1/2} & R(k)^{1/2} & 0 \\ A(k)P(k|k-1)^{1/2} & X(k) & Q_x(k)^{1/2} \end{bmatrix} T_r$$

$$= \begin{bmatrix} R^e(k)^{1/2} & 0 & 0 \\ G(k) & P(k+1|k)^{1/2} & 0 \end{bmatrix}. \tag{5.56}$$

As a result we arrive at the following set of equations:

$$\begin{bmatrix} C(k)\hat{x}(k|k-1) - y(k) \\ B(k)u(k) + A(k)\hat{x}(k|k-1) \end{bmatrix}$$

$$= \begin{bmatrix} 0 \\ I_n \end{bmatrix} x(k+1) + \begin{bmatrix} R^e(k)^{1/2} & 0 & 0 \\ G(k) & P(k+1|k)^{1/2} & 0 \end{bmatrix} \begin{bmatrix} \nu(k) \\ \tilde{x}(k+1) \\ \tilde{w}'(k) \end{bmatrix}, \tag{5.57}$$

and the generalized covariance expression for the state $x(k+1)$ is

$$A(k)\hat{x}(k|k-1) + B(k)u(k) + G(k)R^e(k)^{-1/2}\Big(y(k) - C(k)\hat{x}(k|k-1)\Big)$$

$$= x(k+1) + P(k+1|k)^{1/2}\tilde{x}(k+1). \tag{5.58}$$

The proof that the orthogonal transformation T_r in Equation (5.56) indeeds reveals a square root of the minimal-covariance matrix $P(k+1|k)$ and that the left-hand side of Equation (5.58) equals $\hat{x}(k+1|k)$ can be retrieved following the outline given in Theorems 5.2 and 5.3. This is left as an exercise for the interested reader.

Equation (5.56) together with Equation (5.58) defines the combined time and measurement update of the square-root covariance filter.

Summary of the square-root covariance filter

Given: the system matrices $A(k)$, $B(k)$, and $C(k)$ of a linear system and the covariance matrices

$$\begin{bmatrix} v(k) \\ w(k) \end{bmatrix} = \begin{bmatrix} R(k)^{1/2} & 0 \\ X(k) & Q_x(k)^{1/2} \end{bmatrix} \begin{bmatrix} \widetilde{v}(k) \\ \widetilde{w}(k) \end{bmatrix}, \quad \text{with} \begin{bmatrix} \widetilde{v}(k) \\ \widetilde{w}(k) \end{bmatrix} \sim (0, I_{\ell+n}),$$

for time instants $k = 0, 1, \ldots, N - 1$, and the initial state $\widehat{x}(0| - 1)$, and the square root of its covariance matrix $P(0| - 1)^{1/2}$.

For: $k = 0, 1, 2, \ldots, N - 1$.

Apply the orthogonal transformation T_r such that

$$\begin{bmatrix} C(k)P(k|k-1)^{1/2} & R(k)^{1/2} & 0 \\ A(k)P(k|k-1)^{1/2} & X(k) & Q_x(k)^{1/2} \end{bmatrix} T_r$$

$$= \begin{bmatrix} R^e(k)^{1/2} & 0 & 0 \\ G(k) & P(k+1|k)^{1/2} & 0 \end{bmatrix}.$$

Update the one-step-ahead prediction as

$$\widehat{x}(k+1|k) = A(k)\widehat{x}(k|k-1) + B(k)u(k)$$
$$+ G(k)R^e(k)^{-1/2}\Big(y(k) - C(k)\widehat{x}(k|k-1)\Big), \quad (5.59)$$

and, if desired, its covariance matrix,

$$E\Big[\big(x(k+1) - \widehat{x}(k+1|k)\big)\big((x(k+1) - \widehat{x}(k+1|k)\big)^{\mathrm{T}}\Big]$$
$$= P(k+1|k)^{1/2}P(k+1|k)^{\mathrm{T}/2}.$$

5.5.5 *The innovation form representation*

Many different models may be proposed that generate the same sequence $y(k)$ as in

$$x(k+1) = A(k)x(k) + B(k)u(k) + w(k),$$
$$y(k) = C(k)x(k) + v(k).$$

One such model that is of interest to system identification is the so-called *innovation form representation*. This name stems from the fact that the representation has the innovation sequence

$$e(k) = y(k) - C(k)\widehat{x}(k|k-1)$$

as a stochastic input. The stochastic properties of the innovation sequence are easily derived from Equation (5.57) and are given by the following generalized covariance representation:

$$C(k)\widehat{x}(k|k-1) - y(k) = R^e(k)^{1/2}\nu(k), \quad \nu(k) \sim (0, I_\ell).$$

Therefore, the mean and covariance matrix of the innovation sequence are

$$E[C(k)\widehat{x}(k|k-1) - y(k)] = 0,$$

$$E\left[\Big(C(k)\widehat{x}(k|k-1) - y(k)\Big)\Big(C(j)\widehat{x}(j|j-1) - y(j)\Big)^{\mathrm{T}}\right]$$
$$= R^e(k)\Delta(k-j).$$

Thus, the innovation signal $e(k)$ is a zero-mean white-noise sequence.

If we define the *Kalman gain* $K(k)$ as $K(k) = G(k)R^e(k)^{-1/2}$, then the state-update equation (5.59) and the output $y(k)$ can be written as

$$\widehat{x}(k+1|k) = A(k)\widehat{x}(k|k-1) + B(k)u(k) + K(k)e(k), \quad (5.60)$$
$$y(k) = C(k)\widehat{x}(k|k-1) + e(k). \quad (5.61)$$

This state-space system is the so-called *innovation representation*.

Example 5.2 (SRCF for a double-tank process) Consider again the double-tank process of Example 5.1 on page 130 depicted in Figure 5.1. The discrete-time model of this process is extended with "artificial" process and measurement noise as follows:

$$x(k+1) = \begin{bmatrix} 0.9512 & 0 \\ 0.0476 & 0.9512 \end{bmatrix} x(k) + \begin{bmatrix} 0.0975 \\ 0.0024 \end{bmatrix} u(k)$$

$$+ \begin{bmatrix} 0.0975 & 0 \\ 0.0024 & 0.0975 \end{bmatrix} w(k),$$

$$y(k) = \begin{bmatrix} 0 & 1 \end{bmatrix} x(k) + \underbrace{\begin{bmatrix} 0 & 0.5 \end{bmatrix} w(k) + \overline{v}(k)}_{v(k)},$$

with $w(k)$ and $v(k)$ zero-mean white-noise sequences with covariance matrix

$$E\begin{bmatrix} v(k) \\ w(k) \end{bmatrix}\begin{bmatrix} v(j)^{\mathrm{T}} | w(j)^{\mathrm{T}} \end{bmatrix} = \begin{bmatrix} 0.0125 & 0 & 0.005 \\ 0 & 0.01 & 0 \\ 0.005 & 0 & 0.01 \end{bmatrix}\Delta(k-j).$$

The input sequence $u(k)$ is the same block sequence as defined in Example 5.1. Figure 5.4 shows the noisy output signal. The goal is to estimate $x_1(k)$ and $x_2(k)$ on the basis of the noisy measurements of $y(k)$.

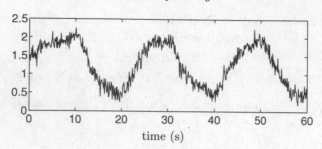

time (s)

Fig. 5.4. The noisy output signal of the double-tank process in Example 5.2.

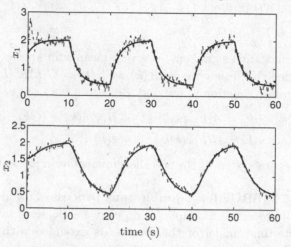

time (s)

Fig. 5.5. True (solid line) and reconstructed (broken line) state-vector sequences of the double-tank process in Example 5.2 with the asymptotic observer of Example 5.1 using noisy measurements.

First we use the asymptotic observer with poles at 0.7 and 0.8 designed in Example 5.1 to reconstruct the state sequence. The true and reconstructed state-vector sequences are plotted in Figure 5.5. On comparing this with Figure 5.3 on page 132 in Example 5.1, we see that, due to the noise, the reconstruction of the states has deteriorated.

Next we use the SRCF implementation of the Kalman filter, which takes the noise into account. The reconstructed state variables are compared with the true ones in Figure 5.6. Clearly, the error variance of the state estimates computed by the SRCF is much smaller than the error variance of the state estimates computed by the asymptotic observer. This is what we expect since the SRCF yields the minimum-error variance estimates.

So far we have inspected the performance of the observer and the Kalman filter by comparing the reconstructed states with the true ones. However, the true states are not available in practice. Under realistic

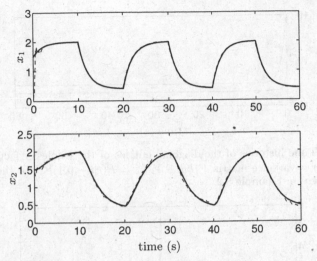

Fig. 5.6. True (solid line) and reconstructed (broken line) state-vector sequences of the double-tank process in Example 5.2 with the SRCF using noisy measurements.

circumstances the optimality of the Kalman filter estimates has to be judged in other ways. One way is to inspect the innovation sequence. From the derivation of the innovation form representation in Section 5.5.5, we know that the standard deviation of this sequence is given by the number $|R^e(k)^{1/2}|$. For $k = 600$ this value is 0.114. If we compute a sample estimate of the standard deviation making use of the single realization of the sequence $y(k) - C\widehat{x}(k|k-1)$ via the formula

$$\widehat{\sigma}_e = \sqrt{\frac{1}{N} \sum_{k=1}^{N} \Big(y(k) - C\widehat{x}(k|k-1)\Big)^2},$$

we obtain the value 0.121. This value is higher than that "predicted" by its theoretical value given by $|R^e(k)^{1/2}|$. This difference follows from the fact that the SRCF is a time-varying filter even when the underlying system is time-invariant. Therefore the innovation sequence, which is the output of that filter, is nonstationary, and we cannot derive a sample-mean estimate of the standard deviation of the innovation from a single realization.

If we inspect the time histories of the diagonal entries of the matrix $P(k|k-1)^{1/2} = P(t|t-0.1)^{1/2}$ (sampling time 0.1 s), given in Figure 5.7 on page 156, we observe that the Kalman filter converges to a time-invariant filter after approximately 5 s.

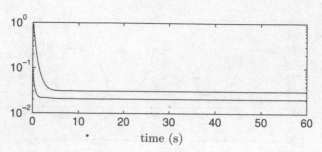

Fig. 5.7. Time histories of the diagonal entries of the Cholesky factor of the state error covariance matrix $P(k|k-1)^{1/2} = P(t|t-0.1)^{1/2}$ for the double-tank process in Example 5.2.

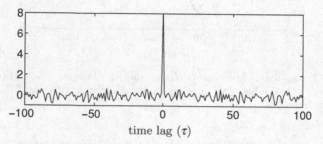

Fig. 5.8. The estimated auto-covariance function $C_e(\tau)$ of the innovation sequence in Example 5.2.

On the basis of this observation we again compute $\hat{\sigma}_e$ based on the last 550 samples only. The value we now get is 0.116, which is in close correspondence to the one predicted by $|R^e(600)^{1/2}|$.

A property of the innovation sequence is that it is white. To check this property, we plot the estimate of the auto-covariance function $C_e(\tau)$ in Figure 5.8. This graph shows that the innovation is indeed close to a white-noise sequence.

Example 5.3 (Minimum-variance property) The goal of this example is to visualize the minimum-variance property of the Kalman filter. The minimum-variance property of the Kalman filter states that the covariance matrix of the estimation error

$$E\left[\left(x(k) - \hat{x}(k)\right)\left(x(k) - \hat{x}(k)\right)^{\mathrm{T}}\right]$$

between the true state $x(k)$ and the Kalman-filter estimate $\hat{x}(k)$ is smaller than the covariance of the estimation error of any other (unbiased) linear estimator.

We use the linearized model of the double-tank process of Example 5.2. With this model 100 realizations of $y(k)$ containing 20 samples are

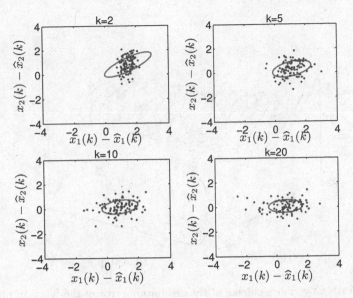

Fig. 5.9. Estimation error of the Kalman filter for different realizations of the output $y(k)$ at four different time steps. The ellipses indicate the covariance of the error.

created (using different realizations of the noise sequences $w(k)$ and $v(k)$). For each of these realizations the state has been estimated both with the Kalman filter and with the pole-placement observer of Example 5.2. In Figure 5.9 the estimation error of the Kalman filter for the first state has been plotted against the estimation error of the second state for the 100 different realizations at four different time steps. Each point corresponds to one realization of $y(k)$. The covariance of the simulation error has been computed using a sample average and it is indicated in the figure by an ellipse. Note that, as the time step k increases, the center of the ellipse converges to the origin.

Figure 5.10 shows the covariance ellipses of the Kalman filter for $k = 1, 2, \ldots, 20$. So, exactly the same ellipse as in Figure 5.9 can be found in Figure 5.10 at $k = 2, 5, 10, 20$. The centers of the ellipses are connected by a line. In Figure 5.11 the covariance ellipses for the pole-placement observer are shown.

On comparing Figure 5.10 with Figure 5.11, we see that at each time step the ellipses for the observer have a larger volume than the ellipses for the Kalman filter. The volume of the ellipses is equal to the determinant of the covariance matrix. Thus the increase in volume is an indication that an (arbitrary) observer reconstructs the state with a larger covariance matrix of the state error vector.

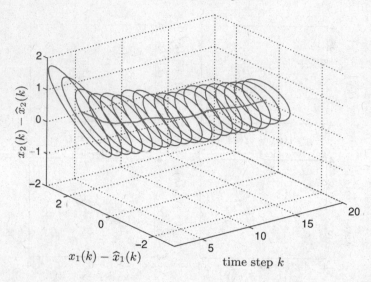

Fig. 5.10. Covariance ellipses of the estimation error of the Kalman filter.

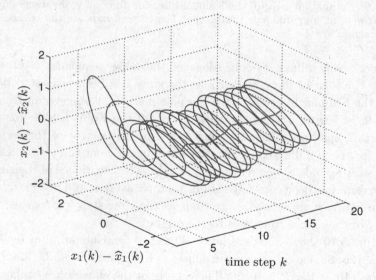

Fig. 5.11. Covariance ellipses of the estimation error of the pole-placement observer.

5.6 Fixed-interval smoothing

The weighted least-squares problem formulation treated in Section 5.5 allows us to present and solve other state-reconstruction problems. In this section, it is outlined how to obtain the so-called smoothed state estimate. We start by formulating the related weighted least-squares problem and briefly discuss its solution.

Combining the data equation of the initial state estimate, denoted by $\widehat{x}(0)$, and the data equation (5.16) on page 135 with $L(k)$ replaced by the square-root expression given in (5.26) on page 141 yields for $k > 0$ the following equations about the unknown state vector sequence:

$$\widehat{x}_0 = x(0) + P_0^{1/2}\widetilde{x}(0), \tag{5.62}$$

$$y(k) = C(k)x(k) + R(k)^{1/2}\widetilde{v}(k), \tag{5.63}$$

$$-B(k)u(k) = A(k)x(k) - x(k+1) + X(k)\widetilde{v}(k) + Q_x(k)^{1/2}\widetilde{w}(k). \tag{5.64}$$

Given measurements of $u(k)$ and $y(k)$ for $k = 0, 1, \ldots, N-1$, we can formulate the following set of equations:

$$
\begin{bmatrix}
\widehat{x}_0 \\
y(0) \\
-B(0)u(0) \\
y(1) \\
-B(1)u(1) \\
\vdots \\
-B(N-2)u(N-2) \\
y(N-1)
\end{bmatrix}
$$

$$
=
\begin{bmatrix}
I_n & 0 & 0 & 0 & \cdots & & 0 \\
C(0) & 0 & 0 & 0 & \cdots & & 0 \\
A(0) & -I_n & 0 & 0 & \cdots & & 0 \\
0 & C(1) & 0 & 0 & \cdots & & 0 \\
0 & A(1) & -I_n & 0 & \cdots & & 0 \\
\vdots & & & & \ddots & & \vdots \\
0 & \cdots & 0 & & A(N-2) & -I_n \\
0 & \cdots & 0 & & 0 & C(N-1)
\end{bmatrix}
\begin{bmatrix}
x(0) \\
x(1) \\
x(2) \\
\vdots \\
x(N-1)
\end{bmatrix}
$$

$$
+
\begin{bmatrix}
P_0^{1/2} & 0 & 0 & \cdots & & 0 \\
0 & R(0)^{1/2} & 0 & \cdots & & 0 \\
0 & X(0) & Q_x(0)^{1/2} & \cdots & & 0 \\
\vdots & & & \ddots & & \vdots \\
0 & \cdots & & R(N-1)^{1/2} & & 0 \\
0 & \cdots & & X(N-1) & Q_x(N-1)^{1/2}
\end{bmatrix}
\begin{bmatrix}
\widetilde{x}(0) \\
\widetilde{v}(0) \\
\widetilde{w}(0) \\
\vdots \\
\widetilde{v}(N-1) \\
\widetilde{w}(N-1)
\end{bmatrix},
$$

where all random variables $\widetilde{x}(0)$, and $\widetilde{v}(0), \widetilde{v}(1), \ldots, \widetilde{v}(N-1)$, and also $\widetilde{w}(0), \widetilde{w}(1), \ldots, \widetilde{w}(N-1)$ are uncorrelated, and have mean zero and unit covariance matrix. Obviously, this set of equations can be denoted by

$$y_N = F_N x_N + L_N \mu_N, \quad \mu_N \sim (0, I), \tag{5.65}$$

with the appropriate definitions of the F_N and L_N matrices and the zero-mean, unit-covariance-matrix stochastic variable μ_N. The data equation (5.65) contains the measurements in the time interval $[0, N-1]$ as well as the prior statistical information about $x(0)$. Thus this equation can be used in a weighted least-squares problem to estimate the random variable x_N, as discussed in detail in Section 4.5.4.

The weighted least-squares problem for finding the smoothed state estimates is defined as

$$\min_{x_N} \mu_N^{\mathrm{T}} \mu_N \quad \text{subject to } y_N = F_N x_N + L_N \mu_N, \quad \mu_N \sim (0, I). \tag{5.66}$$

Under the assumption that the matrix L_N is invertible, the weighted least-squares problem is equivalent to the following minimization problem:

$$\min_{x_N} \| L_N^{-1} F_N x_N - L_N^{-1} y_N \|_2^2. \tag{5.67}$$

Since the matrix L_N is lower-triangular, its inverse can be determined analytically and we get

$$L_N^{-1} F_N$$

$$= \begin{bmatrix} P_0^{-1/2} & 0 & \cdots & 0 \\ R(0)^{-1/2}C(0) & 0 & & 0 \\ -Q_x(0)^{-1/2}\big(X(0)R(0)^{-1/2}C(0) - A(0)\big) & -Q_x(0)^{-1/2} & & 0 \\ 0 & R(1)^{-1/2}C(1) & \cdots & 0 \\ \vdots & \vdots & & \ddots \end{bmatrix},$$

$$L_N^{-1} y_N = \begin{bmatrix} P_0^{-1/2}\widehat{x}_0 \\ R(0)^{-1/2}y(0) \\ -Q_x(0)^{-1/2}\big(X(0)R(0)^{-1/2}y(k) + B(0)u(0)\big) \\ R(1)^{-1/2}y(1) \\ \vdots \end{bmatrix}.$$

We minimize (5.67) using the QR factorization (see Section 2.5) of $L_N^{-1}F_N$ denoted by

$$L_N^{-1}F_N = T_N \begin{bmatrix} \overline{R}(0) & \overline{G}(0) & 0 & \cdots & & 0 \\ 0 & \overline{R}(1) & \overline{G}(1) & & & 0 \\ \vdots & & \ddots & \ddots & & \\ \vdots & & & & \overline{R}(N-2) & \overline{G}(N-2) \\ 0 & 0 & \cdots & & 0 & \overline{R}(N-1) \\ 0 & 0 & \cdots & & 0 & 0 \end{bmatrix}, \quad (5.68)$$

where T_N is an orthogonal matrix (playing the role of the Q matrix of the QR factorization). Applying the transpose of the transformation to $L_N^{-1}y_N$ yields

$$T_N^{\mathrm{T}}L_N^{-1}y_N(k) = \begin{bmatrix} \overline{c}(0) \\ \overline{c}(1) \\ \vdots \\ \overline{c}(N-2) \\ \overline{c}(N-1) \\ \varepsilon(N) \end{bmatrix}. \quad (5.69)$$

Combining the results of the QR factorization in (5.68) and (5.69) transforms the problem in (5.67) into

$$\min_{x_N} \left\| \begin{bmatrix} \overline{R}(0) & \overline{G}(0) & 0 & \cdots & & 0 \\ 0 & \overline{R}(1) & \overline{G}(1) & & & 0 \\ \vdots & & \ddots & \ddots & & \\ \vdots & & & \overline{R}(N-2) & \overline{G}(N-2) \\ 0 & 0 & \cdots & 0 & \overline{R}(N-1) \\ 0 & 0 & \cdots & 0 & 0 \end{bmatrix} x_N - \begin{bmatrix} \overline{c}(0) \\ \overline{c}(1) \\ \vdots \\ \overline{c}(N-2) \\ \overline{c}(N-1) \\ \varepsilon(N) \end{bmatrix} \right\|_2^2 .$$

The solution to this least-squares problem is called the fixed-interval *smoothed state estimate*. It is the state estimate that is obtained by using all the measurements over a fixed time interval of $N-1$ samples. It is denoted by $\widehat{x}(k|N-1)$ and can be obtained by back substitution. We have $\widehat{x}(N|N-1) = \overline{R}(N-1)^{-1}\overline{c}(N-1)$ and, for $k < N$,

$$\widehat{x}(k|N-1) = \overline{R}(k)^{-1}\left(\overline{c}(k) - \overline{G}(k)\widehat{x}(k+1|N-1)\right).$$

Note that we are in fact running the Kalman filter backward in time. This backward filtering is called smoothing. In the end it updates the initial state $\widehat{x}(0| -1)$ that we started off with. Various recursive smoothing

algorithms have appeared in the literature, for example the square-root
algorithm of Paige and Saunders (1977) and the well-known Bryson–
Frazier algorithm (Kailath *et al.*, 2000).

5.7 The Kalman filter for LTI systems

As we have seen in Example 5.2 on page 153, the recursions for $P(k|k - 1)^{1/2}$ converge to a stationary (constant) value when the linear system
involved is time-invariant. The exact conditions for such a stationary
solution of the Kalman-filter problem are summarized in the following
theorem.

Theorem 5.4 *(Anderson and Moore, 1979) Consider the linear time-
invariant system*

$$x(k + 1) = Ax(k) + Bu(k) + w(k), \tag{5.70}$$
$$y(k) = Cx(k) + v(k), \tag{5.71}$$

with $w(k)$ and $v(k)$ zero-mean random sequences with covariance matrix

$$E \begin{bmatrix} w(k) \\ v(k) \end{bmatrix} [w(j)^{\mathrm{T}} \quad v(j)^{\mathrm{T}}] = \begin{bmatrix} Q & S \\ S^{\mathrm{T}} & R \end{bmatrix} \Delta(k - j) \tag{5.72}$$

such that

$$\begin{bmatrix} Q & S \\ S^{\mathrm{T}} & R \end{bmatrix} \geq 0, \quad and \quad R > 0.$$

*If the pair (A, C) is observable and the pair $(A, Q^{1/2})$ is reachable,
then*

$$P(k|k - 1) = E\left[\left(x(k) - \widehat{x}(k|k - 1) \right) \left((x(k) - \widehat{x}(k|k - 1))^{\mathrm{T}} \right) \right],$$

with $\widehat{x}(k|k - 1) = E[x(k)]$, satisfies

$$\lim_{k \to \infty} P(k|k - 1) = P > 0$$

for any symmetric initial condition $P(0| -1) > 0$, where P satisfies

$$P = APA^{\mathrm{T}} + Q - (S + APC^{\mathrm{T}})(CPC^{\mathrm{T}} + R)^{-1}(S + APC^{\mathrm{T}})^{\mathrm{T}}. \tag{5.73}$$

*Moreover, such a P is unique. If this matrix P is used to define the
Kalman-gain matrix K as*

$$K = (S + APC^{\mathrm{T}})(CPC^{\mathrm{T}} + R)^{-1}, \tag{5.74}$$

then the matrix $A - KC$ is asymptotically stable.

The condition $R > 0$ in this theorem guarantees that the matrix $CPC^{T} + R$ is nonsingular. There exist several refinements of this theorem in which the observability and reachability requirements are replaced by weaker notions of detectability and stabilizability, as pointed out in Anderson and Moore (1979) and Kailath *et al.* (2000). A discussion of these refinements is outside the scope of this book.

Equation (5.73) in Theorem 5.4 is called a discrete algebraic Riccati equation (DARE). It is a steady-state version of Equation (5.25) on page 140. It can have several solutions P (this is illustrated in Example 5.5), but we are interested only in the solution that is positive-definite. From the theorem it follows that the positive-definite solution is unique and ensures asymptotic stability of $A - KC$. For a discussion on computing a solution to the DARE, using the so-called invariant-subspace method, we refer to Kailath *et al.* (2000). Under the conditions given in Theorem 5.4, the innovation form representation (5.60)–(5.61) on page 153 converges to the following time-invariant state-space system:

$$\hat{x}(k+1|k) = A\hat{x}(k|k-1) + Bu(k) + Ke(k), \tag{5.75}$$
$$y(k) = C\hat{x}(k|k-1) + e(k). \tag{5.76}$$

Since $v(k)$ is a zero-mean white-noise sequence, both $x(k)$ and $\hat{x}(k|k-1)$ are uncorrelated with $v(k)$. Therefore, the covariance matrix of the asymptotic innovation sequence $e(k)$ equals

$$E[e(k)e(k)^{T}] = E\left[C\left(x(k) - \hat{x}(k|k-1)\right) + v(k)\right]$$
$$\times \left[C\left(x(k) - \hat{x}(k|k-1)\right) + v(k)\right]^{T}$$
$$= CPC^{T} + R. \tag{5.77}$$

The update equation for the one-step-ahead prediction of the state given by Equation (5.59) on page 152 becomes

$$\hat{x}(k+1) = A\hat{x}(k) + Bu(k) + K\left(y(k) - C\hat{x}(k)\right)$$
$$= (A - KC)\hat{x}(k) + Bu(k) + Ky(k), \tag{5.78}$$
$$\hat{y}(k) = C\hat{x}(k). \tag{5.79}$$

These two equations are called a *predictor model* or *innovation predictor model*. An important property of this model is that it is asymptotically stable if the Kalman gain is computed from the positive-definite solution of the DARE, as indicated by Theorem 5.4.

Example 5.4 (Double-tank process) Consider the double-tank process and the experiment outlined in Example 5.2 on page 153. For this example the positive-definite solution to Equation (5.73) on page 162 equals

$$P = 10^{-3} \begin{bmatrix} 0.9462 & 0.2707 \\ 0.2707 & 0.5041 \end{bmatrix}.$$

It can be computed with the function *dare* in the Matlab Control Toolbox (MathWorks, 2000c). This solution equals $P(k|k-1)^{1/2}P(k|k-1)^{T/2}$ for $k = 600$ up to 16 digits. This example seems to suggest that the Kalman-filter solution via the Riccati equation is equivalent to that obtained via the SRCF. However, as shown in Verhaegen and Van Dooren (1986), the contrary is the case. For example, the recursion via Equation (5.25) on page 140 can diverge in the sense that the numerically stored covariance matrix (which in theory needs to be symmetric) loses its symmetry.

The standard deviation of the innovation sequence given by the quantity $R^e(600)^{1/2}$ also corresponds up to 16 digits to the square root of the analytically derived variance given by Equation (5.77) on page 163. The reader is left to check the asymptotic stability of the Kalman filter (5.78)–(5.79) on page 163 with the Kalman gain derived from the computed matrix P.

Example 5.5 (Estimating a constant) Consider the estimation of the ("almost" constant) temperature $x(k)$, modeled by Equations (5.1) and (5.2) on page 127,

$$x(k+1) = x(k) + w(k),$$
$$y(k) = x(k) + v(k).$$

If we use a stationary Kalman filter to estimate $x(k)$, the minimal obtainable error variance of the estimated temperature is the positive-definite solution of the DARE (5.73) on page 162. If we assume that the noise sequences $v(k)$ and $w(k)$ are uncorrelated ($S = 0$), the DARE reduces to the following *quadratic* equation:

$$P = P + Q - P(P + R)^{-1}P.$$

Since P, Q, and R are all scalars in this case, we get

$$P^2 - QP - QR = 0. \tag{5.80}$$

The two solutions to this equation are

$$P^+ = \frac{Q + \sqrt{Q^2 + 4QR}}{2}, \qquad P^- = \frac{Q - \sqrt{Q^2 + 4QR}}{2}.$$

From this expression a number of observations can be made.

(i) If Q is zero, which indicates that the model assumes the temperature to be constant, the stationary value of the variance P will become zero. The Kalman gain also becomes zero, since $K = P(R + P)^{-1}$. The Kalman-filter update equation (5.78) in that case becomes

$$\hat{x}(k + 1|k) = \hat{x}(k|k - 1),$$

and no update takes place. Therefore, to find the right constant temperature we have to use the recursive Kalman filter, which, even for the constant-signal-generation model, is a time-varying filter of the following form:

$$\hat{x}(k + 1|k) = \hat{x}(k|k - 1) + K(k)\Big(y(k) - \hat{x}(k|k - 1)\Big).$$

It is, however, remarked that, since $P(k|k - 1)$ approaches 0 for $k \to \infty$, $K(k)$ also goes to 0.

(ii) Taking $Q \neq 0$, we can express the (lack of) confidence that we have in the temperature being constant. In this case the stationary Kalman filter has a gain K different from zero.

(iii) Taking $Q \neq 0$, we are left with two choices for P. The positive root P^+ of Equation (5.80) is strictly smaller than R, provided that

$$R > 2Q.$$

The error in the estimate of the state obtained from the stationary Kalman filter will have a smaller variance (equal to P^+) than the variance of the measurement errors on $y(k)$ (equal to R) only when $R > 2Q$. Therefore, we can find circumstances under which the use of a Kalman filter does not produce more accurate estimates of the temperature than the measurements themselves. This is, for example, the case when $Q \geq R/2$. Such a situation may occur if the temperature is changing rapidly in an unknown fashion. Using the positive root, we also observe that the stationary Kalman filter is asymptotically stable. This observation follows directly from the stationary Kalman-filter-update equations:

$$\widehat{x}(k+1|k) = (1 - K)\,\widehat{x}(k|k-1) + Ky(k)$$

$$= \left(1 - \frac{P^+}{P^+ + R}\right)\widehat{x}(k|k-1) + \frac{P^+}{P^+ + R}y(k).$$

This filter is indeed asymptotically stable, since

$$0 < 1 - \frac{P^+}{P^+ + R} < 1.$$

If we take the negative root P^- of Equation (5.80) then the stationary Kalman filter becomes unstable, since

$$1 - \frac{P^-}{P^- + R} > 1.$$

The previous example has illustrated that the algebraic Riccati equation (5.73) on page 162 can have more than one solution. However, the maximal and positive-definite solution $P_{\max}$, in the sense that

$$P_{\max} \geq P$$

for every solution P of Equation (5.73), guarantees that the stationary Kalman filter is asymptotically stable (see Theorem 5.4 on page 162).

5.8 The Kalman filter for estimating unknown inputs

In this section we present an application of Kalman filtering. We discuss how Kalman filtering can be used to estimate unknown inputs of a linear dynamical system.

We start with a rather crude problem formulation. Consider the time-invariant state-space system

$$x(k+1) = Ax(k) + Bu(k) + w(k), \tag{5.81}$$
$$y(k) = Cx(k) + v(k), \tag{5.82}$$

and let measurements of the output sequence $y(k)$ be given for $k = 1, 2, \ldots, N$. The problem is to determine an estimate of the input sequence, $\widehat{u}(k)$ for $k = 1, 2, \ldots, N$, such that the corresponding output $\widehat{y}(k)$ closely approximates $y(k)$.

A more precise problem formulation requires the definition of a model that describes how the input $u(k)$ is generated. An example of such a model can follow from knowledge of the class of signals to which $u(k)$

belongs, as illustrated in Exercise 5.3 on page 173. In this section we assume the so-called *random-walk* process as a model for the class of inputs we consider:

$$u(k+1) = u(k) + w_u(k), \tag{5.83}$$

with $w_u(k)$ a white-noise sequence that is uncorrelated with $w(k)$ and $v(k)$ in (5.81)–(5.82) on page 166, and has the covariance representation:

$$w_u(k) = Q_u^{1/2} \tilde{w}_u(k), \quad \tilde{w}_u(k) \sim (0, I_m).$$

For such input signals, the problem is to determine the input covariance matrix Q_u and a realization of the input sequence $u(k)$ for $k = 1, 2, \ldots, N$, such that the output $\hat{y}(k)$ is a minimum-error variance approximation of $y(k)$.

The combination of the time-invariant model (5.81)–(5.82) on page 166 and the model representing the class of input signals (5.83) results in the following augmented state-space model:

$$\begin{bmatrix} x(k+1) \\ u(k+1) \end{bmatrix} = \begin{bmatrix} A & B \\ 0 & I \end{bmatrix} \begin{bmatrix} x(k) \\ u(k) \end{bmatrix} + \begin{bmatrix} w(k) \\ w_u(k) \end{bmatrix}, \tag{5.84}$$

$$y(k) = \begin{bmatrix} C & 0 \end{bmatrix} \begin{bmatrix} x(k) \\ u(k) \end{bmatrix} + v(k), \tag{5.85}$$

with the process and measurement noise having a covariance matrix

$$E \begin{bmatrix} w(k) \\ w_u(k) \\ v(k) \end{bmatrix} \begin{bmatrix} w^T(k) & w_u^T(k) & v^T(k) \end{bmatrix} = \begin{bmatrix} Q & 0 & S \\ 0 & Q_u & 0 \\ S^T & 0 & R \end{bmatrix}.$$

An important design variable for the Kalman filter of the augmented state-space model is the covariance matrix Q_u. The tuning of this parameter will be illustrated in Example 5.6 on page 168.

The augmented state-space model (5.84)–(5.85) has no measurable input sequence. A bounded solution for the state-error covariance matrix of the Kalman filter of the extended state-space model (5.84)–(5.85) requires that the pair

$$\left(\begin{bmatrix} A & B \\ 0 & I \end{bmatrix}, \begin{bmatrix} C & 0 \end{bmatrix} \right)$$

be observable (see Lemma 5.1 on page 130 and Exercise 5.5 on page 174).

The conditions under which the observability of the original pair (A, C) is preserved are given in the following lemma.

Lemma 5.2 *The pair*

$$\left(\begin{bmatrix} A & B \\ 0 & I \end{bmatrix}, \begin{bmatrix} C & 0 \end{bmatrix}\right)$$

is observable if the pair (A, C) is observable and, for any $\zeta \in \mathbb{R}^m$, $C(A-I)^{-1}B\zeta = 0$ implies $\zeta = 0$.

Proof By the Popov–Belevitch–Hautus test for checking observability (Lemma 3.4 on page 69) we have to prove that, for all eigenvectors v of the augmented system matrix

$$\begin{bmatrix} A & B \\ 0 & I \end{bmatrix},$$

the condition

$$\begin{bmatrix} C & 0 \end{bmatrix} v = 0 \tag{5.86}$$

holds only if $v = 0$. If we partition the eigenvector v as

$$\begin{bmatrix} \eta \\ \zeta \end{bmatrix}$$

then

$$\begin{bmatrix} A & B \\ 0 & I \end{bmatrix}\begin{bmatrix} \eta \\ \zeta \end{bmatrix} = \lambda \begin{bmatrix} \eta \\ \zeta \end{bmatrix}, \tag{5.87}$$

with λ the corresponding eigenvalue. It follows from the lower part of this Equation that $\zeta = 0$ or $\lambda = 1$.

With $\zeta = 0$ we read from Equation (5.87) that $A\eta = \lambda\eta$. Since the pair (A, C) is observable, application of the Popov–Belevitch–Hautus test shows that $C\eta$ can be zero only provided that η is zero.

With $\lambda = 1$, the top row of Equation (5.87) reads $(A - I)\eta = -B\zeta$. Hence, $C\eta = 0$ implies $C(A - I)^{-1}B\zeta = 0$, but this holds only if $\zeta = 0$. $\square$

The condition in Lemma 5.2 on ζ is, for single-input, single-output LTI systems, equivalent to the fact that the original system (A, B, C), does not have zeros at the point $z = 1$ of the complex plane (Kailath, 1980). For multivariable systems the condition corresponds to the original system having no so-called *transmission zeros* (Kailath, 1980) at the point $z = 1$.

Example 5.6 (Estimating an unknown input signal) In this example we are going to use a Kalman filter to estimate the wind speed that

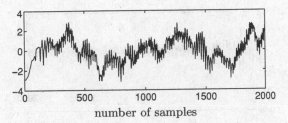

number of samples

Fig. 5.12. A wind-speed signal acting on a wind turbine (the mean has been subtracted).

acts on a wind turbine with a horizontal axis. Wind turbines are widely used for generation of electrical energy. A wind turbine consists of a rotor on top of a tower. The purpose of the rotor is to convert the linear motion of the wind into rotational energy: the rotor shaft is set into motion, which is used to drive a generator, where electrical energy is generated. The rotor consists of a number of rotor blades that are attached to the rotor shaft with a blade-pitch mechanism. This mechanism can be used to change the pitch angle of the blades. It is used to control the rotor speed. For efficient operation it is desirable to keep the rotor speed close to a certain nominal value irrespective of the intensity of the wind acting on the rotor. For this purpose, the wind turbine has a feedback controller that uses the measurement of the rotor speed to control the pitch angle of the blades.

In a typical wind turbine, the speed of the wind acting on the rotor cannot be measured accurately, because the rotor is disturbing it. We use a realistic simulation model of a closed-loop wind-turbine system to show that the Kalman filter can be used to estimate the unknown wind-speed signal. The input to this model is the wind-speed signal that needs to be estimated. The model has three outputs: the magnitude of the tower vibrations in the so-called nodding direction, the magnitude of the tower vibrations in the so-called naying direction, and the difference between the nominal value of the rotor speed and its actual value. For simulation, the wind signal shown in Figure 5.12 was used. The corresponding output signals are shown in Figure 5.13. In the simulations no process noise was used and a measurement noise with covariance matrix $R = 10^{-6}I_\ell$ was added to the outputs.

As discussed above, the unknown wind input signal is modeled as a random-walk process:

$$u(k + 1) = u(k) + w_u(k).$$

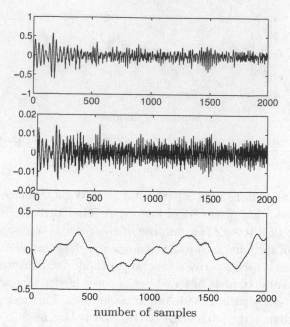

number of samples

Fig. 5.13. Output signals of the wind-turbine model. From top to bottom: tower acceleration in the nodding direction, tower acceleration in the naying direction, and the difference between the nominal rotor speed and its actual value.

An augmented state-space model was created. A Kalman filter, implemented as an SRCF, was used to estimate the states and the input from 2000 samples of the output signals. Since there was no process noise, the covariance of the noise acting on the states was taken to be very small: $Q = 10^{-20} I_n$. Although it may seem strange to model the wind signal as a random-walk process, the simulations presented below show that adequate reconstruction results can be obtained by selecting an appropriate value for Q_u. Four different values of the covariance matrix Q_u are selected: $Q_u = 10^{-12}$, $Q_u = 10^{-8}$, $Q_u = 10^{-4}$, and $Q_u = 1$, respectively. The results are presented in Figure 5.14. By setting $Q_u = 10^{-12}$, we assume that the input $u(k)$ is "almost" constant. This assumption is not correct, because the wind signal is far from constant. As can be seen from Figure 5.14(a), the real input signal and the estimated input signal barely look alike. By increasing the matrix Q_u we allow for more variation in the input signal. Figure 5.14(b) shows that the choice $Q_u = 10^{-8}$ results in more variation in the reconstructed input signal. On increasing the matrix Q_u further to $Q_u = 10^{-4}$, the reconstructed

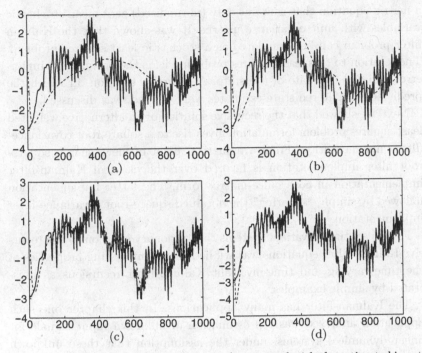

Fig. 5.14. The true input signal (solid line) compared with the estimated input signals (dashed line) for four choices of the matrix Q_u: (a) 10^{-12}, (b) 10^{-8}, (c) 10^{-4}, and (d) 1.

input signal better resembles the form of the original input signal. This is illustrated in Figure 5.14(c). For the final choice of $Q_u = 1$, shown in Figure 5.14(d), the reconstructed input and the original input are almost equal, except for some small initial deviation. We can conclude that, by carefully choosing the matrix Q_u, a high-quality reconstruction of the input signal can be obtained using a Kalman filter.

5.9 Summary

This chapter discussed the reconstruction of the state of a linear time-varying dynamic system, given a known state-space model. The construction of a filter that estimates the state, in the presence of noise, is known as the Kalman-filter problem. The formulation of the problem and its solution are presented within the framework of the linear least-squares method. The key to this formulation is the general representation of the mean and the covariance matrix of a random variable in a matrix equation that contains only (normalized) zero-mean stochastic variables with unit covariance matrix. It was shown that the Kalman filter problem can be formulated as a stochastic least-squares problem. The solution to this least-squares problem yields the optimal minimum-error variance estimate of the state. A recursive solution, in which the predicted and filtered state estimates were defined, was discussed.

Next, we showed that the recursive solution of an alternative weighted least-squares problem formulation gives rise to a square-root covariance-filter implementation. From a numerical point of view, such a square-root filter implementation is favored over the classical Kalman-filter implementation in covariance-matrix form. The latter implementation followed by simply "squaring" the obtained square-root covariance-filter implementation.

For linear, time-invariant systems the stationary solution to the recursive Kalman-filter equations was studied. Some interesting properties of the time-varying and time-invariant Kalman-filter recursions as illustrated by simple examples.

The Kalman-filter has many applications. In this chapter one such application analyzed was the estimation of unknown input signals for linear dynamical systems, under the assumption that these unknown input signals belong to a particular class of signals, for example the class of random-walk signals.

Exercises

5.1   Derive the conventional Kalman filter for the case in which the system model is represented by

$$x(k+1) = Ax(k) + Bu(k) + w(k), \quad x(0) = x_0,$$
$$y(k) = Cx(k) + Du(k) + v(k),$$

instead of (5.8)–(5.9). Accomplish this derivation in the following steps.

(a) Derive the modification caused by this new signal-generation model to the data equation (5.16) on page 135.

(b) Show the modifications necessary for the measurement-update equations (5.17) and (5.18) on page 136.

(c) Show the modifications for the time-update equations (5.19) and (5.20) on page 137.

5.2   Verify that the representation

$$\begin{bmatrix} v(k) \\ w(k) \end{bmatrix} = \begin{bmatrix} R(k)^{1/2} & 0 \\ X(k) & Q_x(k)^{1/2} \end{bmatrix} \begin{bmatrix} \widetilde{v}(k) \\ \widetilde{w}(k) \end{bmatrix},$$

with

$$\begin{bmatrix} \widetilde{v}(k) \\ \widetilde{w}(k) \end{bmatrix} \sim (0, I_{\ell+n}),$$

and with the matrices $X(k)$ and $Q_x(k)$ satisfying

$$X(k) = S(k)R(k)^{-T/2},$$
$$Q_x(k) = Q(k) - S(k)R(k)^{-1}S(k)^{T},$$

represents the covariance matrix of $v(k)$ and $w(k)$ given by

$$E \begin{bmatrix} v(k) \\ w(k) \end{bmatrix} \begin{bmatrix} v(k-j)^{T} & w(k-j)^{T} \end{bmatrix} = \begin{bmatrix} R(k) & S(k)^{T} \\ S(k) & Q(k) \end{bmatrix} \Delta(j).$$

5.3   Consider the following LTI system:

$$x(k+1) = Ax(k) + Bu(k) + w(k), \quad x(0) = x_0,$$
$$y(k) = Cx(k) + Du(k) + v(k),$$

with the system matrices A, B, C, D given and with $w(k)$ and $v(k)$ mutually correlated zero-mean white-noise sequences with known covariance matrix

$$E \left[\begin{bmatrix} w(k) \\ v(k) \end{bmatrix} \begin{bmatrix} w(j)^{T} & v(j)^{T} \end{bmatrix} \right] = \begin{bmatrix} Q & S \\ S^{T} & R \end{bmatrix} \Delta(k-j).$$

The goal is to estimate both the input $u(k)$ and the state $x(k)$ using measurements of the output sequence $y(k)$. It is assumed that the input $u(k)$ belongs to the class of harmonic sequences of the form

$$u(k) = \alpha \cos(\omega k + \phi),$$

for $\alpha, \omega, \phi \in \mathbb{R}$ and ω given.

(a) Determine the system matrices Φ and Γ and the initial condition z_0 of the autonomous system

$$z(k+1) = \Phi z(k), \quad z(0) = z_0,$$
$$u(k) = \Gamma z(k),$$

such that $u(k) = \alpha \cos(\omega k + \phi)$.
(*Hint: formulate the input $u(k)$ as the solution to an autonomous difference equation.*)

(b) Under the assumption that the pair (A, C) is observable, determine the condition under which the augmented pair

$$\left(\begin{bmatrix} A & B\Gamma \\ 0 & \Phi \end{bmatrix}, \begin{bmatrix} C & D\Gamma \end{bmatrix} \right)$$

is observable.

(c) Take the system matrices A, B, C, D given in Example 5.2 on page 153. and verify part (b) of this exercise.

(d) Take the system matrices A, B, C, D and the noise covariance matrices Q, R, and S equal to those given in Example 5.2 on page 153. Show for $\omega = 0.5$ by means of a Matlab experiment that the stationary Kalman filter for the augmented system referred to in (b) allows us to determine an input such that the output $y(k)$ *asymptotically tracks* the given trajectory $d(k) = 2\cos(0.5k + 0.1)$.

5.4   Consider the SISO autonomous system:

$$x(k+1) = Ax(k), \quad x(0) = x_0,$$
$$y(k) = Cx(k),$$

with $x(k) \in \mathbb{R}^n$.

(a) Prove that the canonical form of the pair (A, C) given by

$$A = \begin{bmatrix} 0 & 0 & \cdots & 0 & -a_0 \\ 1 & 0 & \cdots & 0 & -a_1 \\ 0 & 1 & \cdots & 0 & -a_2 \\ \vdots & \vdots & \ddots & & \vdots \\ 0 & 0 & \cdots & 1 & -a_{n-1} \end{bmatrix}, \quad C = \begin{bmatrix} 0 & \cdots & 0 & 1 \end{bmatrix},$$

is always observable.

(b) Prove that for the above pair (A, C) it is always possible to find a vector $K \in \mathbb{R}^n$ such that the eigenvalues of the matrix $A - KC$ have an arbitrary location in the complex plane, and thus can always be located inside the unit disk.

5.5   Consider the LTI system

$$x(k+1) = Ax(k) + Bu(k) + w(k), \quad x(0) = x_0,$$
$$y(k) = Cx(k),$$

with the system matrices A, B, and C given and with $w(k)$ a zero-mean white-noise sequence with known covariance matrix

$E\big[w(k)w(j)^{\mathrm{T}}\big] = Q\Delta(k-j)$. To reconstruct the state of this system making use of the observations $u(k)$ and $y(k)$, we use an observer of the form

$$\widehat{x}(k+1) = A\widehat{x}(k) + Bu(k) + K\Big(y(k) - C\widehat{x}(k)\Big), \quad \widehat{x}(0) = \widehat{x}_0,$$

$$\widehat{y}(k) = C\widehat{x}(k),$$

Assume that (1) the pair (A, C) is observable, (2) a matrix K has been determined such that the matrix $A - KC$ is asymptotically stable (see Exercise 5.4), and (3) the covariance matrix,

$$E\left[\Big(x(0) - \widehat{x}_0\Big)\Big(x(0) - \widehat{x}_0\Big)^{\mathrm{T}}\right]$$

is bounded, then prove that the covariance matrix of the state error $x(k) - \widehat{x}(k)$ remains bounded for increasing time k.

5.6   Consider the LTI system

$$x(k+1) = Ax(k) + w(k),$$

$$y(k) = Cx(k),$$

with the system matrices A, B, and C given, with A asymptotically stable, and $w(k)$ a zero-mean white-noise sequence, with known covariance matrix $E\big[w(k)w(j)^{\mathrm{T}}\big] = Q\Delta(k-j)$, and $x(0)$ a random variable that is uncorrelated with $w(k)$ for all $k \in \mathbb{Z}$ and has covariance matrix

$$E\big[x(0)x(0)^{\mathrm{T}}\big] = P(0).$$

(a) Show that $E[x(k)w(j)^{\mathrm{T}}] = 0$ for $j \geq k$.

(b) Show that $P(k) = E[x(k)x(k)^{\mathrm{T}}]$ satisfies

$$P(k+1) = AP(k)A^{\mathrm{T}} + Q.$$

5.7   Consider the LTI system

$$x(k+1) = Ax(k) + Bu(k) + w(k), \qquad \text{(E5.1)}$$

$$y(k) = Cx(k) + v(k), \qquad \text{(E5.2)}$$

with $w(k)$ and $v(k)$ zero-mean random sequences with covariance matrix

$$E\left[\begin{bmatrix} w(k) \\ v(k) \end{bmatrix}\begin{bmatrix} w(j)^{\mathrm{T}} & v(j)^{\mathrm{T}} \end{bmatrix}\right] = \begin{bmatrix} Q & S \\ S^{\mathrm{T}} & R \end{bmatrix}\Delta(k-j),$$

such that

$$\begin{bmatrix} Q & S \\ S^T & R \end{bmatrix} > 0,$$

and with (A, C) observable and $(A, Q^{1/2})$ reachable. Assume that the DARE given by

$$P = APA^T + Q - (S + APC^T)(CPC^T + R)^{-1}(S + APC^T)^T,$$

has a positive-definite solution P that defines the gain matrix K as

$$K = (S + APC^T)(CPC^T + R)^{-1}.$$

Use Theorem 3.5 on page 63 to show that the matrix $A - KC$ is asymptotically stable.

5.8   Let the system $P(z) = C(zI_n - A)^{-1}B$ be given with the pair (A, C) observable and let the system have a single zero at $z = 1$. For this value there exists a vector $y \neq 0$ such that

$$\lim_{z \to 1}\left(C(zI_n - A)^{-1}B\right) y = 0.$$

(a) Show that the matrix C_n in the expression

$$C(zI_n - A)^{-1}B = (z - 1)C_n(zI_n - A)^{-1}B$$

satisfies

$$C_n\begin{bmatrix} (A - I_n) & B \end{bmatrix} = \begin{bmatrix} C & 0 \end{bmatrix}.$$

(b) Show that, if the pair (A, B) is reachable, the matrix

$$\begin{bmatrix} (A - I_n) & B \end{bmatrix}$$

has full row rank.

5.9   Consider the regularized and exponentially weighted least-squares problem

$$\min_x \Bigg[\lambda^N \left(x - \hat{x}(0) \right)^T P(0)^{-T/2} P(0)^{-1/2} \left(x - \hat{x}(0) \right)$$

$$+ \sum_{i=0}^{N-1} \left(y(i) - C(i)x \right)^T R(i)^{-T/2} R(i)^{-1/2} \left(y(i) - C(i)x \right) \Bigg]$$

$$\text{(E5.3)}$$

with $\hat{x}(0) \in \mathbb{R}^n$, $P(0)^{1/2} \in \mathbb{R}^{n \times n}$ invertible, $R(i)^{1/2} \in \mathbb{R}^{\ell \times \ell}$ invertible, $C(i) \in \mathbb{R}^{\ell \times n}$, and $y(i) \in \mathbb{R}^\ell$. The scalar λ is known and is used to "forget old data" (with time index $i < N - 1$); it satisfies $0 \ll \lambda < 1$. Following the steps below, derive a recursive update for the above least-squares estimate similar to the derivation of the measurement update in Section 5.5.2.

(a) Derive an update for the initial estimate $\widehat{x}(0)$ according to

$$\begin{bmatrix} (1/\sqrt{\lambda})C(0)P(0)^{1/2} & R(0)^{1/2} \\ (1/\sqrt{\lambda})P(0)^{1/2} & 0 \end{bmatrix} T_r(0) = x \begin{bmatrix} R^\epsilon(0)^{1/2} & 0 \\ G(0) & P(1)^{1/2} \end{bmatrix},$$

with $T_r(0)$ an orthogonal matrix to make the left matrix array lower-triangular, and

$$\widehat{x}(1) = \widehat{x}(0) + G(0)R^\epsilon(0)^{-1/2}\Big(y(0) - C(0)\widehat{x}(0)\Big).$$

(b) Show that for $i = 0, 1, \ldots, N - 1$ the estimate $\widehat{x}(i)$ is updated according to

$$\begin{bmatrix} (1/\sqrt{\lambda})C(i)P(i)^{1/2} & R(i)^{1/2} \\ (1/\sqrt{\lambda})P(i)^{1/2} & 0 \end{bmatrix} T_r(i) = \begin{bmatrix} R^\epsilon(i)^{1/2} & 0 \\ G(i) & P(i+1)^{1/2} \end{bmatrix},$$

with $T_r(i)$ again an orthogonal matrix, and

$$\widehat{x}(i+1) = \widehat{x}(i) + G(i)R^\epsilon(i)^{-1/2}\Big(y(i) - C(i)\widehat{x}(i)\Big).$$

(c) Show that $\widehat{x}(N)$ obtained by the update of part (b) is a solution to Equation (E5.3).

5.10   Consider the innovation model

$$\widehat{x}(k+1) = A\widehat{x}(k) + Bu(k) + Ke(k),$$
$$y(k) = C\widehat{x}(k) + e(k),$$

as defined in Section 5.7 with predictor

$$\widehat{x}(k+1) = (A - KC)\widehat{x}(k) + Bu(k) + Ky(k),$$
$$\widehat{y}(k) = C\widehat{x}(k).$$

(a) Determine the transfer functions $G(q)$ and $H(q)$ such that

$$y(k) = G(q)u(k) + H(q)e(k).$$

You may assume that the operator $(qI_n - A)$ is invertible.

(b) Let the matrix K be such that $(A - KC)$ is asymptotically stable. Show that

$$\widehat{y}(k) = H(q)^{-1}G(q)u(k) + \Big(I_\ell - H(q)^{-1}\Big)y(k).$$

6

Estimation of spectra and frequency-response functions

After studying this chapter you will be able to

- use the discrete Fourier transform to transform finite-length time signals to the frequency domain;
- describe the properties of the discrete Fourier transform;
- relate the discrete Fourier transform to the discrete-time Fourier transform;
- efficiently compute the discrete Fourier transform using fast-Fourier-transform algorithms;
- estimate spectra from finite-length data sequences;
- reduce the variance of spectral estimates using blocked data processing and windowing techniques;
- estimate the frequency-response function (FRF) and the disturbance spectrum from finite-length data sequences for an LTI system contaminated by output noise; and
- reduce the variance of FRF estimates using windowing techniques.

6.1 Introduction

In this chapter the problem of determining a model from input and output measurements is treated using frequency-domain methods. In the previous chapter we studied the estimation of the state given the system and measurements of its inputs and outputs. In this chapter we

178

do not bother about estimating the state. The models that will be estimated are input–output models, in which the state does not occur. More specifically, we investigate how to obtain in a simple and fast manner an estimate of the dynamic transfer function of an LTI system from recorded input and output data sequences taken from that system. We are interested in estimating the frequency-response function (FRF) that relates the measurable input to the measurable output sequence. The FRF has already been discussed briefly in Section 3.4.4 and its estimation is based on Lemma 4.3 via the estimation of the signal spectra of the recorded input and output data. Special attention is paid to the case of practical interest in which the data records have *finite data length*. The obtained discretization of the FRF estimate is determined by a number of parameters that is of the same order of magnitude as the number of samples used in the estimation process.

FRF models are of interest for a number of reasons. First, for complex engineering problems, such as those involving flexible mechanical components of the International Space Station, that are characterized by a large number of input and output data channels and a large number of data samples, the FRF estimation based on the fast Fourier transform can be computed efficiently. Second, though the model is not parametric, it provides good engineering insight into a number of important (qualitative) properties, such as the location and presence of resonances, the system delay, and the bandwidth. Thus the FRF method is often used as an initial and quick data-analysis procedure to acquire information on the system so that improved identification experiments can be designed subsequently for use in estimating, for example, state-space models.

In addition to the study of the effect of the finite number of samples on the estimation of the FRFs, we explain how to deal with noise on the output signals in this estimation procedure. For simplicity, the discussion will be limited to SISO systems only, although most of the results can readily be extended to the MIMO case.

We start this chapter by introducing the discrete Fourier transform in Section 6.2. The discrete Fourier transform allows us to derive frequency-domain information from finite-duration signals. We show that the discrete Fourier transform can be computed by solving a least-squares problem. We also investigate the relation between the discrete Fourier transform and the discrete-time Fourier transform (see Section 3.3.2). This leads to the discussion of spectral leakage in Section 6.3. Next, in

Section 6.4, we briefly show that the fast Fourier transform is a very fast algorithm for computing the discrete Fourier transform. Section 6.5 explains how to estimate signal spectra from finite-length data sequences using the discrete Fourier transform. Finally, Section 6.6 discusses the estimation of frequency-response functions on the basis of estimates of the input and output spectra.

6.2 The discrete Fourier transform

In Section 3.3.2 we introduced the DTFT. To compute this signal transform, we need information on the signal in the time interval from $-\infty$ to ∞. In practice, however, we have access to only a finite number of data samples. For this class of sequences, the discrete Fourier transform (DFT) is defined as follows.

Definition 6.1 (Discrete Fourier transform) *The discrete Fourier transform (DFT) of a signal $x(k)$ for $k = 0, 1, 2, \ldots, N-1$ is given by*

$$X_N(\omega_n) = \sum_{k=0}^{N-1} x(k) e^{-j\omega_n kT}, \qquad (6.1)$$

for $\omega_n = 2\pi n/(NT)$ radians per second, $n = 0, 1, 2, \ldots, N-1$, and sampling time $T \in \mathbb{R}$.

Therefore, a time sequence of N samples is transformed by the DFT to a sequence of N complex numbers. The DFT is an invertible operation, its inverse can be computed as follows.

Theorem 6.1 (Inverse discrete Fourier transform) *The inverse discrete Fourier transform of a sequence $X_N(\omega_n)$ with $\omega_n = 2\pi n/(NT)$ and $n = 0, 1, 2, \ldots, N-1$ is given by*

$$x(k) = \frac{1}{N} \sum_{n=0}^{N-1} X_N(\omega_n) e^{j\omega_n kT},$$

with time $k = 0, 1, 2, \ldots, N-1$ and sampling time $T \in \mathbb{R}$.

Some important properties of the DFT are listed in Table 6.1. Table 6.2 lists some DFT pairs.

As pointed out by Johansson (1993), the computation of the DFT of a signal can in fact be viewed as a least-squares problem. To illustrate

Table 6.1. *Properties of the discrete Fourier transform*

Property	Time signal	Fourier transform	Conditions
Linearity	$ax(k) + by(k)$	$aX_N(\omega_n) + bY_N(\omega_n)$	$a, b \in \mathbb{C}$
Time–frequency symmetry	$X_N\left(\dfrac{2\pi}{NT}k\right)$	$Nx\left(-\omega_n\dfrac{NT}{2\pi}\right)$	
Multiple shift	$x\left((k-\ell)\bmod N\right)$	$e^{-j\omega_n T\ell}X_N(\omega_n)$	$l \in \mathbb{Z}$
Circular convolution	$\displaystyle\sum_{i=0}^{N-1} x\left((k-i)\bmod N\right)y(i)$	$X_N(\omega_n)Y_N(\omega_n)$	

Table 6.2. *Standard discrete Fourier transform pairs for the time signals defined at the points $k = 1, 2, \ldots, N-1$ and the Fourier-transformed signals defined at the points $\omega_n = 2\pi n/(NT)$, $n = 0, 1, 2, \ldots, N-1$*

Time signal	Fourier transform	Conditions
1	$N\Delta(n)$	
$\Delta(k)$	1	
$\Delta(k - \ell)$	$e^{-j\omega_n \ell T}$	$\ell \in \mathbb{Z}$
a^k	$\dfrac{1 - a^N}{1 - ae^{-j\omega_n T}}$	$a \in \mathbb{C},\ |a| < 1$
$\sin(\omega_\ell k)$	$j\dfrac{N}{2}\Delta\left(n + \dfrac{\ell}{T}\right) - j\dfrac{N}{2}\Delta\left(n - \dfrac{\ell}{T}\right)$	$\omega_l = \dfrac{2\pi\ell}{NT},\ \ell \in \mathbb{Z}$
$\cos(\omega_\ell k)$	$\dfrac{N}{2}\Delta\left(n + \dfrac{\ell}{T}\right) + \dfrac{N}{2}\Delta\left(n - \dfrac{\ell}{T}\right)$	$\omega_l = \dfrac{2\pi\ell}{NT},\ \ell \in \mathbb{Z}$

this, we use the inverse DFT to define

$$\widehat{x}(k) = \frac{1}{N}\sum_{n=0}^{N-1} X_N(\omega_n)e^{j\omega_n kT}.$$

To compute $X_N(\omega_n)$ for $\omega_n = 2\pi n/(NT)$, $n = 0, 1, 2, \ldots, N-1$, we solve the following minimization problem:

$$\min_{X_N(\omega_n)} \sum_{k=0}^{N-1}\left(x(k) - \widehat{x}(k)\right)^2.$$

This problem can be written in matrix form as

$$\min_{\underline{X}_N} \frac{1}{N} \| N\underline{x}_N - \Phi_N \underline{X}_N \|_2^2, \tag{6.2}$$

with

$$\underline{x}_N = \begin{bmatrix} x(0) & x(1) & \cdots & x(N-1) \end{bmatrix}^{\mathrm{T}},$$

$$\underline{X}_N = \begin{bmatrix} X_N(\omega_0) & X_N(\omega_1) & \cdots & X_N(\omega_{N-1}) \end{bmatrix}^{\mathrm{T}},$$

$$\Phi_N = \begin{bmatrix} 1 & 1 & \cdots & 1 \\ e^{j\omega_0 T} & e^{j\omega_1 T} & \cdots & e^{j\omega_{N-1} T} \\ \vdots & \vdots & & \vdots \\ e^{j\omega_0(N-1)T} & e^{j\omega_1(N-1)T} & \cdots & e^{j\omega_{N-1}(N-1)T} \end{bmatrix}.$$

From Section 2.6 we know that the solution follows from

$$N\Phi_N^{\mathrm{H}} \underline{x}_N = \Phi_N^{\mathrm{H}} \Phi_N \underline{X}_N,$$

where Φ_N^{H} denotes the complex-conjugate transpose of the matrix Φ_N. Beware of the presence of the complex-conjugate transpose instead of the transpose since we are dealing with complex-valued matrices. The solution to (6.2) equals

$$\underline{X}_N = N \left(\Phi_N^{\mathrm{H}} \Phi_N \right)^{-1} \Phi_N^{\mathrm{H}} \underline{x}_N.$$

This can be simplified using

$$\sum_{k=0}^{N-1} e^{jw_p Tk} e^{-jw_q Tk} = \sum_{k=0}^{N-1} e^{j2\pi k(p-q)/N}$$

$$= \begin{cases} N, & p = q, \\ \dfrac{1 - e^{j2\pi(p-q)}}{1 - e^{j2\pi(p-q)/N}} = 0, & p \neq q, \end{cases}$$

which leads to

$$\left(\Phi_N^{\mathrm{H}} \Phi_N \right)^{-1} = \frac{1}{N} I_N.$$

Using this fact, we get

$$\underline{X}_N = \Phi_N^{\mathrm{H}} \underline{x}_N,$$

or, equivalently,

$$X_N(\omega_n) = \sum_{k=0}^{N-1} x(k) e^{-j\omega_n kT}.$$

Note that it is not advisable to solve the above-mentioned least-squares problem in order to compute the DFT. Computationally much faster algorithms are available to compute the DFT. One such algorithm is briefly discussed in Section 6.4.

The DFT of a periodic sequence $x(k)$ based on a finite number of samples N can represent the DTFT of $x(k)$ when that number of samples N is an integer multiple of the period length of $x(k)$. This is illustrated in the following example. The DFT of any sequence of finite length can be considered as the DTFT of an infinite-length sequence that is obtained by a periodic repetition of the given finite-length sequence.

Example 6.1 (Relation between DFT and DTFT) We are given the periodic sinusoid

$$x(k) = \cos(\omega_0 kT),$$

for $\omega_0 = 2\pi/(N_0 T)$ and $k = 0, 1, 2, \ldots, N-1$, such that $N = rN_0$ with $r \in \mathbb{Z}, r > 0$. According to Table 3.5 on page 54, the DTFT of $x(k)$ is equal to

$$X(e^{j\omega T}) = \frac{\pi}{T}\delta(\omega + \omega_0) + \frac{\pi}{T}\delta(\omega - \omega_0). \tag{6.3}$$

The DFT of $x(k)$ can be taken as the DTFT of the product $p(k)x(k)$ for all k, with $p(k)$ defined as

$$p(k) = \begin{cases} 1, & k = 0, 1, \ldots, N-1, \\ 0, & \text{otherwise.} \end{cases}$$

Let $y(k) = p(k)x(k)$, then, at the frequency points $\omega = \omega_n$, the DTFT of $y(k)$ equals the DFT of $x(k)$:

$$Y(e^{j\omega_n T}) = \sum_{k=-\infty}^{\infty} p(k)x(k)e^{-j\omega_n kT} = \sum_{k=0}^{N-1} x(k)e^{-j\omega_n kT} = X_N(\omega_n).$$

We first prove that the DTFT of $y(k)$ can be written as the convolution of the DTFT of $p(k)$ with the DTFT of $x(k)$ as follows:

$$Y(e^{j\omega T}) = \frac{T}{2\pi} \int_{-\pi/T}^{\pi/T} P(e^{j(\omega-\lambda)T})X(e^{j\lambda T})d\lambda. \tag{6.4}$$

This expression follows from the derivation below:

$$Y(e^{j\omega T}) = \frac{T}{2\pi} \int_{-\pi/T}^{\pi/T} P(e^{j(\omega-\lambda)T}) X(e^{j\lambda T}) d\lambda$$

$$= \frac{T}{2\pi} \int_{-\pi/T}^{\pi/T} \sum_{k=-\infty}^{\infty} p(k) e^{-j(\omega-\lambda)kT} \sum_{\ell=-\infty}^{\infty} x(\ell) e^{-j\lambda \ell T} d\lambda$$

$$= \frac{T}{2\pi} \sum_{k=-\infty}^{\infty} p(k) e^{-j\omega kT} \sum_{\ell=-\infty}^{\infty} x(\ell) \int_{-\pi/T}^{\pi/T} e^{j\lambda(k-\ell)T} d\lambda$$

$$= \sum_{k=-\infty}^{\infty} p(k) e^{-j\omega kT} \sum_{\ell=-\infty}^{\infty} x(\ell) \Delta(k-\ell)$$

$$= \sum_{k=-\infty}^{\infty} p(k) x(k) e^{-j\omega kT}.$$

With Equation (6.3) we get

$$Y(e^{j\omega T}) = \frac{1}{2} \int_{-\pi/T}^{\pi/T} P(e^{j(\omega-\lambda)T}) \Big(\delta(\lambda+\omega_0) + \delta(\lambda-\omega_0)\Big) d\lambda$$

$$= \frac{1}{2} \Big(P(e^{j(\omega+\omega_0)T}) + P(e^{j(\omega-\omega_0)T}) \Big).$$

With the result from Example 3.8 on page 54 we can evaluate this expression at the points $\omega = \omega_n = 2\pi n/(NT)$. We first evaluate $P(e^{j(\omega_n+\omega_0)})$:

$$P(e^{j(\omega_n+\omega_0)}) = \frac{\sin\left(\dfrac{N}{2}(\omega_n+\omega_0)T\right)}{\sin\left(\dfrac{1}{2}(\omega_n+\omega_0)T\right)} e^{-j(\omega_n+\omega_0)T(N-1)/2}$$

$$= \frac{\sin\left(\dfrac{N}{2}\left(\dfrac{2\pi n}{N}+\dfrac{2\pi r}{N}\right)\right)}{\sin\left(\dfrac{1}{2}\left(\dfrac{2\pi n}{N}+\dfrac{2\pi r}{N}\right)\right)} e^{-j(2\pi n+2\pi r)(N-1)/(2N)}$$

$$= \frac{\sin(\pi(n+r))}{\sin\left(\dfrac{\pi}{N}(n+r)\right)} e^{-j2\pi(n+r)(N-1)/(2N)}$$

$$= \begin{cases} N, & n = N-r, \\ 0, & n = 0,\ldots,N-1-r, N+1-r,\ldots,N-1. \end{cases}$$

In the same way, we get

$$P(e^{j(\omega_n-\omega_0)}) = \begin{cases} N, & n = r, \\ 0, & n = 0,\ldots,r-1, r+1,\ldots,N-1. \end{cases}$$

Therefore,

$$X_N(\omega_n) = Y(e^{j\omega_n T}) = \begin{cases} N/2, & n = r, \\ N/2, & n = N - r, \\ 0, & \text{otherwise}, \end{cases}$$

which can also be written as

$$X_N(\omega_n) = \frac{N}{2}\Delta(n - r) + \frac{N}{2}\Delta(n - N + r).$$

6.3 Spectral leakage

Equation (6.4) on page 183 shows a general relationship between the DTFT and the DFT of a time sequence $x(k)$. Recall this equation:

$$X_N(\omega_n) = \frac{T}{2\pi} \int_{-\pi/T}^{\pi/T} P(e^{j(\omega_n - \lambda)T})X(e^{j\lambda T})d\lambda,$$

with

$$P(e^{j\omega T}) = \frac{\sin\left(\dfrac{N}{2}\omega T\right)}{\sin\left(\dfrac{1}{2}\omega T\right)}e^{-j(N-1)\omega T/2}.$$

This integral in fact represents a convolution between $P(e^{j\omega_n T})$ and the DTFT of $x(k)$. Therefore, in general the DFT will be a distorted version of the DTFT. This distortion is called *spectral leakage*. Figure 6.1 shows the magnitude of $P(e^{j\omega T})$. Because $P(e^{j\omega T})$ is different from zero around $\omega = 0$, a peak in $X(e^{j\omega T})$ at a certain frequency ω_r shows up not only at the frequency ω_r in $X_N(\omega_n)$ but also at neighboring frequencies. Hence, the original frequency content of $X(e^{j\omega T})$ "leaks out" toward other frequencies in $X_N(\omega_n)$. Only for the special case of a periodic sequence $x(k)$ for which the length N of the DFT is an integer multiple of the period, as in Example 6.1, does no leakage occur. When this is not the case, leakage occurs even for periodic sequences, as illustrated in Example 6.2.

Example 6.2 (Spectral leakage) Consider the signal

$$x(k) = \cos\left(\frac{2\pi}{4}k\right),$$

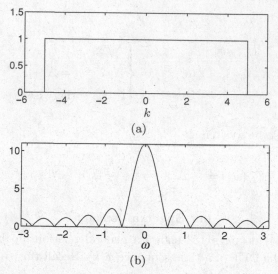

Fig. 6.1. (a) A rectangular window ($N = 11$). (b) The magnitude of the DTFT of this rectangular window ($T = 1$).

with sampling time $T = 1$. It is easy to see that

$$X_N(\omega_n) = \sum_{k=0}^{N-1} \cos\left(\frac{2\pi}{4}k\right) e^{-j2\pi kn/N}$$

$$= \frac{1}{2} \sum_{k=0}^{N-1} \left(e^{j2\pi k/4} + e^{-j2\pi k/4}\right) e^{-j2\pi kn/N}.$$

Figure 6.2(a) shows the magnitude of this DFT for $N = 4$. We see two peaks, as expected. There is no frequency leakage, since N equals the period length of $x(k)$. Figure 6.2(b) shows the magnitude of the DFT for $N = 8$. Again there is no frequency leakage, since N is an integer multiple of the period length of $x(k)$. However, if we take $N = 11$ as shown in Figure 6.2(c), frequency leakage occurs.

To overcome the effect of leakage, the finite-time-length sequence $x(k)$ is usually multiplied by a window that causes the time sequence to undergo smooth transition toward zero at the beginning and the end. The DTFT of this windowed sequence extended with zeros, to make it an infinite sequence, exactly matches the DFT of the finite-length windowed sequence. Hence, no spectral leakage occurs, at the expense of introducing a slight distortion into original time signal. An often-used

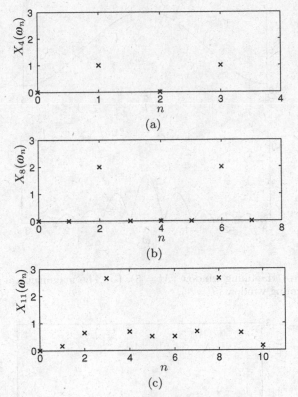

Fig. 6.2. DFTs of the signal $x(k)$ from Example 6.2 for (a) $N = 4$, (b) $N = 8$, and (c) $N = 11$.

window is the Hamming window, defined as

$$w_M(k) = \begin{cases} 0.54 - 0.46\cos\left(\dfrac{\pi}{M}k\right), & -M \leq k \leq M, \\ 0, & \text{otherwise,} \end{cases} \tag{6.5}$$

where M determines the "width" of the window. Figure 6.3(a) shows the Hamming window for $M = 5$. Let $W_M(e^{j\omega T})$ be the DTFT of $w_M(k)$, then

$$X_N(\omega_n) = \frac{T}{2\pi} \int_{-\pi/T}^{\pi/T} W_M(e^{j(\omega_n - \lambda)T})X(e^{j\lambda T})\mathrm{d}\lambda.$$

Figure 6.3(b) shows the magnitude of $W_M(e^{j\omega T})$. We see that the width of the peak around $\omega = 0$ is reduced in comparison with that in Figure 6.1. Therefore, there is less distortion in the DFT.

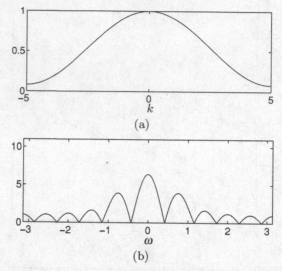

(a)

(b)

Fig. 6.3. (a) A Hamming window ($M = 5$). (b) The magnitude of the DTFT of this Hamming window ($T = 1$).

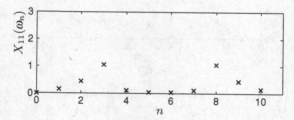

Fig. 6.4. The DFT of the product of a Hamming window with the signal $x(k)$ with $N = 11$ from Example 6.2.

Example 6.3 (Windowing) To overcome spectral leakage, the sequence $x(k)$ from Example 6.2 is multiplied by a Hamming window of width $M = 5$. We take $N = 11$. Figure 6.4 shows the DFT of the product of the Hamming window and the signal $x(k)$. On comparing this with Figure 6.2(c), we see that the distortion in the DFT is reduced. The remaining distortion is a result not of spectral leakage, but of the distortion of the signal in the time domain introduced by the window.

6.4 The FFT algorithm

For a sequence $\{x(k)\}_{k=0}^{N-1}$ given as in Equation (6.1) on page 180, the calculation of the DFT $X(\omega_n)$ at a certain frequency point ω_n requires N complex multiplications and $N - 1$ additions. In total, we therefore need N^2 multiplications and $N(N-1)$ additions to compute the DFT of the sequence $x(k)$ at all frequency points ω_n, $n = 0, 1, \ldots, N-1$. In general, multiplications take much more computer time than additions do; therefore the number of operations needed in order to compute the DFT is said to be of order N^2. If N is large, the computational complexity may hamper, for example, the real-time implementation of the algorithm. There are, however, algorithms that compute the DFT in fewer than N^2 operations. These algorithms are referred to as fast-Fourier-transformation (FFT) algorithms. The Radix-2 FFT algorithm is the most famous one and is summarized here.

The basic idea employed to reduce the computational burden is to divide the total time interval into intervals having a smaller number of points. To simplify the notation, we denote $e^{-j2\pi/N}$ by q_N. Assume that $N > 1$ and thus $q_N \neq 1$, then

$$X_N(\omega_n) = \sum_{k=0}^{N-1} x(k)q_N^{kn}, \quad n = 0, 1, \ldots, N-1.$$

Next, we assume that N is even and divide the calculation of $X(\omega_n)$ into two parts, one part containing the even samples and one part containing the odd samples, like this:

$$\alpha(k) = x(2k), \quad k = 0, 1, \ldots, N/2 - 1,$$
$$\beta(k) = x(2k+1), \quad k = 0, 1, \ldots, N/2 - 1.$$

We have

$$A_{N/2}(\omega_n) = \sum_{k=0}^{N/2-1} \alpha(k)q_{N/2}^{kn}, \quad k = 0, 1, \ldots, N/2 - 1,$$

$$B_{N/2}(\omega_n) = \sum_{k=0}^{N/2-1} \beta(k)q_{N/2}^{kn}, \quad k = 0, 1, \ldots, N/2 - 1.$$

Next, we show that

$$X_N(\omega_n) = A_{N/2}(\omega_n) + q_N^n B_{N/2}(\omega_n), \quad n = 0, 1, \ldots, N/2 - 1,$$
$$(6.6)$$

$$X_N(\omega_{N/2+n}) = A_{N/2}(\omega_n) - q_N^n B_{N/2}(\omega_n), \quad n = 0, 1, \ldots, N/2 - 1.$$
$$(6.7)$$

Since $\alpha(k) = x(2k)$ and $\beta(k) = x(2k+1)$, we have

$$A_{N/2}(\omega_n) + q_N^n B_{N/2}(\omega_n) = \sum_{k=0}^{N/2-1} x(2k)q_{N/2}^{kn} + \sum_{k=0}^{N/2-1} x(2k+1)q_N^n q_{N/2}^{kn}.$$

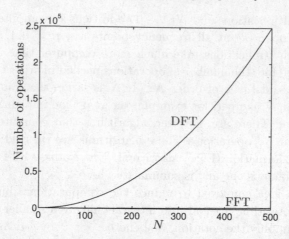

Fig. 6.5. The number of operations required to compute the DFT compared with the number of operations required by the FFT, plotted against N.

Using the properties

$$q_{N/2}^{kn} = q_N^{2kn} \quad \text{and} \quad q_N^n q_{N/2}^{kn} = q_N^{(1+2k)n},$$

we obtain

$$A_{N/2}(\omega_n) + q_N^n B_{N/2}(\omega_n) = \sum_{k=0}^{N/2-1} x(2k)q_N^{2kn} + \sum_{k=0}^{N/2-1} x(2k+1)q_N^{(1+2k)n}$$

$$= \sum_{k=0}^{N-1} x(k)q_N^{kn}.$$

In the same way Equation (6.7) can be proven. Computing both $A_{N/2}(\omega_n)$ and $B_{N/2}(\omega_n)$ takes $2(N/2)^2$ operations, so in total the number of operations needed to compute $X_N(\omega_n)$ is reduced to $(N^2+N)/2$. To compute $A_{N/2}(\omega_n)$ we can, if $N/2$ is even, apply the same trick again, and divide the sequence $\alpha(k)$ into two parts. To compute $B_{N/2}(\omega_n)$ we can then also divide the sequence $\beta(k)$ into two parts. If $N = 2^P$ for some positive integer P, we can repeatedly divide the sequences into two parts until we end up computing two-point DFTs. It can be shown that in this case the complexity has been reduced to

$$\frac{N}{2} \log_2 N.$$

This is an enormous reduction compared with the original N^2 operations, as illustrated by Figure 6.5.

6.5 Estimation of signal spectra

According to Definition 4.14 on page 105, to compute the spectrum of a signal we need an infinite number of samples of this signal. In practice, of course, we have only a finite number of samples available. Therefore, we would like to estimate the signal spectrum using only a finite number of samples. For this we can use the DFT. An estimate of the spectrum of a signal is called the *periodogram*, and is given by

$$\widehat{\Phi}_N^x(\omega_n) = \frac{1}{N}|X_N(\omega_n)|^2,$$

where $\omega_n = 2\pi n/(NT)$ and $n = 0, 1, \ldots, N-1$. The periodogram for the cross-spectrum between $x(k)$ and $y(k)$ equals

$$\widehat{\Phi}_N^{xy}(\omega_n) = \frac{1}{N}X_N(\omega_n)Y_N^*(\omega_n),$$

where $Y_N^*(\omega_n)$ denotes the complex conjugate of $Y_N(\omega_n)$.

The periodogram of a stochastic zero-mean process $x(k)$ with spectral density $\Phi^x(\omega)$ is itself a stochastic process. Its first- and second-order statistical moments are derived in Ljung and Glad (1994) and are given by

$$E\left[\widehat{\Phi}_N^x(\omega_n)\right] = \Phi^x(\omega_n) + R_{1,N}, \tag{6.8}$$

$$E\left[\widehat{\Phi}_N^x(\omega_n) - \Phi^x(\omega_n)\right]^2 = \left(\Phi^x(\omega_n)\right)^2 + R_{2,N}, \tag{6.9}$$

$$E\left[\widehat{\Phi}_N^x(\omega_n) - \Phi^x(\omega_n)\right]\left[\widehat{\Phi}_N^x(\omega_r) - \Phi^x(\omega_r)\right] = R_{3,N},$$

$$\text{for } |\omega_n - \omega_r| \geq \frac{2\pi}{NT}, \tag{6.10}$$

where $\omega_r = 2\pi r/(NT)$, $r = 0, 1, \ldots, N-1$, and

$$\lim_{N\to\infty} R_{i,N} = 0, \quad i = 1, 2, 3.$$

Equation (6.8) shows that the periodogram is asymptotically unbiased at the frequency points ω_n. The periodogram is, however, not a consistent estimate, because the variance of the periodogram does not go to zero as indicated by Equation (6.9). In fact, the variance is as large as that of the original spectrum. This means that the periodogram is highly fluctuating around its mean value. Equation (6.10) shows that the values of the periodogram at neighboring frequencies are asymptotically uncorrelated. This observation gives rise to methods that reduce the variance of the periodogram at the expense of decreasing the *frequency resolution*. The frequency resolution is the fineness of detail that can be distinguished in the periodogram. It is inversely proportional to the frequency separation in the periodogram. Note that for a sequence of N samples the DFT yields frequency values at the points $2\pi n/(NT)$, $n = 0, 1, \ldots, N-1$. Therefore, the known frequency points are separated by $2\pi/(NT)$. Since no information is available between these points, the frequency resolution of the periodogram is $NT/(2\pi)$.

One way to reduce the variance of the periodogram is *blocked-data processing* in which the final periodogram is an average of several periodograms taken from blocks of data. The rationale of this idea is as follows. Let s be a random variable and let N observations of s be collected in the sequence $\{s(k)\}_{k=1}^{N}$. If the random variable s has a mean and a variance given by

$$E[s(k)] = s_0,$$
$$E[s(k) - s_0][s(\ell) - s_0] = \sigma_s^2 \Delta(k - \ell),$$

then the estimate

$$\frac{1}{p} \sum_{\ell=1}^{p} s(\ell)$$

has mean s_0 and variance σ_s^2/p. In Exercise 6.3 on page 204 you are asked to derive this result. Since the periodogram $\widehat{\Phi}_N^x$ is a random variable, the idea of blocked-data processing can be used to reduce the variance of the periodogram. Given a time sequence $\{x(k)\}_{k=1}^{N}$, we split this sequence into p different blocks of data of equal size $N_p = N/p$. On the basis of the periodograms

$$\widehat{\Phi}_{N_p}^{x,i}(\omega_n), \quad i = 1, 2, \ldots, p,$$

we define the averaged periodogram as

$$\widehat{\Phi}_{N_p}^{x}(\omega_n) = \frac{1}{p} \sum_{\ell=1}^{p} \widehat{\Phi}_{N_p}^{x,\ell}(\omega_n),$$

which has the same mean, but a smaller variance than each of the periodograms $\widehat{\Phi}_{N_p}^{x,i}(\omega_n)$. However, in comparison with the periodogram that would have been obtained using the whole time record, the frequency separation is a factor of p larger, that is,

$$\frac{2\pi}{N_p T} = p \frac{2\pi}{NT}.$$

This means that we will lose detail, because it is no longer possible to determine changes of the true spectrum within a frequency interval

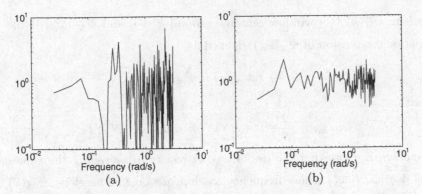

Fig. 6.6. (a) The periodogram of a unit-variance white-noise sequence based on 1000 data points. (b) A block-averaged periodogram based on 10 data blocks, each of 1000 points.

of $p2\pi/(NT)$ rad/s. The frequency resolution decreases. The important conclusion is that choosing the block size is a trade-off between reduced variance and decreased frequency resolution. The right choice depends on the application at hand.

Example 6.4 (Periodogram estimate from blocked-data processing) Let the stochastic process $x(k)$ be a zero-mean white-noise sequence with unit variance. Since the auto-correlation function of this sequence is a unit pulse at time instant zero, we know from Table 3.5 on page 54 that the true spectrum of $x(k)$ equals unity for all frequencies. The estimated spectrum using the periodogram based on $N = 1000$ data points of a realization of $x(k)$ is displayed in Figure 6.6(a). Using $N = 10\,000$ data points, the averaged periodogram computed using 10 blocks, each of 1000 data points, is displayed in Figure 6.6(b). On comparing these two figures, we clearly see that the variance of the block-averaged periodogram is much the smaller.

An alternative to averaging over periodograms and using extensively long data batches is to smooth periodograms by averaging the periodogram over a number of neighboring frequencies using a window(Ljung and Glad, 1994) as follows:

$$\widehat{\Phi}_N^x(\omega_n) = \frac{\sum_{k=-\infty}^{\infty} w_M(k)\widehat{\Phi}_N^{x,p}(\omega_n - \omega_k)}{\sum_{k=-\infty}^{\infty} w_M(k)}, \tag{6.11}$$

where $w_M(k)$ is a window centered around $k = 0$ and $\widehat{\widehat{\Phi}}_N^{x,p}(\omega_n)$ is the periodic extension of $\widehat{\widehat{\Phi}}_N^{x}(\omega_n)$, defined as

$$\widehat{\widehat{\Phi}}_N^{x,p}(\omega_{n+kN}) = \widehat{\widehat{\Phi}}_N^{x}(\omega_n),$$

with

$$\omega_{n+kN} = \frac{2\pi}{NT}(n + kN), \quad n = 0, 1, \ldots, N - 1,$$

and $k \in \mathbb{Z}$. Note that the use of $w_M(k)$ results in averaging the values of $\widehat{\widehat{\Phi}}_N^{x,p}(\omega_n - \omega_k)$ whose frequency ω_k depends on k. Therefore, $w_M(k)$ plays the role of a *frequency window*.

A simple example of a window $w_M(k)$ is the rectangular window, given by

$$w_M(k) = \begin{cases} 1, & \text{for } -M \leq k \leq M, \\ 0, & \text{otherwise.} \end{cases} \tag{6.12}$$

Another example of such a window is the Hamming window given by Equation (6.5) on page 187. The width of the window, M, corresponds to the frequency interval in which the periodogram is smoothed. If in this interval $\Phi^x(\omega)$ remains (nearly) constant, we reduce the variance because we average the periodogram over the length of this interval. Recall that Equation (6.10) teaches us that the periodogram estimated in this interval becomes uncorrelated for $N \to \infty$. However, the window also decreases the frequency resolution, since within the frequency interval that corresponds to the window width it is not possible to distinguish changes in the true spectrum. Hence, the choice of the width of the window is again a trade-off between reduced variance and decreased frequency resolution. For the rectangular window (6.12) it holds that, the larger the width M of the window, the more the variance is reduced and the more the frequency resolution is decreased.

The choice of the width of the window is, of course, very dependent on the total number of available data points N. Therefore, it is customary to work with $\gamma = N/M$, instead of M. Hence, a smaller γ corresponds to a larger width of the window with respect to the total number of data points. The following example illustrates the use of a window for periodogram estimation.

Example 6.5 (Periodogram estimate using a Hamming window) The spectrum of the white-noise process from Example 6.4 is

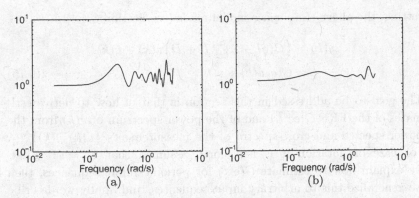

Fig. 6.7. Smoothed periodograms of a unit-variance white-noise sequence based on 1000 data points using a Hamming window with (a) $\gamma = 30$ and (b) $\gamma = 10$.

estimated using windowed periodograms. The window used is a Hamming window given by Equation (6.5) on page 187. Figure 6.7(a) shows the result for $\gamma = 30$ and Figure 6.7(b) shows that for $\gamma = 10$. We clearly observe that, the smaller the value of γ, hence the larger the width of the window, the more the variance is reduced. In this example, a large width of the window can be afforded, because the white-noise signal has a power spectrum that is constant over the whole frequency band. For an unknown signal it cannot be assumed that the power spectrum is constant and the selection of γ will be much more difficult.

6.6 Estimation of FRFs and disturbance spectra

In this section we describe the estimation of the frequency-response function (FRF) of an unknown LTI system from input–output measurements. For that purpose, we consider the following single-input, single-output LTI system:

$$x(k+1) = Ax(k) + Bu(k), \tag{6.13}$$
$$y(k) = Cx(k) + Du(k) + v(k), \tag{6.14}$$

where $x(k) \in \mathbb{R}^n$, $u(k) \in \mathbb{R}$, $y(k) \in \mathbb{R}$, and $v(k) \in \mathbb{R}$. The term $v(k)$ represents an additive unknown perturbation that is assumed to be WSS and uncorrelated with the known input sequence $u(k)$. The system matrix A is assumed to be asymptotically stable. In a transfer-function setting, the above state-space model reads (see Section 3.4.4)

$$y(k) = \left(C(qI - A)^{-1}B + D\right)u(k) + v(k)$$
$$= G(q)u(k) + v(k). \tag{6.15}$$

The goal to be addressed in this section is that of how to derive estimates of the FRF $G(e^{j\omega T})$ and of the power spectrum of $v(k)$ from the power spectra and cross-spectra of the measurements $\{u(k), y(k)\}_{k=1}^{N}$. For ease of notation it is henceforth assumed that $T = 1$. First, we explain how to estimate $G(e^{j\omega})$ for periodic input sequences, then we generalize this to arbitrary input sequences, and finally we describe how to estimate the spectrum of $v(k)$, the so-called disturbance spectrum.

6.6.1 Periodic input sequences

Let the input $u(k)$ be periodic with period N_0, that is,

$$u(k) = u(k + \ell N_0), \quad \text{for } \ell \in \mathbb{Z},$$

then the input–output relationship (6.15) can be written as

$$y(k) \quad = \quad \sum_{\ell=0}^{\infty} g(\ell)u(k - \ell) \tag{6.16}$$

$$\overset{(\ell=r+sN_0)}{=} \quad \sum_{r=0}^{N_0-1} \sum_{s=0}^{\infty} g(r + sN_0)u(k - r - sN_0)$$

$$= \quad \sum_{r=0}^{N_0-1} \left[\sum_{s=0}^{\infty} g(r + sN_0)\right] u(k - r)$$

$$= \quad \sum_{r=0}^{N_0-1} \overline{g}(r)u(k - r), \tag{6.17}$$

where $\overline{g}(r)$ denotes $\sum_{s=0}^{\infty} g(r + sN_0)$. The effect of a periodic input sequence is that the IIR filter (6.16) behaves as an FIR filter (6.17). This feature simplifies the transformation of periodograms greatly.

Given a finite number of samples of the input and output sequences, we seek a relationship between the DFTs of these sequences and the FRF $G(e^{j\omega})$. To derive such a relationship, we first evaluate $Y_{N_0}(\omega_n)$ for the case $v(k) = 0$, with $\omega_n = 2\pi n / N_0$, $n = 0, 1, \ldots, N_0 - 1$:

$$
\begin{aligned}
Y_{N_0}(\omega_n) &= \sum_{k=0}^{N_0-1} y(k)e^{-j\omega_n k} \\
&= \sum_{k=0}^{N_0-1} \sum_{\ell=0}^{N_0-1} \overline{g}(\ell)u(k-\ell)e^{-j\omega_n k} \\
&= \sum_{\ell=0}^{N_0-1} \overline{g}(\ell) \sum_{k=0}^{N_0-1} u(k-\ell)e^{-j\omega_n k} \\
&\overset{(i=k-\ell)}{=} \sum_{\ell=0}^{N_0-1} \overline{g}(\ell)e^{-j\omega_n \ell} \sum_{i=-\ell}^{N_0-1-\ell} u(i)e^{-j\omega_n i} \\
&= \sum_{\ell=0}^{N_0-1} \overline{g}(\ell)e^{-j\omega_n \ell} \sum_{i=0}^{N_0-1} u(i)e^{-j\omega_n i},
\end{aligned}
$$

where the last equation follows from the periodicity of $u(k)$ and the fact that $e^{-j\omega_n i} = e^{-j\omega_n(i+N_0)}$. We have

$$
Y_{N_0}(\omega_n) = \overline{G}_{N_0}(\omega_n)U_{N_0}(\omega_n), \tag{6.18}
$$

with

$$
\begin{aligned}
\overline{G}_{N_0}(\omega_n) &= \sum_{\ell=0}^{N_0-1} \overline{g}(\ell)e^{-j\omega_n \ell} \\
&= \sum_{\ell=0}^{N_0-1} \sum_{i=0}^{\infty} g(\ell + iN_0)e^{-j\omega_n \ell} \\
&\overset{(k=\ell+iN_0)}{=} \sum_{k=0}^{\infty} g(k)e^{-j\omega_n k} \\
&= G(e^{j\omega_n}).
\end{aligned}
$$

We can conclude that, for the general case in which $v(k) \neq 0$, we have

$$
Y_{N_0}(\omega_n) = G(e^{j\omega_n})U_{N_0}(\omega_n) + V_{N_0}(\omega_n). \tag{6.19}
$$

An estimate of the FRF, referred to as the *empirical transfer-function estimate (ETFE)*, can be determined from the foregoing expression as

$$
\widehat{G}_{N_0}(e^{j\omega_n}) = \frac{Y_{N_0}(\omega_n)}{U_{N_0}(\omega_n)}. \tag{6.20}
$$

The statistical properties of this estimate for periodic inputs can easily be deduced from (6.19), namely

$$E\left[\widehat{G}_{N_0}(e^{j\omega_n})\right] = G(e^{j\omega_n}),$$

$$E\left[\widehat{G}_{N_0}(e^{j\omega_n}) - G(e^{j\omega_n})\right]\left[\widehat{G}_{N_0}(e^{-j\omega_r}) - G(e^{-j\omega_r})\right]$$

$$= \frac{E[V_{N_0}(\omega_n)V_{N_0}(-\omega_r)]}{U_{N_0}(\omega_n)U_{N_0}(-\omega_r)},$$

where $\omega_r = 2\pi r/N_0$ and the expectation is with respect to the additive perturbation $v(k)$, since $u(k)$ is assumed to be given. The term

$$\frac{E[V_{N_0}(\omega_n)V_{N_0}(-\omega_r)]}{U_{N_0}(\omega_n)U_{N_0}(-\omega_r)}$$

can be expressed in terms of the spectrum of $\Phi^v(\omega_n)$, similarly to in (6.9) and (6.10) on page 191. According to Lemma 6.1 in Ljung (1999), this expression reads

$$E\left[\widehat{G}_{N_0}(e^{j\omega_n}) - G(e^{j\omega_n})\right]\left[\widehat{G}_{N_0}(e^{-j\omega_r}) - G(e^{-j\omega_r})\right]$$

$$= \begin{cases} \dfrac{1}{|U_{N_0}(\omega_n)|^2}[\Phi_v(\omega_n) + R_{N_0}], & \text{for } \omega_n = \omega_r, \\[2ex] \dfrac{R_{N_0}}{U_{N_0}(\omega_n)U_{N_0}(-\omega_r)}, & \text{for } |\omega_r - \omega_n| = \dfrac{2\pi k}{N_0}, \\ & \qquad k = 1, 2, \ldots, N_0 - 1, \end{cases} \tag{6.21}$$

with $|R_{N_0}| \leq C/N$ for C having some constant value.

Recall that in Example 6.1 on page 183 it was shown that for a periodic sequence $u(k)$ the magnitude of the DFT, $|U_{N_0}(\omega_n)|$, is proportional to N_0 at the frequency points ω_n where $U_{N_0}(\omega_n) \neq 0$, provided that N_0 is an integer multiple of the period length. It has also been shown that the frequency points ω_n where $U_{N_0}(\omega_n) \neq 0$ are fixed and finite in number. These results, together with Equation (6.21), show that the ETFE delivers a consistent estimate of the FRF at a finite number of frequency points where $U_{N_0}(\omega_n) \neq 0$, that is, at these frequency points the variance of the ETFE goes to zero for $N \to \infty$.

6.6.2 General input sequences

For general input sequences, we can define the ETFE, slightly differently from (6.20), as

$$\widehat{G}_N(e^{j\omega_n}) = \frac{Y_N(\omega_n)}{U_N(\omega_n)},$$

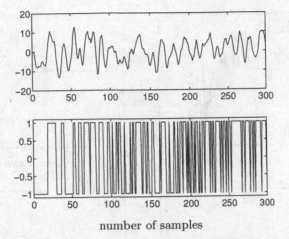

number of samples

Fig. 6.8. Input (bottom) and output (top) data for Example 6.6.

where N denotes the total number of data points available. It can be shown that also for general input signals the ETFE is asymptotically (for $N \to \infty$) an unbiased estimate of the FRF (Ljung, 1999). The variance, in this case, is also described by Equation (6.21), with $\widehat{G}_{N_0}(e^{j\omega_n})$ replaced by $\widehat{\widehat{G}}_N(e^{j\omega_n})$. However, when $u(k)$ is not a periodic signal, $|U_N(\omega_n)|$ is no longer proportional to N and therefore $\widehat{\widehat{G}}_N(e^{j\omega_n})$ is not a consistent estimate. However, the expression (6.21) shows that the ETFE at neighboring frequency points becomes asymptotically uncorrelated. Hence, windowing techniques similar to those discussed in Section 6.5 for averaging periodograms may be employed to reduce the variance of the estimate $\widehat{G}_N(e^{j\omega_n})$. This is illustrated in the next example. In general the use of the ETFE is not without problems, for example its variance can become unbounded (Broersen, 1995).

Example 6.6 (Empirical transfer-function estimate) Let the output of an LTI system be given by

$$y(k) = G(q)u(k) + e(k),$$

with transfer function

$$G(q) = \frac{1}{1 - 1.5q^{-1} + 0.7q^{-2}}$$

and $e(k)$ an unknown zero-mean white-noise sequence. The system is simulated with a pseudo-random binary input sequence for 1000 samples. The first 300 data points of the input and output are shown in Figure 6.8.

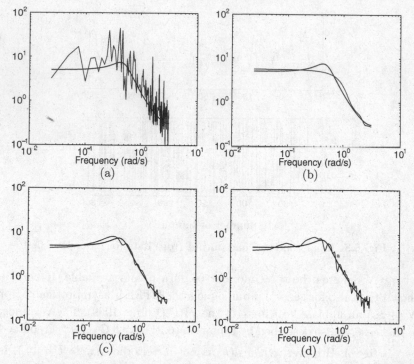

Fig. 6.9. Empirical transfer-function estimates (ETFEs) for Example 6.6: (a) without windowing; (b) with $\gamma = 5$; (c) with $\gamma = 25$; and (d) with $\gamma = 50$. The thick lines represent the true FRF.

Figure 6.9(a) shows the ETFE. The ETFE is smoothed using a Hamming window, Equation (6.5) on page 187. Figures 6.9(b)–(d) show the results for three widths of the Hamming window. It appears that $\gamma = 25$ is a good choice.

6.6.3 Estimating the disturbance spectrum

According to Lemma 4.3 on page 106, the system description

$$y(k) = G(q)u(k) + v(k)$$

gives rise to the following relationships between power spectra and cross-spectra:

$$\Phi^{yu}(\omega) = G(e^{j\omega T})\Phi^u(\omega),$$
$$\Phi^y(\omega) = |G(e^{j\omega T})|^2\Phi^u(\omega) + \Phi^v(\omega).$$

Therefore, if we compute the estimated spectra $\widehat{\Phi}_N^{yu}(\omega_n), \widehat{\Phi}_N^u(\omega_n)$, and $\widehat{\Phi}_N^y(\omega_n)$ from the available data sequences $\{u(k), y(k)\}_{k=1}^N$ as outlined in Section 6.5, assuming $u(k)$ to be a given sequence, an estimate of the power spectrum of the disturbance $v(k)$ is

$$\widehat{\Phi}_N^v(\omega_n) = \widehat{\Phi}_N^y(\omega_n) - \frac{|\widehat{\Phi}_N^{yu}(\omega_n)|^2}{\widehat{\Phi}_N^u(\omega_n)}, \tag{6.22}$$

for the frequency points ω_n for which $\widehat{\Phi}_N^u(\omega_n) \neq 0$. Another expression for this estimate is

$$\widehat{\Phi}_N^v(\omega_n) = \widehat{\Phi}_N^y(\omega_n)\left(1 - \frac{|\widehat{\Phi}_N^{yu}(\omega_n)|^2}{\widehat{\Phi}_N^y(\omega_n)\widehat{\Phi}_N^u(\omega_n)}\right)$$
$$= \widehat{\Phi}_N^y(\omega_n)(1 - \widehat{\kappa}_N^{yu}(\omega_n)), \tag{6.23}$$

for the frequency points ω_n for which $\widehat{\Phi}_N^u(\omega_n) \neq 0$ and, in addition, $\widehat{\Phi}_N^y(\omega_n) \neq 0$. Here $\widehat{\kappa}_N^{yu}(\omega_n)$ is an estimate of the *coherence spectrum*. The coherence spectrum is defined as

$$\kappa^{yu}(\omega) = \frac{|\Phi^{yu}(\omega)|^2}{\Phi^y(\omega)\Phi^u(\omega)}.$$

It expresses the linear correlation in the frequency domain between the sequences $u(k)$ and $y(k)$. This is in analogy with the linear regression coefficient in linear least squares. The coherence spectrum is real-valued and should be as close as possible to unity. Its deviation from unity (in some frequency band) may have the following causes.

(i) The disturbance spectrum $\Phi^v(\omega)$ is large relative to the product $|G(e^{j\omega T})|^2\Phi^u(\omega)$ in a particular frequency band. This is illustrated by the expression

$$\kappa^{yu}(\omega) = 1 - \frac{\Phi^v(\omega)}{|G(e^{j\omega T})|^2\Phi^u(\omega) + \Phi^v(\omega)}.$$

(ii) The output contains a significant response due to nonzero initial conditions of the state of the system (6.13)–(6.14) with transfer function $G(q)$. This effect has been completely neglected in the definition of signal spectra.

(iii) The relationship between $u(k)$ and $y(k)$ has a strong nonlinear component (Schoukens *et al.*, 1998). Hence, the transfer function $G(q)$ is not a good description of the relation between $u(k)$ and $y(k)$.

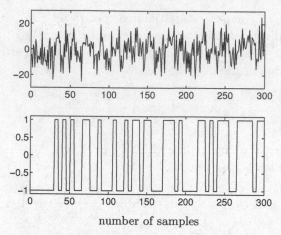

Fig. 6.10. Input (bottom) and output (top) data for Example 6.7.

The above possibilities are by no means the only reasons for the coherence spectrum to differ from unity.

Example 6.7 (Coherence spectrum) Consider the noisy system given by

$$y(k) = G(q)u(k) + H(q)e(k),$$

with $G(q)$ given in Example 6.6, $e(k)$ an unknown zero-mean, unit-variance white-noise sequence, and $H(q)$ given by

$$H(q) = 10 \frac{0.8 - 1.6q^{-1} + 0.8q^{-2}}{1 - 1.6q^{-1} + 0.6q^{-2}}.$$

This system is simulated with a slowly varying pseudo-random binary input sequence for 1000 samples. The first 300 data points of the input and output are shown in Figure 6.10. Figures 6.11(a) and (b) show the periodograms of the input and output sequences, respectively. These periodograms were estimated using a Hamming window with $\gamma = 50$. Figure 6.11(c) shows the noise spectrum estimated using Equation (6.22), and Figure 6.11(d) shows the coherence spectrum. We see that for higher frequencies the coherence spectrum drops below unity. Looking at the periodogram of the input sequence, we conclude that this is due to the fact that the input sequence contains very little energy at high frequencies. We conclude that the high-frequency components present in the output are due to the noise $e(k)$ only.

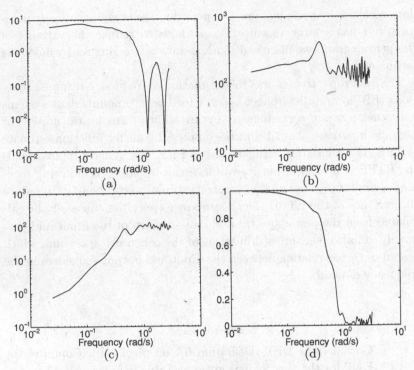

Fig. 6.11. (a) The periodogram of the input used in Example 6.7. (b) The periodogram of the output. (c) The estimated noise spectrum. (d) The estimated coherence spectrum.

6.7 Summary

First, we introduced the discrete Fourier-transform (DFT), which can be used to transform a finite-length sequence into the frequency domain. We showed that the DFT can be computed by solving a least-squares problem. Next, we discussed the relation between the DFT and the DTFT. It turned out that the DFT is a distorted version of the DTFT, because of the phenomenon of spectral leakage. Only for periodic signals are the DFT and DTFT equal, provided that the length used in the DFT is an integer multiple of the period of the signal. We explained how to reduce spectral leakage, using windowing. Next, we showed that the DFT can be efficiently computed using fast Fourier-transform (FFT) algorithms.

For WSS sequences we consider the estimation of signal spectra using finite-length time sequences. The spectral estimate based on finite-length data is called the periodogram. The periodogram is asymptotically unbiased, but has a large variance. As methods to reduce the variance of the periodogram, we discussed blocked-data processing and windowing techniques.

The final topic treated was the estimation of frequency-response functions (FRFs) and disturbance spectra for linear time-invariant systems with their output contaminated by an additive stochastic noise. For periodic inputs we showed that the empirical transfer-function estimate (ETFE) is a consistent estimate of the FRF. For general input signals the ETFE is unbiased, but its variance can be large. Windowing techniques similar to those used in spectral estimation can be used to reduce the variance of the ETFE. The disturbance spectrum can easily be calculated from the power spectra and cross-spectra of the input and output. It can also be expressed in terms of the coherence spectrum, which describes the correlation between the input and output sequences in the frequency domain.

Exercises

6.1   Consider the DFT (Definition 6.1 on page 180). Compute the DFT for the time signals given in Table 6.2 on page 181.

6.2   Given a sequence $\{x(k)\}_{k=1}^N$ with sampling time $T = 1$, for $N = 8$, draw the points $e^{-j\omega_n}$ for which $X_N(\omega_n)$ is defined in the complex plane.

6.3   We are given N observations of a random variable s, collected in the sequence $\{s(k)\}_{k=1}^N$. Let s have the properties

$$E[s(k)] = s_0,$$

$$E\left[\left(s(k) - s_0\right)\left(s(\ell) - s_0\right)\right] = \sigma_s^2 \Delta(k - \ell).$$

Show that the averaged estimate

$$\frac{1}{p} \sum_{\ell=1}^p s(\ell)$$

has mean s_0 and variance σ_s^2/p.

6.4   Given the periodic signal

$$x(k) = \cos\left(\frac{2\pi}{N_1}k\right) + \cos\left(\frac{2\pi}{N_2}k\right),$$

with sampling time $T = 1$,

    (a) determine the length N for the DFT such that we observe an integer multiple of the period of $x(k)$,

    (b) determine $X_N(\omega_n)$ analytically, and

    (c) determine $X_N(\omega_n)$ numerically and illustrate the leakage phenomenon by taking different values for N.

6.5   Let the signal $x(k)$ be a filtered version of the signal $e(k)$,

$$x(k) = G(q)e(k),$$

with the filter given by

$$G(q) = \frac{1}{1 - \sqrt{2}q^{-1} + 0.5q^{-2}},$$

and $e(k)$ an unknown zero-mean, unit-variance, Gaussian white-noise sequence.

    (a) Generate $10\,000$ data points of the signal $x(k)$.

    (b) Numerically estimate the periodogram of $x(k)$ using the first 1000 data points.

    (c) Reduce the variance of the periodogram by block averaging, using 10 blocks of 1000 data points.

    (d) Reduce the variance of the periodogram of $x(k)$ by using a windowed periodogram estimate with a Hamming window. Use only the first 1000 data points of $x(k)$ and determine a suitable width for the Hamming window.

6.6   Let the output of an LTI system be given by

$$y(k) = G(q)u(k) + H(q)e(k),$$

with

$$G(q) = \frac{0.3 + 0.6q^{-1} + 0.3q^{-2}}{1 + 0.2q^{-2}},$$

and

$$H(q) = \frac{0.95 - 2.15q^{-1} + 3.12q^{-2} - 2.15q^{-3} + 0.95q^{-4}}{1 - 2.2q^{-1} + 3.11q^{-2} - 2.1q^{-3} + 0.915q^{-4}}.$$

Take as an input a pseudo-random binary sequence and take the perturbation $e(k)$ equal to a zero-mean, unit-variance, Gaussian white-noise sequence.

(a) Generate 1000 data points of the signal $y(k)$.

(b) Numerically estimate the periodograms of $u(k)$ and $y(k)$.

(c) Numerically estimate the coherence function and the disturbance spectrum.

(d) What can you conclude from the coherence function?

6.7   Let the output of an LTI system be given by

$$y(k) = G(q)u(k) + v(k),$$

with $u(k) = A\sin(\omega_1 k)$, $\omega_1 = 2\pi/N_1$, and $v(k) = B\sin(\omega_2 k)$, $\omega_2 = 2\pi/N_2$. Let N be the greatest common multiple of N_1 and N_2.

(a) Show that, for $N_2 > N_1$,

$$Y_N(e^{j\omega_n}) = G(e^{j\omega_n})U_N(e^{j\omega_n}) + V_N(e^{j\omega_n}),$$

with $\omega_n = 2\pi n/N$ for $n = 0, 1, \ldots, N-1$.

(b) Determine the coherence function $\widehat{\kappa}_N^{yu}$ for $N_2 > N_1$.

(c) Show that $\widehat{\kappa}_N^{yu}$ is defined for $\omega = 2\pi/N$ only if $N_1 = N_2$.

(d) Let $Y_N(e^{j\omega_n})$ and $U_N(e^{j\omega_n})$ be given. Show how to compute the imaginary part of $G(e^{j\omega_n})$ for the special case in which $N_1 = N_2$.

7

Output-error parametric model estimation

After studying this chapter you will be able to

- describe the output-error model-estimation problem;
- parameterize the system matrices of a MIMO LTI state-space model of fixed and known order such that all stable models of that order are presented;
- formulate the estimation of the parameters of a given system parameterization as a nonlinear optimization problem;
- numerically solve a nonlinear optimization problem using gradient-type algorithms;
- evaluate the accuracy of the obtained parameter estimates via their asymptotic variance under the assumption that the signal-generating system belongs to the class of parameterized state-space models; and
- describe two ways for dealing with a nonwhite noise acting on the output of an LTI system when estimating its parameters.

7.1 Introduction

After the treatment of the Kalman filter in Chapter 5 and the estimation of the frequency-response function (FRF) in Chapter 6, we move another step forward in our exploration of how to retrieve information about linear time-invariant (LTI) systems from input and output measurements. The step forward is taken by analyzing how we can estimate (part of) the system matrices of the signal-generating model from acquired input and output data. We first tackle this problem as a complicated estimation problem by attempting to estimate both the state vector and the system matrices. Later on, in Chapter 9, we outline the so-called subspace identification methods that solve such problems by means of linear least-squares problems.

Nonparametric models such as the FRF could also be obtained via the simple least-squares method or the computationally more attractive fast Fourier transform. This was demonstrated in Chapter 6. Though FRF models have proven and still do prove their usefulness in analyzing real-life measurements, such as in modal analysis in the automobile industry (Rao, 1986), other applications require more compact parametric models. One such broad area of application of parametric models is in model-based controller design. Often the starting point in robust controller synthesis methods, such as H_∞-control (Kwakernaak, 1993; Skogestad and Postlethwaite, 1996; Zhou *et al.*, 1996), multi-criteria controller design (Boyd and Baratt, 1991), and model-based predictive control (Clarke *et al.*, 1987; Garcia *et al.*, 1989; Soeterboek, 1992), requires an initial state-space model of the system that needs to be controlled. Another area of application of parametric models is in developing realistic simulators for, for example, airplanes, cars, and virtual surgery (Sorid and Moore, 2000). These simulators are critical in training operators to deal with life-threatening circumstances. Part of the necessary accurate replication of these circumstances is often a parametric mathematical model of the airplane, the car, or the human heart.

This chapter, together with Chapters 8 and 9, presents an introduction to estimating the parameters in a user-defined LTI model. In this chapter we start with the determination of a model to approximate the deterministic relation between measurable input and output sequences. The uncertainties due to noises acting on the system are assumed to be lumped together as an additive perturbation at the output. Therefore, the estimation methods presented in this chapter are referred to as the *output-error methods*. In Chapter 8 we deal with the approximation of both the deterministic and the stochastic parts of the system's response, using an innovation model.

The reason for starting with output-error methods for the analysis of estimating the parameters of a parametric model of an LTI system is twofold. First, in a number of applications, only the deterministic transfer from the measurable input to the output is of interest. An example is identification-based fault diagnosis, in which the estimated parameters

of the deterministic part of the model are compared with their nominal
"fault-free" values (Isermann, 1993; Verdult *et al.*, 2003). Second, the
restriction to the deterministic part simplifies the discussion and allows
us to highlight how the estimation of parameters in an LTI model can
be approached systematically. This systematic approach, which lies at
the heart of many identification methods, is introduced in Section 7.2
and consists of the following four steps. The *first step* is parameterizing the model; that is, the selection of which parameters to estimate
in the model. For MIMO LTI state-space models, some parameterizations and their properties are discussed in Section 7.3. *Step two* consists
of formulating the estimation of the model parameters as an optimization problem. Section 7.4 presents such an optimization problem with
the widely used least-squares cost function. *Step three* is the selection
of a numerical procedure to solve the optimization problem iteratively.
Methods for minimizing a least-squares cost function are presented in
Section 7.5. The *final step* is evaluation of the accuracy of the obtained
estimates via the covariance matrix of the estimates. This is discussed in
Section 7.6. In these four steps it is assumed that the additive error to the
output is a zero-mean white noise. Section 7.7 discusses the treatment
of colored additive noise.

7.2 Problems in estimating parameters of an LTI state-space model

Consider the signal-generating LTI system to be identified, given by

$$y(k) = G(q)u(k) + v(k), \tag{7.1}$$

where $v(k)$ represents measurement noise that is statistically independent from the input $u(k)$. Then a general formulation of the output-error
(OE) model-estimation problem is as follows.

Given a finite number of samples of the input signal $u(k)$ and the
output signal $y(k)$, and the order of the following predictor,

$$\widehat{x}(k+1) = A\widehat{x}(k) + Bu(k), \tag{7.2}$$

$$\widehat{y}(k) = C\widehat{x}(k) + Du(k), \tag{7.3}$$

the goal is to estimate a set of system matrices A, B, C, and D in
this predictor such that the output $\widehat{y}(k)$ approximates the output of
the system (7.1).

First we consider the case in which $v(k)$ is a white-noise sequence. In Section 7.7 and in Chapters 8 and 9 we then consider the more general case in which $v(k)$ is colored noise.

A common way to approach this problem is to assume that the entries of the system matrices depend on a parameter vector θ and to estimate this parameter vector. The parameterized predictor model based on the system (7.2)–(7.3) becomes

$$\widehat{x}(k+1,\theta) = A(\theta)\widehat{x}(k,\theta) + B(\theta)u(k), \tag{7.4}$$

$$\widehat{y}(k,\theta) = C(\theta)\widehat{x}(k,\theta) + D(\theta)u(k). \tag{7.5}$$

The output data $\widehat{y}(k,\theta)$ used in the cost function (7.6) depend not only on the input and the parameters θ used to parameterize the system matrices $A(\theta)$, $B(\theta)$, $C(\theta)$, and $D(\theta)$, but also on the initial state $\widehat{x}(0)$ of the model (7.4)–(7.5). Therefore, the initial state is often also regarded as a parameter and added to the parameter vector θ. The notation $\widehat{x}(0,\theta)$ is used to denote the treatment of the initial state as a part of the parameter vector θ.

The problem of estimating the parameter vector θ can be divided into four parts.

(i) *Determination of a parameterization.* A parameterization of the system (7.4)–(7.5) is the specification of the dependence of the system matrices on the parameter vector θ. One widely used approach to parameterize systems is to use unknown physical constants in a mathematical model derived from the laws of physics, such as Newton's or Kirchoff's laws. An example of such a parameterization is given below in Example 7.1.

(ii) *Selection of a criterion to judge the quality of a particular value of θ.* In this book, we consider a quadratic error criterion of the form

$$\frac{1}{N} \sum_{k=0}^{N-1} \|y(k) - \widehat{y}(k,\theta)\|_2^2, \tag{7.6}$$

with $\widehat{y}(k,\theta)$ given by (7.4) and (7.5). For each particular value of the parameter vector θ, this criterion has a positive value. The optimality may therefore be expressed by selecting that parameter value that yields the minimal value of (7.6). Though such a strategy is a good starting point, a more detailed consideration is generally necessary in order to find the most appropriate model for a particular application. A discussion on this topic of model selection is given in Section 10.4.2.

(iii) *Numerical minimization of the criterion* (7.6). Let the "optimal" parameter vector $\widehat{\theta}_N$ be the argument θ of the cost function (7.6) that minimizes this cost function; this is denoted by

$$\widehat{\theta}_N = \arg\min \frac{1}{N} \sum_{k=0}^{N-1} \|y(k) - \widehat{y}(k, \theta)\|_2^2. \qquad (7.7)$$

As indicated by (7.4) and (7.5), the prediction $\widehat{y}(k, \theta)$ of the output is a filtered version of the input $u(k)$ only. A method that minimizes a criterion of the form (7.6), where $\widehat{y}(k, \theta)$ is based on the input only, belongs to the class of output-error methods (Ljung, 1999). The Kalman filter discussed in Chapter 5 determines a prediction of the output by filtering both the input $u(k)$ and the output $y(k)$. A specific interpretation of the criterion (7.7) will be given when using the Kalman filter to predict the output in Chapter 8.

(iv) *Analysis of the accuracy of the estimate* $\widehat{\theta}_N$. Since the measurements $y(k)$ are assumed to be stochastic processes, the derived parameter estimate $\widehat{\theta}_N$ obtained via optimizing (7.6) will be a random variable. Therefore, a measure of its accuracy could be its bias and (co)variance.

The above four problems, which are analyzed in the listed order in Sections 7.3–7.6, aim, loosely speaking, at determining the "best" predictor such that the difference between the measured and predicted output is made "as small as possible." The output-error approach is illustrated in Figure 7.1.

Example 7.1 (Parameterization of a model of an electrical motor) The electrical–mechanical equations describing a permanent-magnet synchronous motor (PMSM) were derived in Tatematsu *et al.* (2000). These equations are used to obtain a model of a PMSM and summarized below. Figure 7.2 shows a schematic drawing of the PMSM. The magnet, marked with its north and south poles, is turning and along with it is the rotor reference frame indicated by the d-axis and q-axis. In the model the following physical quantities are used:

- (i_d, i_q) are the currents and (v_d, v_q) are the voltages with respect to the rotor reference frame;

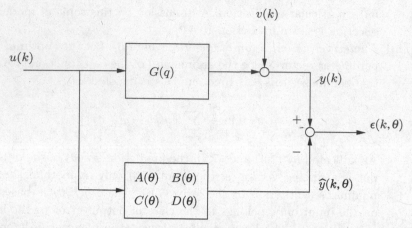

Fig. 7.1. The output-error model-estimation method (the initial state $\widehat{x}(0, \theta)$ of the model has been omitted).

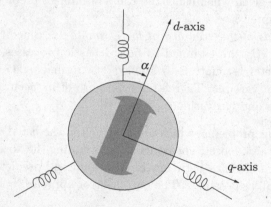

Fig. 7.2. A schematic representation of the permanent-magnet synchronous motor of Example 7.1.

- α is the rotor position and ω its velocity;
- T_L represents the external load;
- N is the number of magnetic pole pairs in the motor;
- R is the phase resistance;
- L_d and L_q are the direct- and quadrature-axis inductances, respectively;
- ϕ_a is the permanent magnetic constant; and
- J is the rotor inertia.

On the basis of these definitions the physical equations describing a PMSM are (Tatematsu *et al.*, 2000)

$$\frac{\mathrm{d}i_d}{\mathrm{d}t} = -\frac{R}{L_d}i_d + \frac{N\omega L_q}{L_d}i_q + \frac{1}{L_d}v_d, \tag{7.8}$$

$$\frac{\mathrm{d}i_q}{\mathrm{d}t} = -\frac{R}{L_q}i_q - \frac{N\omega L_d}{L_q}i_d - \frac{N\phi_\mathrm{a}}{L_q}\omega + \frac{1}{L_q}v_q, \tag{7.9}$$

$$\frac{\mathrm{d}\omega}{\mathrm{d}t} = \frac{N\phi_\mathrm{a}}{J}i_q - \frac{1}{J}T_\mathrm{L}, \tag{7.10}$$

$$\frac{\mathrm{d}\alpha}{\mathrm{d}t} = N\omega. \tag{7.11}$$

The state of this system equals $\begin{bmatrix} i_d & i_q & \omega & \alpha \end{bmatrix}^\mathrm{T}$. The parameters that would allow us to simulate this state, given the (input) sequences T_L, v_d, and v_q, are

$$\{N, R, L_d, L_q, \phi_\mathrm{a}, J\}.$$

Hence, a parameterization of the PMSM model (7.8)–(7.11) corresponds to the mapping from the parameter set $\{N, R, L_d, L_q, \phi_\mathrm{a}, J\}$ to the model description (7.8)–(7.11). Note that a discrete-time model of the PMSM can be obtained by approximating the derivatives in (7.8)–(7.11) by finite differences.

In this chapter we assume that the order of the LTI system, that is, the dimension of the state vector, is known. In practice, this is often not the case. Estimating the order from measurements is discussed in Chapter 10, together with some relevant issues that arise in the practical application of system identification.

7.3 Parameterizing a MIMO LTI state-space model

Finding a model to relate input and output data sequences in the presence of measurement errors and with lack of knowledge about the physical phenomena that relate these data is a highly nonunique, nontrivial problem. To address this problem one specializes to specific *models*, *model sets*, and *parameterizations*. These notions are defined below for MIMO state-space models of finite order given by Equations (7.4) and (7.5).

Let p be the dimension of the parameter vector θ. The set $\Omega \subset \mathbb{R}^p$ that constrains the parameter vector, in order to guarantee that the parameterized models comply with prior knowledge about the system, such as the system's stability or the positiveness of its DC gain, is called the parameter set. By taking different values of θ from the set Ω, we get state-space models of the form (7.4)–(7.5) with different system matrices. A state-space *model set* M is a collection or enumeration of state-space models of the form given by Equations (7.4) and (7.5).

The transfer function of the nth-order system (7.4)–(7.5) is of the form

$$G(q, \theta) = D(\theta) + C(\theta)\big(qI_n - A(\theta)\big)^{-1} B(\theta). \tag{7.12}$$

Thus, for each particular value of θ we get a certain transfer function. From Section 3.4.4 we know that this transfer function is an $\ell \times m$ proper rational function with a degree of at most n. We use $\mathcal{R}_n^{\ell \times m}$ to denote the set of all $\ell \times m$ proper rational transfer functions with real coefficients and a degree of at most n.

A *parameterization* of the nth-order state-space model (7.4)–(7.5) is a mapping from the parameter set $\Omega \in \mathbb{R}^p$ to the space of rational transfer functions $\mathcal{R}_n^{\ell \times m}$. This mapping is called the state-space *model structure* and is denoted by $\mathcal{M} : \Omega \to \mathcal{R}_n^{\ell \times m}$, thus $G(q, \theta) = \mathcal{M}(\theta)$. Since the structure of the transfer function is fixed and given by Equation (7.12), the parameterization defined in this way is nothing but a prescription of how the elements of the system matrices A, B, C, and D are formed from the parameter vector θ.

Before we continue, we recall some properties of a mapping. The map $f : X \to Y$ maps the set X onto the set Y. The set X is called the *domain* of f and Y is called the *range* of f. The map f is called *surjective* if for every $y \in Y$ there exists an $x \in X$ such that $f(x) = y$. In other words, to every point in its range there corresponds at least one point in its domain. It is important to realize that the surjective property of a map depends on the definitions of its domain X and its range Y. The map f is called *injective* if $f(x_1) = f(x_2)$ implies $x_1 = x_2$, that is, to every point in its range there corresponds at most one point in its domain. Finally, if the map f is both surjective and injective, it is called *bijective*.

Since a similarity transformation of the state vector does not alter the transfer function, not all parameterizations need to be injective. A parameterization that is not injective gives rise to a nonunique correspondence between the parameter vector and the transfer function. This is illustrated in the following example.

Example 7.2 (Nonuniqueness in parameterizing a state-space model) Consider the LTI system

$$x(k+1) = \begin{bmatrix} 1.5 & 1 \\ -0.7 & 0 \end{bmatrix} x(k) + \begin{bmatrix} 1 \\ 0.5 \end{bmatrix} u(k),$$

$$y(k) = \begin{bmatrix} 1 & 0 \end{bmatrix} x(k).$$

We parameterize this system using all the entries of the system matrices; this results in the following *parametric* model with $\theta \in \mathbb{R}^8$:

$$\tilde{x}(k+1) = \begin{bmatrix} \theta(1) & \theta(2) \\ \theta(3) & \theta(4) \end{bmatrix} \tilde{x}(k) + \begin{bmatrix} \theta(5) \\ \theta(6) \end{bmatrix} u(k),$$

$$y(k) = \begin{bmatrix} \theta(7) & \theta(8) \end{bmatrix} \tilde{x}(k).$$

However, this parameterization is not injective, since we can find more than one parameter vector θ that results in the same transfer function between the input $u(k)$ and the output $y(k)$. For example, the following two values of the parameter vector θ give rise to the same transfer function:

$$\theta_1^T = \begin{bmatrix} 0 & -0.7 & 1 & 1.5 & 0.5 & 1 & 0 & 1 \end{bmatrix},$$

$$\theta_2^T = \begin{bmatrix} 2.9 & 6.8 & -0.7 & -1.4 & 0 & 0.5 & 1 & 2 \end{bmatrix}.$$

The reason for this nonuniqueness is that the transfer function from input to output remains unchanged when a similarity transformation is applied to the state vector $x(k)$. To obtain the parameter values θ_1, the following similarity transformation of the state vector was used:

$$x(k) = \begin{bmatrix} 0 & 1 \\ 1 & 0 \end{bmatrix} \tilde{x}(k);$$

and for θ_2 we made use of

$$x(k) = \begin{bmatrix} 1 & -2 \\ 0 & 1 \end{bmatrix} \tilde{x}(k).$$

To be able to identify uniquely a model from input and output data requires an injective parameterization. However, often the main objective is to find a state-space model that describes the input and output data, and uniqueness is not needed. In a system-identification context, it is much more important that each transfer function with an order of at most n given by (7.12) can be represented by at least one point in the parameter space Ω. In other words, we need to have a parameterization with domain $\Omega \subset \mathbb{R}^p$ and range $\mathcal{R}_n^{\ell \times m}$ that is surjective. An example of a surjective parameterization results on taking all entries of the system matrices A, B, C, and D as elements of the parameter vector θ as in Example 7.2. This vector then has dimension p equal to

$$p = n^2 + n(\ell + m) + m\ell.$$

Since this number quickly grows with the state dimension n, alternative parameterizations have been developed. For example, for multiple-input, single-output systems, the observable canonical form can be used; it is given (see also Section 3.4.4) by (Ljung, 1999)

$$\widehat{x}(k+1) = \begin{bmatrix} 0 & 0 & \cdots & 0 & -a_0 \\ 1 & 0 & \cdots & 0 & -a_1 \\ 0 & 1 & \cdots & 0 & -a_2 \\ \vdots & \vdots & & \ddots & \vdots \\ 0 & 0 & \cdots & 1 & -a_{n-1} \end{bmatrix} \widehat{x}(k) + \begin{bmatrix} b_{11} & \cdots & b_{1m} \\ b_{21} & \cdots & b_{2m} \\ \vdots & & \\ b_{n1} & \cdots & b_{nm} \end{bmatrix} u(k),$$

(7.13)

$$\widehat{y}(k) = \begin{bmatrix} 0 & 0 & 0 & \cdots & 1 \end{bmatrix} \widehat{x}(k) + \begin{bmatrix} d_{11} & \cdots & d_{1m} \end{bmatrix} u(k). \qquad (7.14)$$

The parameter vector (without incorporating the initial state) is given by

$$\theta^{\mathrm{T}} = \begin{bmatrix} a_0 & \cdots & a_{n-1} & \cdots & b_{11} & \cdots & b_{nm} & d_{11} & \cdots & d_{1m} \end{bmatrix}.$$

The size of θ is

$$p = n + nm + m.$$

This parameterization $\mathcal{M} : \Omega \rightarrow \mathcal{R}_n^{1 \times m}$ is surjective, the reason for this being that, although the observer canonical form is always observable, it can be not reachable. When it is not reachable, it is not minimal and the state dimension can be reduced; the order of the system becomes less than n. For a SISO transfer function it means that roots of the numerator polynomial (the zeros of the system) cancel out those of the denominator (the poles of the system). Different pole–zero cancellations correspond to different parameter values θ that represent the same transfer function, hence the conclusion that the parameterization is surjective.

Apart from the size of the parameter vector θ and the surjective and/or injective property of the mapping $\mathcal{M}(\theta)$, the consequences of selecting a parameterization on the numerical calculations performed with the model need to be considered as well. Some examples of the numerical implications of a parameterization are the following.

(i) In estimating the parameter vector θ by solving the optimization problem indicated in (7.7), it may be required that the mapping is differentiable, such that the Jacobian

$$\frac{\partial y(k, \theta)}{\partial \theta}$$

exists on a subset in $\mathbb{R}^p$.

(ii) In case the mapping is surjective, the parameter optimization (7.7) may suffer from numerical problems due to the redundancy in the entries of the parameter vector. A way to avoid such numerical problems is *regularization* (McKelvey, 1995), which is discussed in Section 7.5.2.

(iii) Restrictions on the set of transfer functions $\mathcal{M}(\theta)$ need to be translated into constraints on the parameter set in $\mathbb{R}^p$. For example, requiring asymptotic stability of the model leads to restrictions on the parameter set. In this respect it may be more difficult to impose such restrictions on one chosen parameterization than on another. Let Ω denote this constraint region in the parameter space, that is, $\Omega \subset \mathbb{R}^p$; then we can formally denote the model set M as

$$M = \{\mathcal{M}(\theta)|\theta \in \Omega\}. \tag{7.15}$$

An example of constraining the parameter space is given below in Example 7.3.

(iv) The numerical sensitivity of the model structure $\mathcal{M}(\theta)$ with respect to the parameter vector θ may vary dramatically between parameterizations. An example of numerical sensitivity is given below in Example 7.4.

Example 7.3 (Imposing stability) Consider the transfer function

$$G(q) = \frac{q+2}{q^2 + a_1 q + a_0}, \tag{7.16}$$

parameterized by $\theta = [a_0, \ a_1]^{\mathrm{T}}$. To impose stability on the transfer function $G(q)$, we need to find a set Ω such that $\theta \in \Omega$ results in a stable transfer function of the form (7.16). In other words, we need to determine a suitable domain for the mapping $\mathcal{M} : \Omega \to U$, with U the set of all stable transfer functions of the form (7.16). For this particular second-order example, the determination of the set Ω is not that difficult and is requested in Exercise 7.2 on page 250. Figure 7.3 shows the set Ω. Every point in the set Ω corresponds uniquely to a point in the set U, and thus the parameterization is injective. The parameterization is

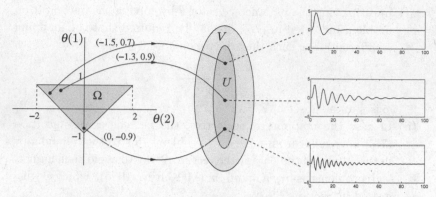

Fig. 7.3. Imposing stability on the second-order transfer function of Example 7.3. The set Ω is mapped onto the set U of all stable second-order transfer functions of the form (7.16). The set V is the set of all stable second-order transfer functions. On the right are the impulse responses for the three indicated points in the parameter space Ω.

bijective with respect to the set U (with the particular choice of zeros in Equation (7.16), no pole–zero cancellation can occur for stable poles), but not with respect to the set V that consists of all stable second-order transfer functions.

Figure 7.3 shows impulse responses of three systems that correspond to three different choices of the parameter θ from the set Ω. These impulse responses are quite different, which illustrates the richness of the set of systems described by Ω.

Example 7.4 (Companion form) The system matrix A in the observer canonical form (7.13)–(7.14) is called a companion matrix (Golub and Van Loan, 1996). A companion matrix is a numerically sensitive representation of the system dynamics; its eigenvalues are very sensitive to small changes in the coefficients $a_0, a_2, \ldots, a_{n-1}$.

We use the observer canonical form (7.13)–(7.14) to represent a system with transfer function

$$G(q) = \frac{1}{q^4 + a_3 q^3 + a_2 q^2 + a_1 q + a_0}.$$

In this case the parameter vector is equal to

$$\theta^{\mathrm{T}} = \begin{bmatrix} a_0 & a_1 & a_2 & a_3 \end{bmatrix}.$$

If we take the parameter vector equal to

$$\theta^{\mathrm{T}} = \begin{bmatrix} 0.915 & -2.1 & 3.11 & -2.2 \end{bmatrix},$$

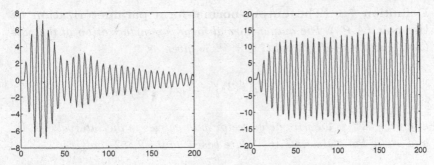

Fig. 7.4. Impulse responses of the stable (left) and the unstable system (right) in Example 7.4.

the matrix A has two eigenvalues with a magnitude equal to 0.9889 up to four digits and two eigenvalues with a magnitude equal to 0.9673 up to four digits. Figure 7.4 shows the impulse response of the system $G(q)$ for this choice of θ.

If we change the parameter $\theta(3) = a_2$ into 3.12, the properties of the system become very different. For this slightly different choice of parameters, the matrix A has two eigenvalues with a magnitude equal to 1.0026 up to four digits and two eigenvalues with a magnitude equal to 0.9541 up to four digits. Hence, even only a small change in the parameter a_2 makes the system unstable. The impulse response of the system with $a_2 = 3.12$ is also shown in Figure 7.4. We clearly see that the impulse response has changed dramatically. It should be remarked that, for systems of larger order, results similar to those illustrated in the example can be obtained with perturbations of magnitude the order of the machine precision of the computer.

In the following subsections, we present two particular parameterizations that are useful for system identification, namely the output normal form and the tridiagonal form.

7.3.1 The output normal form

The output-normal-form parameterization was first introduced for continuous-time state-space models by Hanzon and Ober (1997; 1998), and later extended for MIMO discrete-time state-space models (Hanzon and Peeters, 2000; Hanzon et al., 1999). A big advantage of the output normal form is that the parameterized model is guaranteed to be asymptotically stable without the need for additional constraints on the parameter space. A definition of the output normal parameterization of the pair (A, C) in the case of a state-space model determined by the system matrices A, B, C, and D is as follows.

Definition 7.1 (The output-normal-form parameterization of the pair (A, C)) *The output-normal-form parameterization of the pair (A, C) with $A \in \mathbb{R}^{n \times n}$ and $C \in \mathbb{R}^{\ell \times n}$ is given as*

$$
\begin{bmatrix} C(\theta) \\ A(\theta) \end{bmatrix} = T_1\Big(\theta(1)\Big) T_2\Big(\theta(2)\Big) \cdots T_{n\ell}\Big(\theta(n\ell)\Big) \begin{bmatrix} 0 \\ I_n \end{bmatrix}, \tag{7.17}
$$

where $\theta \in \mathbb{R}^{n\ell}$ is the parameter vector with entries in the interval $[-1, 1]$, and where the matrices $T_i(\theta(i))$ are based on the 2×2 matrix

$$
U(\alpha) = \begin{bmatrix} -\alpha & \sqrt{1-\alpha^2} \\ \sqrt{1-\alpha^2} & \alpha \end{bmatrix},
$$

with $\alpha \in \mathbb{R}$ in the interval $[-1, 1]$; the matrices $T_i(\theta(i)) \in \mathbb{R}^{(n+\ell) \times (n+\ell)}$ are given by

$$
T_1\Big(\theta(1)\Big) = \begin{bmatrix} I_{n-1} & 0 & 0 \\ 0 & U\Big(\theta(1)\Big) & 0 \\ 0 & 0 & I_{\ell-1} \end{bmatrix},
$$

$$
\vdots
$$

$$
T_\ell\Big(\theta(\ell)\Big) = \begin{bmatrix} I_{n+\ell-2} & 0 \\ 0 & U\Big(\theta(\ell)\Big) \end{bmatrix},
$$

$$
T_{\ell+1}\Big(\theta(\ell+1)\Big) = \begin{bmatrix} I_{n-2} & 0 & 0 \\ 0 & U\Big(\theta(\ell+1)\Big) & 0 \\ 0 & 0 & I_\ell \end{bmatrix},
$$

$$
\vdots
$$

$$
T_{2\ell}\Big(\theta(2\ell)\Big) = \begin{bmatrix} I_{n+\ell-3} & 0 & 0 \\ 0 & U\Big(\theta(2\ell)\Big) & 0 \\ 0 & 0 & 1 \end{bmatrix},
$$

$$
\vdots
$$

$$
T_{(n-1)\ell+1}\Big(\theta((n-1)\ell+1)\Big) = \begin{bmatrix} U\Big(\theta((n-1)\ell+1)\Big) & 0 \\ 0 & I_{n+\ell-2} \end{bmatrix},
$$

$$
\vdots
$$

$$
T_{n\ell}\Big(\theta(n\ell)\Big) = \begin{bmatrix} I_{\ell-1} & 0 & 0 \\ 0 & U\Big(\theta(n\ell)\Big) & 0 \\ 0 & 0 & I_{n-1} \end{bmatrix}.
$$

The next lemma shows that the parameterized pair of matrices in Definition 7.1 has the identity matrix as observability Grammian.

Lemma 7.1 *Let an asymptotically stable state-space model be given by*

$$x(k+1) = Ax(k) + Bu(k),$$
$$y(k) = Cx(k) + Du(k),$$

with the pair (A, C) given by the output-normal-form parameterization (7.17) of Definition 7.1, then the observability Grammian Q, defined as the solution of

$$A^{\mathrm{T}}QA + C^{\mathrm{T}}C = Q,$$

is the identity matrix.

Proof The proof follows from the fact that the matrices $U(\alpha)$ satisfy $U(\alpha)^{\mathrm{T}}U(\alpha) = I_2$. $\square$

The output-normal-form parameterization of the pair (A, C) can be used to parameterize any stable state-space model, as shown in the following lemma.

Lemma 7.2 (Output normal form of a state-space model) *Let an asymptotically stable and observable state-space model be given as*

$$\widehat{x}(k+1) = A\widehat{x}(k) + Bu(k), \tag{7.18}$$
$$\widehat{y}(k) = C\widehat{x}(k) + Du(k), \tag{7.19}$$

then a surjective parameterization is obtained by parameterizing the pair (A, C) in the output normal form given in Definition 7.1 with the parameter vector $\theta_{AC} \in \mathbb{R}^{n\ell}$ and parameterizing the pair of matrices (B, D) with the parameter vector $\theta_{BD} \in \mathbb{R}^{m(n+\ell)}$ that contains all the entries of the matrices B and D.

Proof The proof is constructive and consists of showing that any stable, observable state-space system of the form (7.18)–(7.19) can be transformed via a similarity transformation to the proposed parameterization.

Since A is asymptotically stable and since the pair (A, C) is observable, the solution Q to the Lyapunov equation

$$A^{\mathrm{T}}QA + C^{\mathrm{T}}C = Q,$$

is positive-definite. This follows from Lemma 3.8 on page 69. Therefore, a Cholesky factorization can be carried out (see Theorem 2.5 on page 26):

$$Q = T_q T_q^{\mathrm{T}}.$$

The matrix $T_t = T_q^{-\mathrm{T}}$ is the required similarity transformation. Note that T_t exists, because Q is positive-definite. The equivalent matrix pair $(T_t^{-1} A T_t, C T_t) = (A_t, C_t)$ then satisfies

$$A_t^{\mathrm{T}} A_t + C_t^{\mathrm{T}} C_t = I_n.$$

In other words, the columns of the matrix

$$\begin{bmatrix} C_t \\ A_t \end{bmatrix}$$

are orthogonal. To preserve this relationship under a second similarity transformation on the matrices A_t and C_t, this transformation needs to be orthogonal. As revealed by solving Exercise 7.1 on page 249, for any pair (A_t, C_t) there always exists an orthogonal similarity transformation T_{h} such the pair $(T_{\mathrm{h}}^{-1} A_t T_{\mathrm{h}}, C_t T_{\mathrm{h}})$ is in the so-called observer Hessenberg form (Verhaegen, 1985). The observer Hessenberg form has a particular pattern of nonzero entries, which is illustrated below for the case $n = 5$, $\ell = 2$:

$$\begin{bmatrix} C_t T_{\mathrm{h}} \\ T_{\mathrm{h}}^{-1} A_t T_{\mathrm{h}} \end{bmatrix} = \begin{bmatrix} C_{\mathrm{h}} \\ A_{\mathrm{h}} \end{bmatrix} = \begin{bmatrix} \star & 0 & 0 & 0 & 0 \\ \star & \star & 0 & 0 & 0 \\ \star & \star & \star & 0 & 0 \\ \star & \star & \star & \star & 0 \\ \star & \star & \star & \star & \star \\ \star & \star & \star & \star & \star \\ \star & \star & \star & \star & \star \end{bmatrix},$$

with $\star$ denoting a possibly nonzero matrix entry.

The pair $(A_{\mathrm{h}}, C_{\mathrm{h}})$ in observer Hessenberg form can always be represented by a series of real numbers $\theta(i) \in [-1, 1]$ for $i = 1, 2, \ldots, n\ell$ that define an output-normal-form parameterization as in Definition 7.1. This is illustrated for the case $n = 2$ and $\ell = 2$. From the definition (7.17) on page 220 it follows that we need to show that the pair $(A_{\mathrm{h}}, C_{\mathrm{h}})$ satisfies

$$T_{n\ell}^{\mathrm{T}}\big(\theta(n\ell)\big) \cdots T_2^{\mathrm{T}}\big(\theta(2)\big) T_1^{\mathrm{T}}\big(\theta(1)\big) \begin{bmatrix} C_{\mathrm{h}} \\ A_{\mathrm{h}} \end{bmatrix} = \begin{bmatrix} 0 \\ I_n \end{bmatrix}.$$

The first transformation, $T_1^T(\theta(1))$, is applied as

$$
\begin{bmatrix} 1 & 0 & 0 \\ 0 & U^T(\theta(1)) & 0 \\ 0 & 0 & 1 \end{bmatrix} \begin{bmatrix} C_h \\ \hline A_h \end{bmatrix} = \begin{bmatrix} 1 & 0 & 0 \\ 0 & U^T(\theta(1)) & 0 \\ 0 & 0 & 1 \end{bmatrix} \begin{bmatrix} x_{11} & 0 \\ x_{21} & x_{22} \\ x_{31} & x_{32} \\ x_{41} & x_{42} \end{bmatrix}
$$

$$
= \begin{bmatrix} x_{11} & 0 \\ x'_{21} & 0 \\ x'_{31} & x'_{32} \\ x_{41} & x_{42} \end{bmatrix},
$$

with $U(\theta(1))$ such that

$$
U^T(\theta(1)) \begin{bmatrix} x_{22} \\ x_{32} \end{bmatrix} = \begin{bmatrix} 0 \\ x'_{32} \end{bmatrix}
$$

and primes denoting modified entries. The second transformation, $T_2^T(\theta(2))$, yields

$$
\begin{bmatrix} I_2 & 0 \\ 0 & U^T(\theta(2)) \end{bmatrix} \begin{bmatrix} x_{11} & 0 \\ x'_{21} & 0 \\ x'_{31} & x'_{32} \\ x_{41} & x_{42} \end{bmatrix} = \begin{bmatrix} x_{11} & 0 \\ x'_{21} & 0 \\ x''_{31} & 0 \\ x''_{41} & x''_{42} \end{bmatrix},
$$

with double primes denoting modified entries. Since the matrices $U(\theta(1))$ and $U(\theta(2))$ are orthogonal, and the pair (A_h, C_h) satisfies $A_h^T A_h + C_h^T C_h = I_n$, we have

$$
\begin{bmatrix} x_{11} & x'_{21} & x''_{31} & x''_{41} \\ 0 & 0 & 0 & x''_{42} \end{bmatrix} \begin{bmatrix} x_{11} & 0 \\ x'_{21} & 0 \\ x''_{31} & 0 \\ x''_{41} & x''_{42} \end{bmatrix} = I_2.
$$

This implies $x''_{41} = 0$ and $(x''_{42})^2 = 1$. The value of x''_{42} can thus be taken as -1 or 1; in the sequel, the positive value is used. We see that the rightmost column and bottom row of the transformed matrix are already in the correct form. Subsequently, the first column is transformed into the correct form by annihilating the entries x_{11} and x'_{21}. This is done using the orthogonal Givens rotations $U(\theta(3))$ and $U(\theta(4))$.

We obtain

$$
\begin{bmatrix} 1 & 0 & 0 \\ 0 & U^{\mathrm{T}}\big(\theta(4)\big) & 0 \\ 0 & 0 & 1 \end{bmatrix} \begin{bmatrix} U^{\mathrm{T}}\big(\theta(3)\big) & 0 \\ 0 & I_2 \end{bmatrix} \begin{bmatrix} I_2 & 0 \\ 0 & U^{\mathrm{T}}\big(\theta(2)\big) \end{bmatrix}
$$

$$
\times \begin{bmatrix} 1 & 0 & 0 \\ 0 & U^{\mathrm{T}}\big(\theta(1)\big) & 0 \\ 0 & 0 & 1 \end{bmatrix} \begin{bmatrix} x_{11} & 0 \\ x_{21} & x_{22} \\ x_{31} & x_{32} \\ x_{41} & x_{42} \end{bmatrix} = \begin{bmatrix} 0 & 0 \\ 0 & 0 \\ 1 & 0 \\ 0 & 1 \end{bmatrix}.
$$

To complete the parameterization of the state-space system (7.18)–(7.19) the matrices $(B_{\mathrm{h}}, D) = (T_{\mathrm{h}}^{-1} T_t^{-1} B, D)$ of the transformed state-space system are parameterized by all their entries. This completes the proof. $\qquad\square$

The total number of parameters for the output normal parameterization of the state-space model (7.18)–(7.19) is

$$
p = n\ell + nm + m\ell.
$$

Example 7.5 (Output-normal-form parameterization) Consider a second-order state-space model with system matrices

$$
A = \begin{bmatrix} 1.5 & -0.7 \\ 1 & 0 \end{bmatrix}, \qquad B = \begin{bmatrix} 1 \\ 0 \end{bmatrix}, \qquad C = \begin{bmatrix} 1 & 0.5 \end{bmatrix}, \qquad D = 0.
$$

Since A is asymptotically stable and the pair (A, C) is observable, we can apply Lemma 7.2. We start by finding a similarity transformation T_t such that the transformed pair $\big(T_t^{-1} A T_t, C T_t\big) = (A_t, C_t)$ satisfies $A_t^{\mathrm{T}} A_t + C_t^{\mathrm{T}} C_t = I$. Since the pair (A, C) is observable and the system matrix A is asymptotically stable, the solution Q of the Lyapunov equation

$$
A^{\mathrm{T}} Q A + C^{\mathrm{T}} C = Q
$$

is positive-definite. Therefore the matrix Q has a Cholesky factorization $T_q T_q^{\mathrm{T}}$ that defines the necessary similarity transformation $T_t = T_q^{-\mathrm{T}}$

$$
T_q = \begin{bmatrix} 4.3451 & 0 \\ -2.6161 & 1.6302 \end{bmatrix}, \qquad T_t = \begin{bmatrix} 0.2301 & 0.3693 \\ 0 & 0.6134 \end{bmatrix}.
$$

By applying the transformation T_t to the quartet of system matrices we obtain a similarly equivalent quartet. The pair (A_t, C_t) of this quartet reads

$$\begin{bmatrix} C_t \\ \hline A_t \end{bmatrix} = \begin{bmatrix} 0.2301 & 0.6760 \\ \hline 0.8979 & -0.4248 \\ 0.3752 & 0.6021 \end{bmatrix}.$$

This pair (A_t, C_t) already satisfies $A_t^\mathrm{T} A_t + C_t^\mathrm{T} C_t = I_2$. However, to obtain the factorization in (7.17), we have to perform a number of additional transformations. First we perform an orthogonal similarity transformation T_h such that

$$\begin{bmatrix} C_t \\ \hline C_t A_t \end{bmatrix} T_\mathrm{h}$$

is lower triangular. This transformation can be derived from the Q factor of the RQ factorization of the matrix

$$\begin{bmatrix} C_t \\ \hline C_t A_t \end{bmatrix}.$$

It follows that

$$T_\mathrm{h} = \begin{bmatrix} 0.3233 & 0.9466 \\ 0.9466 & -0.3233 \end{bmatrix}.$$

Applying the similarity transformation T_h yields the following transformed pair:

$$\begin{bmatrix} C_\mathrm{h} \\ \hline A_\mathrm{h} \end{bmatrix} = \begin{bmatrix} 0.7141 & 0 \\ \hline 0.6176 & 0.4706 \\ -0.3294 & 0.8824 \end{bmatrix}.$$

To yield the factorization (7.17), we search for a transformation T_1 such that

$$T_1^\mathrm{T} \begin{bmatrix} C_\mathrm{h} \\ \hline A_\mathrm{h} \end{bmatrix} = \begin{bmatrix} \star & 0 \\ \star & 0 \\ 0 & 1 \end{bmatrix},$$

where the $\star$ indicate a number not of interest in this particular step. The required transformation T_1 is based on the Givens rotation (given by the matrix $U(\alpha)$ in Definition 7.1 on page 220) that transforms the lower-right elements $[0.4706, 0.8823]^\mathrm{T}$ into $[0, 1]^\mathrm{T}$, and is given by

$$T_1 = \begin{bmatrix} 1 & 0 & 0 \\ 0 & -0.8824 & 0.4706 \\ 0 & 0.4706 & 0.8824 \end{bmatrix}.$$

defining the parameter $\theta(1)$ equal to 0.8824 and yielding

$$T_1^T \begin{bmatrix} C_h \\ \overline{A_h} \end{bmatrix} = \begin{bmatrix} 0.7141 & 0 \\ -0.7 & 0 \\ 0 & 1 \end{bmatrix}.$$

Finally, the matrix T_2^T transforms the upper-left elements $[0.7141, -0.7]^T$ into $[0, 1]^T$ and again is based on a Givens rotation. The transformation T_2 is given by

$$T_2 = \begin{bmatrix} 0.7 & 0.7141 & 0 \\ 0.7141 & -0.7 & 0 \\ 0 & 0 & 1 \end{bmatrix},$$

defining $\theta(2)$ equal to -0.7.

The parameter vector $\theta_{AC} = \theta$ to parameterize the transformed pair (A, C) then equals

$$\theta_{AC} = \begin{bmatrix} 0.8824 \\ -0.7 \end{bmatrix}.$$

To complete the parameterization in output normal form, the vector θ_{BD} is defined equal to

$$\theta_{BD} = \begin{bmatrix} T_h^{-1} T_t^{-1} B \\ 0 \end{bmatrix} = \begin{bmatrix} 1.4003 \\ 4.1133 \\ 0 \end{bmatrix}.$$

7.3.2 The tridiagonal form

The tridiagonal parameterization exploits the numerical property that for every square matrix A there exists a (nonsingular) similarity transformation T, such that $T^{-1}AT$ is a tridiagonal matrix (Golub and Van Loan, 1996). A tridiagonal matrix has nonzero entries only on the diagonal and one layer above and below the diagonal. An illustration of the form is given for $n = 4$:

$$A(\theta) = \begin{bmatrix} \theta(1) & \theta(2) & 0 & 0 \\ \theta(3) & \theta(4) & \theta(5) & 0 \\ 0 & \theta(6) & \theta(7) & \theta(8) \\ 0 & 0 & \theta(9) & \theta(10) \end{bmatrix}.$$

To complete the parameterization of the LTI system (7.4)–(7.5), we add the entries of the matrices B, C, and D. The total number of parameters equals in this case

$$p = 3n - 2 + n(m + \ell) + m\ell,$$

which is an excess of $3n - 2$ parameters compared with the number of parameters required in Section 7.3.1. The surjective property of this parameterization requires that special care is taken during the numerical search for the parameter vector θ (McKelvey, 1995). This special care is called *regularization* and is discussed in Section 7.5.2.

7.4 The output-error cost function

As stated in Section 7.2 to estimate a state-space model of the form (7.4)–(7.5) from input and output data we consider the quadratic cost function

$$J_N(\theta) = \frac{1}{N} \sum_{k=0}^{N-1} \|y(k) - \widehat{y}(k,\theta)\|_2^2, \tag{7.20}$$

where $y(k)$ is the measured output signal, and $\widehat{y}(k,\theta)$ is the output signal of the model (7.4)–(7.5). The cost function $J_N(\theta)$ is scalar-valued and depends on the parameter vector θ. In mathematical terms it is a *functional* (Rudin, 1986). Taking the constraints on the parameter vector θ into account, we denote the optimization problem as

$$\min_{\theta} J_N(\theta) \quad \text{subject to } \theta \in \Omega \subset \mathbb{R}^p \text{ and } (7.4)\text{--}(7.5). \tag{7.21}$$

Properties such as convexity of the functional $J_N(\theta)$ have a great influence on the numerical way of finding the optimum of (7.21). In general we are able to find only a local minimum and finding the global minimum, when it exists, requires either special properties of $J_N(\theta)$ or an immense computational burden.

For state-space models, a more specific form of $J_N(\theta)$, including the effect of the initial state (as discussed in Section 7.2), is given in the following theorem.

Theorem 7.1 *For the state-space model (7.4)–(7.5), the functional $J_N(\theta)$ can be written as*

$$J_N(\theta_{AC}, \theta_{BD}) = \frac{1}{N} \sum_{k=0}^{N-1} \|y(k) - \phi(k, \theta_{AC})\theta_{BD}\|_2^2, \tag{7.22}$$

with θ_{AC} the parameters necessary to parameterize the pair (A, C) and

$$\theta_{BD} = \begin{bmatrix} \widehat{x}(0) \\ \text{vec}(B) \\ \text{vec}(D) \end{bmatrix}.$$

The matrix $\phi(k, \theta_{AC}) \in \mathbb{R}^{\ell \times (n+m(\ell+n))}$ *is explicitly given as*

$$\phi(k, \theta_{AC})$$

$$= \left[C(\theta_{AC})A(\theta_{AC})^k, \sum_{\tau=0}^{k-1} u^{\mathrm{T}}(\tau) \otimes C(\theta_{AC})A(\theta_{AC})^{k-1-\tau}, u^{\mathrm{T}}(k) \otimes I_\ell \right].$$

Proof The parameterized state-space model (7.4)–(7.5) is given by

$$\widehat{x}(k+1, \theta_{AC}, \theta_{BD}) = A(\theta_{AC})\widehat{x}(k, \theta_{AC}, \theta_{BD}) + B(\theta_{BD})u(k),$$
$$\widehat{y}(k, \theta_{AC}, \theta_{BD}) = C(\theta_{AC})\widehat{x}(k, \theta_{AC}, \theta_{BD}) + D(\theta_{BD})u(k).$$

The output of this state-space model can explicitly be written in terms of the input and the initial state $\widehat{x}(0, \theta_{BD})$ as (see Section 3.4.2)

$$\widehat{y}(k, \theta_{AC}, \theta_{BD}) = C(\theta_{AC})A(\theta_{AC})^k \widehat{x}(0, \theta_{BD})$$

$$+ \sum_{\tau=0}^{k-1} C(\theta_{AC})A(\theta_{AC})^{k-1-\tau} B(\theta_{BD})u(\tau)$$

$$+ D(\theta_{BD})u(k).$$

Application of the property that $\mathrm{vec}(XYZ) = (Z^{\mathrm{T}} \otimes X)\mathrm{vec}(Y)$ (see Section 2.3) and writing down the resulting equation for $k = 1, 2, \ldots, N$ completes the proof. $\qquad \square$

The parameter vector θ in the original state-space model (7.4)–(7.5) could be constructed by simply stacking the vectors θ_{AC} and θ_{BD} of Theorem 7.1 as

$$\theta = \begin{bmatrix} \theta_{AC} \\ \theta_{BD} \end{bmatrix}.$$

The output normal form presented in Lemma 7.2 will give rise to the formulation of the functional as expressed in Theorem 7.1. If the parameters θ_{AC} are fixed, the cost function (7.22) is linear in the parameters θ_{BD}. This fact can be exploited by applying the principle of separable least squares (Golub and Pereyra, 1973) in the search for the minimum of the cost function. Separable least squares first eliminates the parameters θ_{BD} from the cost function and searches for a minimum with respect to the parameters θ_{AC} only. Once the optimal value of the parameter vector θ_{AC} has been found, the parameter values θ_{BD} are derived by simply solving a linear least-squares problem. The critical requirement is that there are no parameters in common between those contained in

θ_{AC} and θ_{BD}. This is the case for the output normal form, defined in Section 7.3.1, but not for the tridiagonal form of Section 7.3.2. The application of separable least squares for the identification of LTI state-space models is discussed by Bruls *et al.* (1999) and Haverkamp (2000).

The influence of the choice of the parameterization on the shape of the cost function $J_N(\theta)$, and therefore on the numerical optimization process (7.21), is illustrated in the example below.

Example 7.6 (Shape of the cost function) Consider the state-space system from Example 7.5 on page 224. We demonstrate that the shape of the cost function $J_N(\theta)$ depends on the parameterization of the state-space system. We consider three cases.

- The system is converted into observer canonical form. For this particular system we just have to switch the two states to arrive at

$$A = \begin{bmatrix} 0 & -a_0 \\ 1 & -a_1 \end{bmatrix}, \qquad B = \begin{bmatrix} 0.5 \\ 1 \end{bmatrix}, \qquad C = \begin{bmatrix} 0 & 1 \end{bmatrix},$$

  where $a_0 = 0.7$ and $a_1 = -1.5$. We parameterize the system with the parameter vector $\theta = [a_0, a_1]^T$. Figure 7.5 shows how the cost function varies with the parameter vector θ. The minimum value of the cost function occurs for $\theta = [0.7, -1.5]$. This function is clearly nonlinear, it has several local minima. Varying the parameters can make the system unstable (see also Example 7.3 on page 217); this results in the "steep walls" displayed in Figure 7.5.
- We take again the observer canonical form, but now take the parameter vector θ equal to $[a_0/a_1, a_1]^T$. This means that we parameterize the A matrix as follows:

$$A = \begin{bmatrix} 0 & \theta(1)\theta(2) \\ 1 & -\theta(2) \end{bmatrix}.$$

  Figure 7.6 shows how the cost function varies with the parameter vector θ. The minimum value of the cost function occurs for $\theta \approx [0.47, -1.5]$.
- The system is converted to the output normal form, as explained in Example 7.5. We vary the two parameters that parameterize

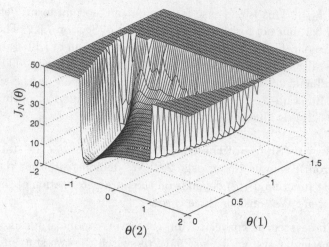

Fig. 7.5. The cost function $J_N(\theta)$ as a function of the parameters $\theta(1)$ and $\theta(2)$ for the system of Example 7.6 in observer canonical form with parameter vector $\theta = [a_0, a_1]^{\mathrm{T}}$.

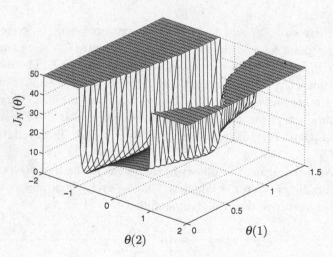

Fig. 7.6. The cost function $J_N(\theta)$ as a function of the parameters $\theta(1)$ and $\theta(2)$ for the system of Example 7.6 in observer canonical form with parameter vector $\theta = [a_0/a_1, a_1]^{\mathrm{T}}$.

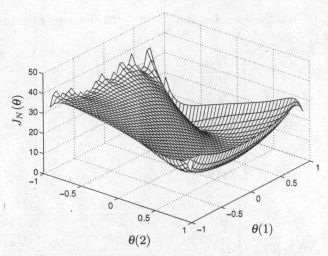

Fig. 7.7. The cost function $J_N(\theta)$ as a function of the parameters $\theta(1)$ and $\theta(2)$ for the system of Example 7.6 in the output normal form with as parameter vector the parameters that describe the matrices A and C.

the matrices A and C. The minimum value of the cost function occurs for $\theta \approx [0.8824, -0.7]$. The cost function is displayed in Figure 7.7. Again we see that the cost function is nonlinear. Unlike in the previous cases, it always remains bounded, since with the output-normal parameterization the system can never become unstable. However, we still observe that the cost function is nonconvex.

7.5 Numerical parameter estimation

To determine a numerical solution to the parameter-optimization problem (Equation (7.21) on page 221) of the previous section, the cost function $J_N(\theta)$ is expanded in a Taylor series around a given point $\theta^{(i)}$ in the parameter space Ω. This point $\theta^{(i)}$ may be the starting point of the optimization process or an intermediate estimate obtained during the search for the minimum of $J_N(\theta)$. The Taylor-series expansion is given by

$$
J_N(\theta) = J_N(\theta^{(i)}) + \left(J'_N(\theta^{(i)}) \right)^{\mathrm{T}} \left(\theta - \theta^{(i)} \right)
$$
$$
+ \frac{1}{2} \left(\theta - \theta^{(i)} \right)^{\mathrm{T}} J''_N(\theta^{(i)}) \left(\theta - \theta^{(i)} \right)
$$
$$
+ \text{ higher-order terms,}
$$

where $J'_N(\theta^{(i)})$ is the Jacobian and $J''_N(\theta^{(i)})$ the Hessian of the functional $J_N(\theta)$ at $\theta^{(i)}$, given by

$$
J'_N(\theta) = \frac{\partial J_N(\theta)}{\partial \theta} = \begin{bmatrix} \dfrac{\partial J_N(\theta)}{\partial \theta(1)} \\[2mm] \dfrac{\partial J_N(\theta)}{\partial \theta(2)} \\[2mm] \vdots \\[2mm] \dfrac{\partial J_N(\theta)}{\partial \theta(p)} \end{bmatrix},
$$

$$
J''_N(\theta) = \frac{\partial^2 J_N(\theta)}{\partial \theta \, \partial \theta^T} = \begin{bmatrix} \dfrac{\partial J_N(\theta)}{\partial \theta(1)\partial \theta(1)} & \dfrac{\partial J_N(\theta)}{\partial \theta(1)\partial \theta(2)} & \cdots & \dfrac{\partial J_N(\theta)}{\partial \theta(1)\partial \theta(p)} \\[2mm] \dfrac{\partial J_N(\theta)}{\partial \theta(2)\partial \theta(1)} & \dfrac{\partial J_N(\theta)}{\partial \theta(2)\partial \theta(2)} & \cdots & \dfrac{\partial J_N(\theta)}{\partial \theta(2)\partial \theta(p)} \\[2mm] \vdots & \vdots & \ddots & \vdots \\[2mm] \dfrac{\partial J_N(\theta)}{\partial \theta(p)\partial \theta(1)} & \dfrac{\partial J_N(\theta)}{\partial \theta(p)\partial \theta(2)} & \cdots & \dfrac{\partial J_N(\theta)}{\partial \theta(p)\partial \theta(p)} \end{bmatrix}.
$$

We approximate $J_N(\theta)$ as

$$
J_N(\theta) \approx J_N(\theta^{(i)}) + \left(J'_N(\theta^{(i)})\right)^T \left(\theta - \theta^{(i)}\right)
$$
$$
+ \frac{1}{2}\left(\theta - \theta^{(i)}\right)^T J''_N(\theta^{(i)})\left(\theta - \theta^{(i)}\right). \tag{7.23}
$$

The necessary condition for minimizing this approximation of $J_N(\theta)$ becomes

$$
J'_N(\theta^{(i)}) + J''_N(\theta^{(i)})\left(\theta - \theta^{(i)}\right) = 0.
$$

Therefore, provided that the Hessian at $\theta^{(i)}$ is invertible, we can update the parameter vector $\theta^{(i)}$ to θ by the update equation

$$
\theta = \theta^{(i)} - J''_N(\theta^{(i)})^{-1} J'_N(\theta^{(i)}). \tag{7.24}
$$

This type of parameter update is called the Newton method. To arrive at explicit expressions for $J'_N(\theta)$ and $J''_N(\theta)$, we introduce the error vector

$$
E_N(\theta) = \begin{bmatrix} \epsilon(0,\theta) \\ \epsilon(1,\theta) \\ \vdots \\ \epsilon(N-1,\theta) \end{bmatrix},
$$

with $\epsilon(k, \theta) = y(k) - \widehat{y}(k, \theta)$. We can denote the cost function $J_N(\theta)$ as

$$J_N(\theta) = \frac{1}{N} \sum_{k=0}^{N-1} \|y(k) - \widehat{y}(k, \theta)\|_2^2 = \frac{1}{N} E_N^T(\theta) E_N(\theta). \tag{7.25}$$

Using the calculus of differentiating functionals outlined in Brewer (1978), and using the notation

$$\Psi_N(\theta) = \frac{\partial E_N(\theta)}{\partial \theta^T}, \tag{7.26}$$

the Jacobian and Hessian of $J_N(\theta)$ can be expressed as

$$
\begin{aligned}
J_N'(\theta) &= \frac{1}{N} \frac{\partial E_N^T(\theta)}{\partial \theta} E_N(\theta) + \frac{1}{N} \left(I_p \otimes E_N^T(\theta) \right) \frac{\partial E_N(\theta)}{\partial \theta} \\
&= \frac{2}{N} \frac{\partial E_N^T(\theta)}{\partial \theta} E_N(\theta) \\
&= \frac{2}{N} \left(\frac{\partial E_N(\theta)}{\partial \theta^T} \right)^T E_N(\theta) \\
&= \frac{2}{N} \Psi_N^T(\theta) E_N(\theta), \tag{7.27}
\end{aligned}
$$

$$
\begin{aligned}
J_N''(\theta) &= \frac{2}{N} \frac{\partial^2 E_N^T(\theta)}{\partial \theta^T \partial \theta} \left(I_p \otimes E_N(\theta) \right) + \frac{2}{N} \frac{\partial E_N^T(\theta)}{\partial \theta} \frac{\partial E_N(\theta)}{\partial \theta^T} \\
&= \frac{2}{N} \frac{\partial^2 E_N^T(\theta)}{\partial \theta^T \partial \theta} \left(I_p \otimes E_N(\theta) \right) + \frac{2}{N} \left(\frac{\partial E_N(\theta)}{\partial \theta^T} \right)^T \frac{\partial E_N(\theta)}{\partial \theta^T} \\
&= \frac{2}{N} \frac{\partial^2 E_N^T(\theta)}{\partial \theta^T \partial \theta} \left(I_p \otimes E_N(\theta) \right) + \frac{2}{N} \Psi_N^T(\theta) \Psi_N(\theta). \tag{7.28}
\end{aligned}
$$

7.5.1 The Gauss–Newton method

The Gauss–Newton method consists of approximating the Hessian $J_N''(\theta^{(i)})$ by the matrix $H_N(\theta^{(i)})$:

$$H_N(\theta^{(i)}) = \frac{2}{N} \Psi_N^T(\theta) \Psi_N(\theta).$$

Such an approximation of the Hessian holds in the neighborhood of the optimum where the second derivative of the error and the error itself are weakly correlated. In that case the first term of Equation (7.28) can be neglected. This results in considerable computational savings. When the matrix $H_N(\theta^{(i)})$ is invertible, we can write the parameter update equation for the Gauss–Newton method as

$$\theta^{(i+1)} = \theta^{(i)} - H_N(\theta^{(i)})^{-1} J_N'(\theta^{(i)}). \tag{7.29}$$

A different way to derive this update equation is by using a Taylor-series expansion on $E_N(\theta)$ in the neighborhood of $\theta^{(i)}$ as follows:

$$J_N(\theta^{(i)} + \delta\theta^{(i)}) = \frac{1}{N}\|E_N(\theta^{(i)} + \delta\theta)\|_2^2 \approx \frac{1}{N}\|E_N(\theta^{(i)}) + \Psi_N(\theta^{(i)})\delta\theta^{(i)}\|_2^2,$$

(7.30)

where $\Psi_N^{\mathrm{T}}(\theta)$ is given by (7.26). The parameter update $\delta\theta^{(i)} = \theta^{(i+1)} - \theta^{(i)}$ follows on solving the following linear least-squares problem:

$$\min_{\delta\theta^{(i)}} \frac{1}{N}\|E_N(\theta^{(i)}) + \Psi_N(\theta^{(i)})\delta\theta^{(i)}\|_2^2,$$

and we get

$$\theta^{(i+1)} = \theta^{(i)} - \left(\Psi_N(\theta^{(i)})^{\mathrm{T}}\Psi_N(\theta^{(i)})\right)^{-1}\Psi_N(\theta^{(i)})^{\mathrm{T}}E_N(\theta^{(i)})$$

$$= \theta^{(i)} - H_N(\theta^{(i)})^{-1}J'_N(\theta^{(i)}),$$

(7.31)

which equals Equation (7.29).

According to Equation (7.29), at every iteration we need to calculate the approximate Hessian $H_N(\theta^{(i)})$ and the Jacobian $J'_N(\theta^{(i)})$. To ease the computational burden it is important to have an efficient way of calculating these quantities. Equations (7.27) and (7.28) show that in fact we need calculate only $E_N(\theta)$ and $\Psi_N(\theta)$. To compute $E_N(\theta)$ we need to compute $\widehat{y}(k,\theta)$ for $k = 1, 2, \ldots, N$. This can be done efficiently by simulating the following system:

$$\widehat{x}(k+1, \theta) = A(\theta)\widehat{x}(k, \theta) + B(\theta)u(k),$$

(7.32)

$$\widehat{y}(k, \theta) = C(\theta)\widehat{x}(k, \theta) + D(\theta)u(k).$$

(7.33)

This will also yield the signal $\widehat{x}(k, \theta)$ which we need to compute $\Psi_N(\theta)$, as explained below. Note that $\Psi_N(\theta)$ is given by

$$\Psi_N(\theta) = \begin{bmatrix} \dfrac{\partial\epsilon(0, \theta)}{\partial\theta^{\mathrm{T}}} \\[2mm] \dfrac{\partial\epsilon(1, \theta)}{\partial\theta^{\mathrm{T}}} \\[2mm] \vdots \\[2mm] \dfrac{\partial\epsilon(N-1, \theta)}{\partial\theta^{\mathrm{T}}} \end{bmatrix} = -\begin{bmatrix} \dfrac{\partial\widehat{y}(0, \theta)}{\partial\theta^{\mathrm{T}}} \\[2mm] \dfrac{\partial\widehat{y}(1, \theta)}{\partial\theta^{\mathrm{T}}} \\[2mm] \vdots \\[2mm] \dfrac{\partial\widehat{y}(N-1, \theta)}{\partial\theta^{\mathrm{T}}} \end{bmatrix},$$

and that

$$\frac{\partial\widehat{y}(k, \theta)}{\partial\theta^{\mathrm{T}}} = \begin{bmatrix} \dfrac{\partial\widehat{y}(k, \theta)}{\partial\theta(1)} & \dfrac{\partial\widehat{y}(k, \theta)}{\partial\theta(2)} & \cdots & \dfrac{\partial\widehat{y}(k, \theta)}{\partial\theta(p)} \end{bmatrix},$$

where $\theta(i)$ denotes the ith entry of the vector θ. It is easy to see that for every parameter $\theta(i)$ we have

$$\frac{\partial \widehat{x}(k+1,\theta)}{\partial \theta(i)} = A(\theta)\frac{\partial \widehat{x}(k,\theta)}{\partial \theta(i)} + \frac{\partial A(\theta)}{\partial \theta(i)}\widehat{x}(k,\theta) + \frac{\partial B(\theta)}{\partial \theta(i)}u(k),$$

$$\frac{\partial \widehat{y}(k,\theta)}{\partial \theta(i)} = C(\theta)\frac{\partial \widehat{x}(k,\theta)}{\partial \theta(i)} + \frac{\partial C(\theta)}{\partial \theta(i)}\widehat{x}(k,\theta) + \frac{\partial D(\theta)}{\partial \theta(i)}u(k).$$

On taking $X_i(k,\theta) = \partial \widehat{x}(k,\theta)/\partial \theta(i)$, this becomes

$$X_i(k+1,\theta) = A(\theta)X_i(k,\theta) + \frac{\partial A(\theta)}{\partial \theta(i)}\widehat{x}(k,\theta) + \frac{\partial B(\theta)}{\partial \theta(i)}u(k), \quad (7.34)$$

$$\frac{\partial \widehat{y}(k,\theta)}{\partial \theta(i)} = C(\theta)X_i(k,\theta) + \frac{\partial C(\theta)}{\partial \theta(i)}\widehat{x}(k,\theta) + \frac{\partial D(\theta)}{\partial \theta(i)}u(k). \quad (7.35)$$

The previous two equations show that the derivative of $\widehat{y}(k,\theta)$ with respect to $\theta(i)$ can be obtained by simulating a linear system with state $X_i(k,\theta)$ and inputs $\widehat{x}(k,\theta)$ and $u(k)$. Note that the matrices

$$\frac{\partial A(\theta)}{\partial \theta(i)}, \quad \frac{\partial B(\theta)}{\partial \theta(i)}, \quad \frac{\partial C(\theta)}{\partial \theta(i)}, \quad \frac{\partial D(\theta)}{\partial \theta(i)}$$

are fixed and depend only on the particular parameterization that is used to describe the system. We conclude that the calculation of $\Psi_N(\theta)$ boils down to simulating a linear system for every element of the parameter vector θ. Therefore, if θ contains p parameters, we need to simulate $p+1$ linear systems in order to compute both $E_N(\theta)$ and $\Psi_N(\theta)$.

Example 7.7 (Minimizing a quadratic cost function) Let the model output be given by $\widehat{y}(k,\theta) = \phi(k)^{\mathrm{T}}\theta$, with $y(k) \in \mathbb{R}$ and $\phi(k) \in \mathbb{R}^p$; then the cost function $J_N(\theta)$ is

$$J_N(\theta) = \frac{1}{N}\sum_{k=0}^{N-1}(y(k) - \phi(k)^{\mathrm{T}}\theta)^2, \quad (7.36)$$

and the vector $E_N(\theta)$ is

$$E_N(\theta) = \begin{bmatrix} y(0) - \phi(0)^{\mathrm{T}}\theta \\ y(1) - \phi(1)^{\mathrm{T}}\theta \\ \vdots \\ y(N-1) - \phi(N-1)^{\mathrm{T}}\theta \end{bmatrix}.$$

Let $\phi_i(j)$ denote the ith entry of the vector $\phi(j)$, then

$$\frac{\partial E_N^{\mathrm{T}}(\theta)}{\partial \theta(i)} = -\begin{bmatrix} \phi_i(0) & \phi_i(1) & \cdots & \phi_i(N-1) \end{bmatrix}.$$

Hence

$$\Psi_N(\theta)^{\mathrm{T}} = -[\phi(0) \quad \phi(1) \quad \cdots \quad \phi(N-1)],$$

$$\left(\frac{\partial E_N(\theta)}{\partial \theta^{\mathrm{T}}}\right)^{\mathrm{T}} E_N(\theta) = -[\phi(0) \quad \phi(1) \quad \cdots \quad \phi(N-1)]$$

$$\times \begin{bmatrix} y(0) - \phi(0)^{\mathrm{T}}\theta \\ y(1) - \phi(1)^{\mathrm{T}}\theta \\ \vdots \\ y(N-1) - \phi(N-1)^{\mathrm{T}}\theta \end{bmatrix}$$

$$= -\Phi_N^{\mathrm{T}}(Y_N - \Phi_N\theta),$$

with

$$Y_N = [y(0) \quad y(1) \quad \cdots \quad y(N-1)]^{\mathrm{T}},$$

$$\Phi_N = [\phi(0) \quad \phi(1) \quad \cdots \quad \phi(N-1)]^{\mathrm{T}}.$$

Assuming that the matrix $\Phi_N^{\mathrm{T}}\Phi_N/N$ is invertible, we can write the parameter update equation (7.31) as

$$\theta^{(i+1)} = \theta^{(i)} + \left(\frac{1}{N}\Phi_N^{\mathrm{T}}\Phi_N\right)^{-1}\frac{1}{N}\Phi_N^{\mathrm{T}}(Y_N - \Phi_N\theta^{(i)})$$

$$= \left(\frac{1}{N}\Phi_N^{\mathrm{T}}\Phi_N\right)^{-1}\frac{1}{N}\Phi_N^{\mathrm{T}}Y_N. \tag{7.37}$$

The assumed invertibility condition depends on the vector time sequence $\phi(k)$. A systematic framework has been developed to relate this invertibility condition to the notion of persistency of excitation of the time sequence (Ljung, 1999). An analysis of this notion in the context of designing a system-identification experiment is presented in Section 10.2.4.

The updated parameter vector $\theta^{(i+1)}$ becomes independent from the initial one $\theta^{(i)}$. Therefore the iterative parameter-update rule (7.37) can be stopped after one iteration (one cycle) and the estimate becomes

$$\widehat{\theta}_N = \left(\frac{1}{N}\Phi_N^{\mathrm{T}}\Phi_N\right)^{-1}\frac{1}{N}\Phi_N^{\mathrm{T}}Y_N. \tag{7.38}$$

The underlying reason for this is that the functional (7.36) is quadratic in θ. The latter is a consequence of the model output $\phi(k)^{\mathrm{T}}\theta$ being *linear* in the unknown parameter vector θ.

Note that the derived solution of the quadratic cost function (7.37) equals the one obtained by solving the normal equations for a linear least-squares problem (see Section 2.6).

7.5.2 Regularization in the Gauss–Newton method

The matrix $H_N(\theta^{(i)})$ used in the Gauss–Newton update equation (7.29) on page 233 to approximate the Hessian may be singular. This will, for example, be the case when the parameterization selected is non-injective; different sets of parameters yield the same value of the cost function $J_N(\theta)$ and thus the θ that minimizes $J_N(\theta)$ no longer need be unique. One possible means of rescue to cope with this singularity is via *regularization*, which leads to a numerically more attractive variant of the Gauss–Newton method. In regularization a penalty term is added to the cost function to overcome the nonuniqueness of the minimizing θ. Instead of just minimizing $J_N(\theta)$, the minimization problem becomes

$$\min_{\theta} J_N(\theta) + \lambda||\theta||_2^2.$$

The real number λ is positive and has to be selected by the user. Using the same approximation of the cost function $J_N(\theta)$ as in Equation (7.30) on page 234, the regularized Gauss–Newton update can be derived as

$$\theta^{(i+1)} = \theta^{(i)} - \left(H_N(\theta^{(i)}) + \lambda I_p\right)^{-1} J'_N(\theta^{(i)}).$$

By adding λI_p to $H_N(\theta^{(i)})$, the matrix $H_N(\theta^{(i)}) + \lambda I_p$ is made nonsingular for $\lambda > 0$. However, the selection of the regularization parameter λ is far from trivial. A systematic approach that is widely used is known as the *Levenberg–Marquardt* method (Moré, 1978).

7.5.3 The steepest descent method

The steepest-descent method does not compute or approximate the Hessian, it just changes the parameters into the direction of the largest decrease of the cost function. This direction is, of course, given by the Jacobian. Hence, the steepest-descent algorithm updates the parameters as follows:

$$\theta^{(i+1)}(\mu) = \theta^{(i)} - \mu J'_N(\theta^{(i)}), \tag{7.39}$$

where an additional step size $\mu \in [0,1]$ is introduced. This step size is usually determined via the additional *scalar* optimization problem,

$$\widehat{\theta}^{(i+1)} = \arg \min_{\mu \in [0,1]} J_N\left(\theta^{(i+1)}(\mu)\right).$$

In general, the iteration process of the steepest-descent algorithm has a lower convergence speed than that of the iteration in the Gauss–Newton method. However, the steepest-descent algorithm results in considerable computational savings in each individual iteration step. This is due to the fact that, to compute $J'_N(\theta)$, we compute the product $\Psi_N^\mathrm{T}(\theta)E_N(\theta)$ directly, without computing $\Psi_N(\theta)$ and $E_N(\theta)$ separately. This requires only two simulations of an nth-order system, as explained below. Recall that

$$\Psi_N^\mathrm{T}(\theta)E_N(\theta) = \sum_{k=0}^{N-1} \left(-\frac{\partial \widehat{y}(k,\theta)}{\partial \theta}\right)^\mathrm{T} \epsilon(k,\theta).$$

Using Equation (7.35) on page 235, we can write the right-hand side as

$$\sum_{k=0}^{N-1} \left(\frac{\partial \widehat{y}(k,\theta)}{\partial \theta(i)}\right)^\mathrm{T} \epsilon(k,\theta) = \sum_{k=0}^{N-1} X_i(k,\theta)^\mathrm{T} C(\theta)^\mathrm{T} \epsilon(k,\theta)$$

$$+ \sum_{k=0}^{N-1} \widehat{x}(k,\theta)^\mathrm{T} \left(\frac{\partial C(\theta)}{\partial \theta(i)}\right)^\mathrm{T} \epsilon(k,\theta)$$

$$+ \sum_{k=0}^{N-1} u(k)^\mathrm{T} \left(\frac{\partial D(\theta)}{\partial \theta(i)}\right)^\mathrm{T} \epsilon(k,\theta).$$

To obtain $\widehat{x}(k,\theta)$, one simulation of the state equation (7.32) on page 234 is required. From the discussion in Section 7.5.1, it follows that, to compute $X_i(k,\theta)$, the p systems defined by (7.34) and (7.35) need to be simulated. However, for the steepest-descent method $X_i(k,\theta)$ is not needed; only the sum

$$\sum_{k=0}^{N-1} X_i(k,\theta)^\mathrm{T} C(\theta)^\mathrm{T} \epsilon(k,\theta)$$

is needed. This sum can be computed by just one (backward) simulation of the system

$$\overline{X}(k-1,\theta) = A(\theta)^\mathrm{T}\overline{X}(k,\theta) + C(\theta)^\mathrm{T}\epsilon(k,\theta), \tag{7.40}$$

involving the *adjoint* state $\overline{X}(k,\theta)$, because

$$\sum_{k=0}^{N-1} X_i(k,\theta)^\mathrm{T} C(\theta)^\mathrm{T} \epsilon(k,\theta) = \sum_{k=0}^{N-1} W_i(k,\theta)^\mathrm{T}\overline{X}(k,\theta), \tag{7.41}$$

where

$$W_i(k,\theta) = \frac{\partial A(\theta)}{\partial \theta(i)}\,\widehat{x}(k,\theta) + \frac{\partial B(\theta)}{\partial \theta(i)}\,u(k).$$

The equality (7.41) can be derived by writing Equation (7.34) on page 235 as

$$X_i(k+1, \theta) = A(\theta)X_i(k, \theta) + W_i(k, \theta).$$

Taking $X_i(0, \theta) = 0$, we can write

$$
\begin{bmatrix}
X_i(0, \theta) \\
X_i(1, \theta) \\
X_i(2, \theta) \\
\vdots \\
X_i(N-1, \theta)
\end{bmatrix}
=
\begin{bmatrix}
0 & 0 & & \cdots & 0 \\
I_n & 0 & & \cdots & 0 \\
A & I_n & 0 & \cdots & 0 \\
\vdots & & \ddots & \ddots & \vdots \\
A^{N-2} & \cdots & A & I_n & 0
\end{bmatrix}
\begin{bmatrix}
W_i(0, \theta) \\
W_i(1, \theta) \\
W_i(2, \theta) \\
\vdots \\
W_i(N-1, \theta)
\end{bmatrix}. \quad (7.42)
$$

For the adjoint state $\overline{X}(N-1, \theta) = 0$ we have

$$
\begin{bmatrix}
\overline{X}(0, \theta) \\
\overline{X}(1, \theta) \\
\overline{X}(2, \theta) \\
\vdots \\
\overline{X}(N-1, \theta)
\end{bmatrix}
=
\begin{bmatrix}
0 & I_n & A^{\mathrm{T}} & \cdots & (A^{\mathrm{T}})^{N-2} \\
0 & 0 & I_n & & \vdots \\
 & & 0 & \ddots & A^{\mathrm{T}} \\
\vdots & \vdots & & \ddots & I_n \\
0 & 0 & 0 & \cdots & 0
\end{bmatrix}
\begin{bmatrix}
C(\theta)^{\mathrm{T}}\epsilon(0, \theta) \\
C(\theta)^{\mathrm{T}}\epsilon(1, \theta) \\
C(\theta)^{\mathrm{T}}\epsilon(2, \theta) \\
\vdots \\
C(\theta)^{\mathrm{T}}\epsilon(N-1, \theta)
\end{bmatrix}.
$$

$$(7.43)$$

On combining Equations (7.42) and (7.43), it is easy to see that Equation (7.41) holds. We can conclude that only *two simulations* of an nth-order system are required for the steepest-descent method, instead of $p+1$ simulations.

7.5.4 Gradient projection

When a chosen parameterization is non-injective, the Hessian needs to be regularized as discussed in Section 7.5.2. For the special case when the surjective parameterization consists of taking all entries of the system matrices A, B, C, and D, the singularity of the Hessian due to similarity transformations of the state-space system can be dealt with in another way. This parameterization that has all the entries of the system matrices in the parameter vector θ is called the *full parameterization*.

Consider the system given by the matrices $\overline{A}$, $\overline{B}$, $\overline{C}$, and $\overline{D}$ obtained by applying a similarity transformation $T \in \mathbb{R}^{n \times n}$ to the matrices A, B, C, and D as

$$
\begin{bmatrix}
\overline{A} & \overline{B} \\
\overline{C} & \overline{D}
\end{bmatrix}
=
\begin{bmatrix}
T^{-1}AT & T^{-1}B \\
CT & D
\end{bmatrix}. \quad (7.44)
$$

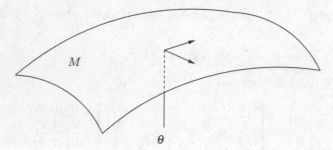

Fig. 7.8. A schematic representation of the manifold M of similar systems and the directions that span the tangent plane at the point θ.

The system given by $\overline{A}$, $\overline{B}$, $\overline{C}$, and $\overline{D}$ has the same transfer function, and thus the same input–output behavior, as the system defined by A, B, C, and D.

By taking all possible nonsingular similarity transformations T, we obtain a set of systems that have the same input–output behavior, and can thus not be distinguished on the basis of input and output data. This set of similar systems forms a manifold M in the parameter space θ, as pictured schematically in Figure 7.8. By changing the parameters along the manifold M we do not change the input–output behavior of the system and we therefore do not change the value of the cost function $J_N(\theta)$.

To avoid problems with the numerical parameter update in minimizing $J_N(\theta)$, we should avoid modifying the parameters such that they stay on this manifold. This idea has been put forward by McKelvey and Helmersson (1997) and by Lee and Poolla (1999). At a certain point θ on the manifold M we can determine the tangent plane (see Figure 7.8). The tangent plane contains the directions in the parameter space along which an update of the parameters does not change the cost function $J_N(\theta)$. The tangent plane of the manifold is determined by considering similar systems for small perturbations of the similarity transformation around the identity matrix, that is $T = I_n + \Delta T$. A first-order approximation of similarly equivalent systems is then (see Exercise 7.8 on page 253) given by

$$\begin{bmatrix} \overline{A} & \overline{B} \\ \overline{C} & \overline{D} \end{bmatrix} = \begin{bmatrix} T^{-1}AT & T^{-1}B \\ CT & D \end{bmatrix} \approx \begin{bmatrix} A & B \\ C & D \end{bmatrix} + \begin{bmatrix} A\,\Delta T - \Delta T A & -\Delta T B \\ C\Delta T & 0 \end{bmatrix}.$$

$$(7.45)$$

If the entries of the system matrices are stacked in column vectors as

$$\theta = \begin{bmatrix} \text{vec}(A) \\ \text{vec}(B) \\ \text{vec}(C) \\ \text{vec}(D) \end{bmatrix}, \qquad \overline{\theta} = \begin{bmatrix} \text{vec}(\overline{A}) \\ \text{vec}(\overline{B}) \\ \text{vec}(\overline{C}) \\ \text{vec}(\overline{D}) \end{bmatrix},$$

applying the vec operator to Equation (7.45) and using the relation $\text{vec}(XYZ) = (Z^{\mathrm{T}} \otimes X)\text{vec}(Y)$ (see Section 2.3) shows that the parameters of the similar systems are related as

$$\overline{\theta} = \theta + Q(\theta)\text{vec}(\Delta T), \tag{7.46}$$

with the matrix $Q(\theta)$ defined by

$$Q(\theta) = \begin{bmatrix} I_n \otimes A - A^{\mathrm{T}} \otimes I_n \\ -B^{\mathrm{T}} \otimes I_n \\ I_n \otimes C \\ 0 \end{bmatrix}.$$

The matrix Q depends on θ, since θ contains the entries of the system matrices A, B, C, and D. Equation (7.46) shows that the columns of the matrix $Q(\theta)$ span the tangent plane at the point θ on the manifold of similar systems. If we update the parameters θ along the directions of the orthogonal complement of the matrix $Q(\theta)$, we will avoid the criterion that we do not change the cost function $J_N(\theta)$. The orthogonal complement of $Q(\theta)$ follows from an SVD of the matrix $Q(\theta)$:

$$Q(\theta) = \begin{bmatrix} U(\theta) & U_\perp(\theta) \end{bmatrix} \begin{bmatrix} \Sigma(\theta) & 0 \\ 0 & 0 \end{bmatrix} \begin{bmatrix} V_1(\theta)^{\mathrm{T}} \\ V_2(\theta)^{\mathrm{T}} \end{bmatrix},$$

with $\Sigma(\theta) > 0$ and $U_\perp(\theta) \in \mathbb{R}^{p \times p - r}$, with $p = n^2 + n(\ell + m) + \ell m$ and $r = \text{rank}(Q(\theta))$. The columns of the matrix $U(\theta)$ form a basis for the column space of $Q(\theta)$; the columns of the matrix $U_\perp(\theta)$ form a basis for the orthogonal complement of the column space of $Q(\theta)$. The matrices $U(\theta)$ and $U_\perp(\theta)$ can be used to decompose the parameter vector θ into two components:

$$\theta = U(\theta)U(\theta)^{\mathrm{T}}\theta + U_\perp(\theta)U_\perp(\theta)^{\mathrm{T}}\theta, \tag{7.47}$$

where the first component corresponds to directions that do not influence the cost function (the column space of Q) and the second component to the directions that change the value of the cost function (the orthogonal complement of the column space of Q).

In solving the optimization problem (7.21) on page 227 the parameters θ are updated according to the rule

$$\theta^{(i+1)} = \theta^{(i)} + \delta\theta^{(i)},$$

where $\delta\theta^{(i)}$ is the update. For the steepest-descent method (7.39) on page 237 this update equals $\delta\theta^{(i)} = -\mu J_N'(\theta^{(i)})$. Preventing an update of the parameters in directions that do not change the cost function is achieved by decomposing $\delta\theta^{(i)}$ similarly to in Equation (7.47) and discarding the first component. On the basis of this observation, the parameter update of the steepest-descent method (7.39) becomes

$$\theta^{(i+1)} = \theta^{(i)} - \mu U_\perp(\theta^{(i)}) U_\perp(\theta^{(i)})^T J_N'(\theta^{(i)}),$$

and the update of the Gauss–Newton method (7.29) on page 233, which is implemented by imposing an update in the direction of the range space of $U_\perp(\theta^{(i)})$ only, is given by

$$\theta^{(i+1)} = \theta^{(i)} - \mu U_\perp(\theta^{(i)}) \Big(U_\perp(\theta^{(i)})^T H_N(\theta^{(i)}) U_\perp(\theta^{(i)}) \Big)^{-1}$$
$$\times U_\perp(\theta^{(i)})^T J_N'(\theta^{(i)}).$$

This insight can be obtained by solving Exercise 7.10 on page 253.

7.6 Analyzing the accuracy of the estimates

The result of the numerical optimization procedure described in the previous section is

$$\widehat{\theta}_N = \arg\min \frac{1}{N} \sum_{k=0}^{N-1} \|y(k) - \widehat{y}(k,\theta)\|_2^2.$$

A possible way to characterize the accuracy of the estimate $\widehat{\theta}_N$ is via an expression for its mean and covariance matrix. In this section we derive this covariance matrix for the case that the system to be identified belongs to the model class. This means that $G(q)$ of the system

$$y(k) = G(q)u(k) + v(k)$$

belongs to the parameterized model set $\mathcal{M}(\theta)$.

The Gauss–Newton optimization method approximates the cost function as in Equation (7.23) on page 232. This approximation holds exactly in the special case of a model output that is linear in the parameters as treated in Example 7.7 on page 235. Therefore, we study the asymptotic variance *first* for the special case when $J_N(\theta)$ is given by

$$J_N(\theta) = \frac{1}{N} \sum_{k=0}^{N-1} (y(k) - \phi(k)^T \theta)^2. \tag{7.48}$$

We assume that the system is in the model class, thus the measured output $y(k)$ is assumed to be generated by the system

$$y(k) = \phi(k)^T \theta_0 + e(k), \tag{7.49}$$

where θ_0 are the true parameter values, and $e(k)$ is a zero-mean white-noise sequence with variance σ_e^2 that is statistically independent from $\phi(k)$.

Expanding the cost function (7.48) and using the expression for $y(k)$ yields

$$J_N(\theta) = \frac{1}{N} \sum_{k=0}^{N-1} e(k)^2 + \frac{1}{N} \sum_{k=0}^{N-1} e(k)\phi(k)^T(\theta_0 - \theta)$$

$$+ \frac{1}{N} \sum_{k=0}^{N-1} (\theta_0 - \theta)^T \phi(k)\phi(k)^T (\theta_0 - \theta),$$

which is exactly the right-hand side of Equation (7.23) on page 232. The parameter vector $\widehat{\theta}_N$ that minimizes this criterion for $\mu = 1$ was derived in Example 7.7 and equals

$$\widehat{\theta}_N = \left(\frac{1}{N} \Phi_N^T \Phi_N \right)^{-1} \frac{1}{N} \Phi_N Y_N$$

$$= \left[\frac{1}{N} \sum_{k=0}^{N-1} \phi(k)\phi(k)^T \right]^{-1} \left[\frac{1}{N} \sum_{k=0}^{N-1} \phi(k)y(k) \right].$$

Again using the expression for $y(k)$, we get

$$\widehat{\theta}_N - \theta_0 = \left[\frac{1}{N} \sum_{k=0}^{N-1} \phi(k)\phi(k)^T \right]^{-1} \left[\frac{1}{N} \sum_{k=0}^{N-1} \phi(k)e(k) \right].$$

Since $e(k)$ and $\phi(k)$ are independent, $E[\widehat{\theta}_N - \theta_0] = 0$ and thus the estimated parameters $\widehat{\theta}_N$ are unbiased. The covariance matrix of $\widehat{\theta}_N - \theta_0$

equals

$$E\left[[\widehat{\theta}_N - \theta_0][\widehat{\theta}_N - \theta_0]^T\right]$$

$$= \left[\frac{1}{N}\sum_{k=0}^{N-1}\phi(k)\phi(k)^T\right]^{-1} E\left[\frac{1}{N^2}\sum_{k=0}^{N-1}\phi(k)e(k)\sum_{j=0}^{N-1}\phi(j)^Te(j)\right]$$

$$\times\left[\frac{1}{N}\sum_{k=0}^{N-1}\phi(k)\phi(k)^T\right]^{-1}$$

$$= \left[\frac{1}{N}\sum_{k=0}^{N-1}\phi(k)\phi(k)^T\right]^{-1} \frac{1}{N^2}\sum_{k=0}^{N-1}\phi(k)\phi(k)^T\sigma_e^2$$

$$\times\left[\frac{1}{N}\sum_{k=0}^{N-1}\phi(k)\phi(k)^T\right]^{-1}$$

$$= \frac{\sigma_e^2}{N}\left[\frac{1}{N}\sum_{k=0}^{N-1}\phi(k)\phi(k)^T\right]^{-1}.$$

When the matrix

$$\left[\frac{1}{N}\sum_{k=0}^{N-1}\phi(k)\phi(k)^T\right]^{-1}$$

converges to a constant bounded matrix Σ_ϕ, the last equation shows that the covariance matrix of $\widehat{\theta}_N$ goes to zero asymptotically (as $N \to \infty$). In this case the estimate is called *consistent*. The fact that $y(k)$ is given by Equation (7.49) indicates that the system used in optimizing (7.48) is in the model set. In this case the output-error method is able to find the unbiased and consistent estimates of the parameter vector θ.

Now, we take a look at the more general case in which the cost function is given by

$$J_N(\theta_{AC}, \theta_{BD}) = \frac{1}{N}\sum_{k=0}^{N-1}\|y(k) - \phi(k, \theta_{AC})\theta_{BD}\|_2^2,$$

as in Theorem 7.1 on page 227. We assume again that the system to be identified is in the model class; that is, the system to be identified can be described by the parameters $\theta_{AC,0}$ and $\theta_{BD,0}$ such that the measured output satisfies

$$y(k) = \phi(k, \theta_{AC,0})\theta_{BD,0} + e(k),$$

where $e(k)$ is a zero-mean white-noise sequence with variance σ_e^2 that is statistically independent from $\phi(k, \theta_{AC})$. Denoting the true parameters by

$$\theta_0 = \begin{bmatrix} \theta_{AC,0} \\ \theta_{BD,0} \end{bmatrix},$$

and the estimated parameters obtained from the output-error method by $\widehat{\theta}_N$, it can again be shown that $E[\widehat{\theta}_N - \theta_0] = 0$ (Ljung, 1999) and thus the estimated parameters $\widehat{\theta}_N$ are unbiased. The covariance matrix of this unbiased estimate is (Ljung, 1999)

$$E[\widehat{\theta}_N - \theta_0][\widehat{\theta}_N - \theta_0]^{\mathrm{T}} = \frac{\sigma_e^2}{N}[J''(\theta_0)]^{-1},$$

and it can be approximated as

$$E[\widehat{\theta}_N - \theta_0][\widehat{\theta}_N - \theta_0]^{\mathrm{T}} \approx \frac{1}{N}\sum_{k=0}^{N-1} \epsilon(k, \widehat{\theta}_N)^2 \left[\frac{1}{N}\sum_{k=0}^{N-1} \psi(k, \widehat{\theta}_N)\psi(k, \widehat{\theta}_N)^{\mathrm{T}}\right]^{-1},$$

$$(7.50)$$

with

$$\epsilon(k, \widehat{\theta}_N) = y(k) - \phi(k, \theta_{AC})\theta_{BD},$$

$$\psi(k, \widehat{\theta}_N) = -\left.\frac{\partial \epsilon(k, \theta)^{\mathrm{T}}}{\partial \theta}\right|_{\theta = \widehat{\theta}_N}.$$

The approximation of the covariance matrix of the estimated parameters holds only asymptotically in N. This needs to be taken into account when using the approximation to describe the model error.

7.7 Dealing with colored measurement noise

At the beginning of this chapter, we considered the signal model

$$y(k) = G(q)u(k) + v(k),\tag{7.51}$$

where $v(k)$ is a white-noise sequence. In this section we investigate the more general case in which $v(k)$ is nonwhite noise. Consider the cost function

$$J_N(\theta) = \frac{1}{N}\sum_{k=0}^{N-1} \|y(k) - \widehat{y}(k, \theta)\|_2^2 = \frac{1}{N}\sum_{k=0}^{N-1} \|\epsilon(k, \theta)\|_2^2 = \frac{1}{N}E_N^{\mathrm{T}}E_N.$$

$$(7.52)$$

If v_k is a white-noise sequence, the residual vector $\epsilon(k, \theta)$ will also be a white-noise sequence if the following two conditions are satisfied: (1) the transfer function $G(q)$ of Equation (7.51) belongs to the parameterized model set $\mathcal{M}(\theta)$; and (2) the estimate $\widehat{\theta}$ is the global minimizing argument of (7.52) in the limit $N \to \infty$. In this case, all temporal information has been modeled; there is no correlation between different samples of error $\epsilon(k, \theta)$. If the output measurements are perturbed by colored noise, the error $\epsilon(k, \theta)$ can never become a white-noise sequence. The consequence is that, although the estimated parameter θ can still be unbiased, it no longer has minimum variance. This is illustrated in the following example.

Example 7.8 (Quadratic cost function and minimum variance)
Consider the quadratic cost function of Example 7.7 on page 235 given by

$$J_N(\theta) = \frac{1}{N} \sum_{k=0}^{N-1} (y(k) - \phi(k)^{\mathrm{T}}\theta)^2. \tag{7.53}$$

We assume that the system is in the model class, thus the measured output $y(k)$ is assumed to be generated by the system

$$y(k) = \phi(k)^{\mathrm{T}}\theta_0 + v(k), \tag{7.54}$$

where θ_0 are the true parameter values, and $v(k)$ is a zero-mean random sequence that is statistically independent from $\phi(k)$.

Adopting the notation of Example 7.7, we can write the minimization of $J_N(\theta)$ as the least-squares problem

$$\min_{\theta} V_N^{\mathrm{T}} V_N \quad \text{subject to } Y_N = \Phi_N \theta + V_N, \tag{7.55}$$

where

$$V_N = \begin{bmatrix} v(0) & v(1) & \cdots & v(N-1) \end{bmatrix}^{\mathrm{T}}.$$

We know from Section 4.5.2 that, to obtain a minimum-variance estimate of θ, we have to solve the weighted least-squares problem

$$\min_{\theta} E_N^{\mathrm{T}} E_N \quad \text{subject to } Y_N = \Phi_N \theta + L E_N,$$

where $E(E_N E_N^{\mathrm{T}}) = I_N$. On comparing this with Equation (7.55), we see that, to obtain a minimum-variance estimate, we need to have $L E_N = V_N$ with $L = \Sigma_v^{1/2}$ such that $E(V_N V_N^{\mathrm{T}}) = \Sigma_v$. If no information about $v(k)$ is available, this is not possible. From Theorem 4.2 on page 112 it follows that simply setting $L = I$ will lead to a minimum-variance estimate only if $v(k)$ is white noise; for colored noise $v(k)$ the minimum variance is obtained for $L = \Sigma_v^{1/2}$.

7.7.1 Weighted least squares

One way to obtain a minimum-variance parameter estimate when the additive noise $v(k)$ at the output in (7.1) is nonwhite requires that we know its covariance matrix. Let the required covariance matrix be denoted by Σ_v and equal to

$$
\Sigma_v = E \begin{bmatrix} v(0) \\ v(1) \\ \vdots \\ v(N-1) \end{bmatrix} \begin{bmatrix} v(0) & v(1) & \cdots & v(N-1) \end{bmatrix}.
$$

Then, if we assume that $\Sigma_v > 0$, we adapt the cost function (7.25) to the following weighted least-squares sum:

$$
J_N(\theta, \Sigma_v) = \frac{1}{N} E_N^{\mathrm{T}} \Sigma_v^{-1} E_N = \frac{1}{N} (\Sigma_v^{-\mathrm{T}/2} E_N)^{\mathrm{T}} (\Sigma_v^{-\mathrm{T}/2} E_N). \tag{7.56}
$$

The numerical methods outlined in Section 7.5 can be adapted in a straightforward manner by replacing E_N by $\Sigma_v^{-\mathrm{T}/2} E_N$ and Ψ_N by $\Sigma_v^{-\mathrm{T}/2} \Psi_N$.

In general, the covariance matrix is a full $N\ell \times N\ell$ matrix, and, therefore, for large N its formation and inversion requires a prohibitive amount of memory. However, recent work by David (2001) provides a way to circumvent this problem, by employing an analytic and sparse expression for the inverse covariance matrix based on the Gohberg–Heinig inversion theorem. This sparsity can be taken into account to derive computationally efficient methods (Bergboer *et al.*, 2002).

A practical procedure for applying the weighting discussed above is the following.

(i) Minimize the output-error cost function (7.52) and compute the corresponding residual vector E_N for the optimum.

(ii) Use the residual vector from the previous step to estimate a multivariable AR model of the noise, and use that model to compute the Cholesky factor of the inverse covariance matrix as described by David (2001).

(iii) Minimize the weighted cost function (7.56).

After step (iii), again the residual vector E_N can be computed, and steps (ii) and (iii) can be repeated. This can be done several times, but in our experience two iterations are usually sufficient, which corresponds to the observations made by David and Bastin (2001).

7.7.2 Prediction-error methods

Another way to improve the accuracy of the estimates of a parametric model of $G(q)$ in Equation (7.51) when the perturbation $v(k)$ is nonwhite noise consists of incorporating a model of this noise into the estimation procedure. We assume that $v(k)$ can be described by a filtered white-noise sequence $e(k)$, such that

$$y(k) = G(q)u(k) + H(q)e(k).$$

Prediction-error methods (PEM) aim at finding parameters of a model that models both of the transfer functions $G(q)$ and $H(q)$. Making use of the Kalman-filter theory of Section 5.5.5, the above transfer-function model can be described *together* with the following innovation state-space model:

$$\widehat{x}(k+1) = A\widehat{x}(k) + Bu(k) + Ke(k),$$
$$y(k) = C\widehat{x}(k) + Du(k) + e(k),$$

where $e(k)$ is a white-noise sequence. Note that, in general, the dimension of the state vector can be larger than the order n of the transfer function $G(q)$, to incorporate the dynamics of $H(q)$; the dimension equals n only in the special case in which $G(q)$ and $H(q)$ have the same system poles.

From Chapter 5 we recall the one-step-ahead predictor of the innovation representation,

$$\widehat{x}(k+1|k) = (A - KC)x(k|k-1) + (B - KD)u(k) + Ky(k),$$
$$\widehat{y}(k|k-1) = Cx(k|k-1) + Du(k).$$

If we can parameterize this predictor by the parameter vector θ, we are able to use a number of the instruments outlined in this chapter to estimate these parameters by means of minimizing a cost function based on the one-step-ahead prediction error

$$J_N(\theta) = \frac{1}{N} \sum_{k=0}^{N-1} \|y(k) - \widehat{y}(k|k-1, \theta)\|_2^2.$$

The resulting prediction-error methods are widely used and so important that we will devote the next chapter to them.

7.8 Summary

In this chapter we discussed the identification of an LTI state-space model based on a finite number of input and output measurements. We assume that the order of the system is given and that the disturbances can be modeled as an additive white-noise signal to the output.

The first step in estimating the parameters is the determination of a parameterization of the LTI state-space system. A parameterization is a mapping from the space of parameters to the space of rational transfer functions that describe the LTI system. We discuss injective, surjective, and bijective properties of parameterizations and highlight the numerical sensitivity of certain parameterizations. We describe the output-normal parameterization and the tridiagonal parameterization in detail.

For the estimation of the parameters, we need a criterion to judge the quality of a particular value of the parameters. We introduce the output-error cost function for this purpose and show that the properties of this cost function depend on the particular parameterization that is used. For most parameterizations considered in this chapter, the cost function is non-convex and has multiple local minima.

To obtain the optimal values of the parameters with respect to the output-error cost function, we numerically minimize this cost function. We discuss the Gauss–Newton, regularized Gauss–Newton, and steepest-descent methods. In addition, we present an alternative approach called the gradient-projection method that can be used to deal with full parameterizations. These numerical procedures are guaranteed only to find local minima of the cost function.

To analyze the accuracy of the estimates obtained by minimizing the output-error cost function, we derived an expression for the covariance matrix of the error between the true and the estimated parameters.

If the additive disturbance to the output is a colored, nonwhite noise, then the output-error method does not yield the minimum-variance estimates of the parameters. To deal with this problem, we discussed two approaches. The first approach is to apply a weighting with the inverse of the covariance matrix of the additive disturbance in the output-error cost function. The second approach is to optimize the prediction error instead of the output error. The prediction-error methods will be discussed in greater detail in the next chapter.

Exercises

7.1 For a given vector $y \in \mathbb{R}^n$, there always exists an orthogonal Householder transformation Q such that (Golub and Van Loan, 1996)

$$Qy = \begin{bmatrix} \xi \\ 0 \\ \vdots \\ 0 \end{bmatrix},$$

with $\xi = \pm \|y\|_2$. Use this transformation to show that, for any pair of matrices $A \in \mathbb{R}^{n \times n}$ and $C \in \mathbb{R}^{\ell \times n}$, there exists an orthogonal transformation T_h such that the entries above the main diagonal of the matrix

$$\begin{bmatrix} CT_h \\ T_h^{-1} A T_h \end{bmatrix}$$

are zero. An illustration for $n = 5$ and $\ell = 2$ is given in the proof of Lemma 7.2 on page 221.

7.2   Consider a parameterized model with parameters a_0, a_1, b_0, and b_1; and a transfer function given by

$$H(q, a_0, a_1, b_0, b_1) = \frac{b_1 q + b_0}{q^2 + a_1 q + a_0}.$$

For which values of the parameters a_0 and a_1 is this transfer function stable?

7.3   Consider the following single-input, multiple-output system:

$$y(k) = \begin{bmatrix} \dfrac{1 + aq^{-1}}{(1 + aq^{-1})(1 + bq^{-1})} \\ \dfrac{1 + bq^{-1}}{(1 + aq^{-1})(1 + bq^{-1})} \end{bmatrix} u(k).$$

(a) Determine a state-space model of this system such that the C matrix of this state-space model equals the identity matrix.

(b) Denote the state-space model derived above by

$$x(k+1) = Ax(k) + Bu(k),$$
$$y(k) = x(k).$$

Show that the matrices A and B of this state-space model can be determined from a finite number of input and output measurements by solving a linear least-squares problem.

7.4   Consider the predictor model

$$\hat{y}(k, \theta) = \frac{b_1 q^{-1} + b_2 q^{-2}}{1 + a_1 q^{-1} + a_2 q^{-2}} u(k),$$

for $k \geq 2$, with unknown initial conditions $\hat{y}(0)$ and $\hat{y}(1)$. Show that, for

$$A = \begin{bmatrix} -a_1 & 1 \\ -a_2 & 0 \end{bmatrix}; \qquad C = \begin{bmatrix} 1 & 0 \end{bmatrix},$$

the predictor can be written in the following form:

$$\widehat{y}(k,\theta) = \phi(k,a_1,a_2)\begin{bmatrix} \widehat{y}(0) \\ \widehat{y}(1) \\ b_1 \\ b_2 \end{bmatrix},$$

with $\phi(k,a_1,a_2)$ given by

$$\phi(k,a_1,a_2)$$

$$= \left[CA^k \begin{bmatrix} 1 & 0 \\ a_1 & 1 \end{bmatrix} \quad \sum_{\tau=0}^{k-1} CA^{k-1-\tau}\begin{bmatrix}1\\0\end{bmatrix}u(\tau) \quad \sum_{\tau=0}^{k-1} CA^{k-1-\tau}\begin{bmatrix}0\\1\end{bmatrix}u(\tau) \right],$$

for $k \geq 2$.

7.5 Consider the predictor model

$$\widehat{x}(k+1,\theta) = A(\theta)\widehat{x}(k,\theta) + B(\theta)u(k),$$
$$\widehat{y}(k,\theta) = C(\theta)\widehat{x}(k,\theta) + D(\theta)u(k),$$

in observer canonical form with system matrices

$$A = \begin{bmatrix} 0 & -a_0 \\ 1 & -a_1 \end{bmatrix}, \qquad B = \begin{bmatrix} b_0 \\ b_1 \end{bmatrix}, \qquad C = \begin{bmatrix} 0 & 1 \end{bmatrix}, \qquad D = 0,$$

so that the parameter vector equals

$$\theta = \begin{bmatrix} a_0 & a_1 & b_0 & b_1 \end{bmatrix}.$$

(a) Determine for this parameterization the system matrices

$$\frac{\partial A(\theta)}{\partial\theta(i)}, \quad \frac{\partial B(\theta)}{\partial\theta(i)}, \quad \frac{\partial C(\theta)}{\partial\theta(i)}, \quad \frac{\partial D(\theta))}{\partial\theta(i)},$$

for $i = 1,2,3,4$, which are needed to compute the Jacobian of the output-error cost function using Equations (7.34) and (7.35) on page 235.

(b) Determine the conditions on the parameter vector θ such that the combination of the above predictor model with the dynamic equations (7.34) and (7.35) on page 235 is asymptotically stable.

7.6 Consider the predictor model

$$\widehat{x}(k+1,\theta) = A(\theta)\widehat{x}(k,\theta) + B(\theta)u(k),$$
$$\widehat{y}(k,\theta) = C(\theta)\widehat{x}(k,\theta) + D(\theta)u(k),$$

with system matrices

$$A = \begin{bmatrix} 0 & 1 & 0 & \cdots & 0 \\ 0 & 0 & 1 & & 0 \\ \vdots & \vdots & & \ddots & \vdots \\ 0 & 0 & 0 & \cdots & 0 \end{bmatrix}, \qquad B = \begin{bmatrix} b_1 \\ b_2 \\ \vdots \\ b_n \end{bmatrix},$$

$$C = \begin{bmatrix} 1 & 0 & 0 & \cdots & 0 \end{bmatrix}, \qquad D = 0,$$

and parameter vector $\theta = [b_1, b_1, \ldots, b_n]$.

(a) Show that the predictor model can be written as

$$\widehat{y}(k, \theta) = (b_1 q^{-1} + b_2 q^{-2} + \cdots b_n q^{-n}) u(k).$$

(b) Show that the gradients

$$\frac{\partial \widehat{y}(k, \theta)}{\partial \theta_i}, \quad i = 1, 2, \ldots, n,$$

are equal to their finite-difference approximations given by

$$\frac{y(k, \theta) - y(k, \theta + \Delta e_i)}{\Delta}, \quad i = 1, 2, \ldots, n,$$

with $\Delta \in \mathbb{R}$ and $e_i \in \mathbb{R}^n$ a vector with the ith entry equal to 1 and the other entries equal to zero.

(c) Determine the adjoint state-space equation (7.40) on page 238 and evaluate Equation (7.41) on page 238.

7.7   We are given the system described by

$$y(k) = (b_0 + b_1 q^{-1}) u(k) + e(k),$$

with $u(k)$ and $e(k)$ ergodic, zero-mean, and statistically independent stochastic sequences. The sequence $u(k)$ satisfies

$$E[u(k)^2] = \sigma_u^2, \qquad E[u(k)u(k-1)] = \gamma,$$

where $\gamma \in \mathbb{R}$ and $e(k)$ is a white-noise sequence with variance σ_e^2. Using input–output measurements of this system, we attempt to estimate the unknown coefficient b of the output predictor given by

$$\widehat{y}(k, b) = bu(k-1).$$

(a) Determine a closed-form expression for the prediction-error criterion for $N \to \infty$, given by

$$\overline{J}(b) = \lim_{N \to \infty} \frac{1}{N} \sum_{k=0}^{N-1} (y(k) - \widehat{y}(k))^2,$$

in terms of the unknown parameter b.

(b) Determine the parameter value of $\widehat{b}$ that satisfies

$$\widehat{b} = \arg\min \overline{J}(b).$$

(c) Use the expression derived for $\widehat{b}$ to determine conditions on the input $u(k)$ such that $\widehat{b} = b_1$.

7.8   Show that, for $X \in \mathbb{R}^{n \times n}$,

$$(I_n + X)^{-1} = I_n - X + X^2 - X^3 + \cdots + (-1)^n X^n (I_n + X)^{-1},$$

and thus that a first-order approximation of $(I_n + X)^{-1}$ equals $I_n - X$.

7.9   Given the matrices

$$A = \begin{bmatrix} 1.5 & 1 \\ -0.7 & 0 \end{bmatrix}, \qquad \overline{A} = \begin{bmatrix} 1.5 & 1 \\ -\alpha^2 + 1.5\alpha - 0.7 & \alpha \end{bmatrix},$$

with $\alpha \in \mathbb{R}$,

(a) determine a similarity transformation such that $\overline{A} = T^{-1}AT$.

(b) Approximate the similarity transformation as $I_n + \Delta T$ and determine ΔT as in Section 7.5.4.

7.10   Consider the constrained least-squares problem

$$\min_{\theta \in \text{range}(U)} \|Y - \Phi\theta\|_2^2, \tag{E7.1}$$

with the matrices $\Phi \in \mathbb{R}^{N \times n}$ $(n < N)$, $Y \in \mathbb{R}^N$, and $\theta \in \mathbb{R}^n$, and with the matrix $U \in \mathbb{R}^{n \times p}$ $(p < n)$ of full column rank. Show that, if the product ΦU has full column rank, the solution to Equation (E7.1) satisfies

$$\widehat{\theta} = U(U^T \Phi^T \Phi U)^{-1} U^T \Phi^T Y.$$

8

Prediction-error parametric
model estimation

After studying this chapter you will be able to

- describe the prediction-error model-estimation problem;
- parameterize the system matrices of a Kalman filter of fixed and known order such that all stable MIMO Kalman filters of that order are presented;
- formulate the estimation of the parameters of a given Kalman-filter parameterization via the solution of a nonlinear optimization problem;
- evaluate qualitatively the bias in parameter estimation for specific SISO parametric models, such as ARX, ARMAX, output-error, and Box–Jenkins models, under the assumption that the signal-generating system does not belong to the class of parameterized Kalman filters; and
- describe the problems that may occur in parameter estimation when using data generated in closed-loop operation of the signal-generating system.

8.1 Introduction

This chapter continues the discussion started in Chapter 7, on estimating the parameters in an LTI state-space model. It addresses the determination of a model of both the deterministic and the stochastic part of an LTI model.

The objective is to determine, from a finite number of measurements of the input and output sequences, a one-step-ahead predictor given by the stationary Kalman filter without using knowledge of the system and covariance matrices of the stochastic disturbances. In fact, these system and covariance matrices (or alternatively the Kalman gain) need to be estimated from the input and output measurements. Note the difference from the approach followed in Chapter 5, where knowledge of these matrix quantities was used. The restriction imposed on the derivation of a Kalman filter from the data is the assumption of a stationary one-step-ahead predictor of a known order. The estimation of a Kalman filter from input and output data is of interest in problems where predictions of the output or the state of the system into the future are needed. Such predictions are necessary in model-based control methodologies such as predictive control (Clarke *et al.*, 1987; Garcia *et al.*, 1989; Soeterboek, 1992). Predictions can be made from state-space models or from transfer-function models. The estimation problems related to both model classes are treated in this chapter.

We start in Section 8.2 with the estimation of the parameters in a state-space model of the one-step-ahead predictor given by a stationary Kalman filter. As in Chapter 7, we address the four steps of the systematic approach to estimating the parameters in a state-space model, but now for the case in which this state-space model is a Kalman filter. Although the output-error model can be considered as a special case of the Kalman filter, it will be shown that a lot of insight about parameterizations, numerical optimization, and analysis of the accuracy of the estimates acquired in Chapter 7 can be reused here.

In Section 8.3 specific and widely used SISO transfer-function models, such as ARMAX, ARX, output-error, and Box–Jenkins, are introduced as special parameterizations of the innovation state-space model introduced in Chapter 5. This relationship with the Kalman-filter theory is used to derive the one-step-ahead predictors for each of these specific classical transfer-function models.

When the signal-generating system does not belong to the class of parameterized models, the predicted output has a *systematic error* or *bias* even when the number of observations goes to infinity. Section 8.4 presents, for several specific SISO parameterizations of the Kalman filter given in Section 8.3, a qualitative analysis of this bias. A typical example of a case in which the signal-generating system does not belong to the model class is when the signal-generating system is of higher order than the parameterized model. The bias analysis presented here is based on the work of Ljung (1978) and Wahlberg and Ljung (1986).

We conclude this chapter in Section 8.5 by illustrating points of caution when using output-error or prediction-error methods with input and output measurements recorded in a feedback experiment. Such closed-loop data experiments in general require additional algorithmic operations to get consistent estimates, compared with the case in which the data are recorded in open-loop mode. The characteristics of a number of situations advocate the need to conduct parameter estimation with data acquired in a feedback experiment. An example is the identification of an F-16 fighter aircraft that is unstable without a feedback control system. In addition to this imposed need for closed-loop system identification, it has been shown that models identified with closed-loop data may result in improved feedback controller designs (Gevers, 1993; van den Hof and Schrama, 1994; De Bruyne and Gevers, 1994). The dominant plant dynamics in closed-loop mode are more relevant to designing an improved controller than the open-loop dynamics are.

8.2 Prediction-error methods for estimating state-space models

In Section 7.7 we briefly introduced prediction-error methods. When the output of an LTI system is disturbed by additive colored measurement noise, the estimates of the parameters describing the system obtained by an output-error method do not have minimum variance. The second alternative presented in that section as a means by which to obtain minimum-variance estimates was the use of prediction-error methods.

The signal-generating system that is considered in this chapter represents the colored-noise perturbation as a filtered white-noise sequence. Thus the input–output data to be used for identification are assumed to be generated in the following way:

$$y(k) = G(q)u(k) + H(q)e(k), \qquad (8.1)$$

where $e(k)$ is a zero-mean white-noise sequence that is statistically independent from $u(k)$, and $G(q)$ represents the deterministic part and $H(q)$ the stochastic part of the system. If we assume a set of input–output data sequences on a finite time interval then a general formulation of the prediction-error model-estimation problem is as follows.

Given a finite number of samples of the input signal $u(k)$ and the output signal $y(k)$, and the order of the predictor

$$\widehat{x}(k+1) = A\widehat{x}(k) + Bu(k) + K(y(k) - C\widehat{x}(k) - Du(k)), \quad (8.2)$$
$$\widehat{y}(k) = C\widehat{x}(k) + Du(k), \quad (8.3)$$

the goal is to estimate the system matrices A, B, C, D, and K in this predictor such that the output $\widehat{y}(k)$ approximates the output of (8.1).

Recall from Section 5.5.5 that the postulated model (8.2)–(8.3) represents a stationary Kalman filter. If we assume that the entries of the system matrices of this filter depend on the parameter vector θ, then we can define the underlying innovation model as

$$\widehat{x}(k+1|k,\theta) = A(\theta)\widehat{x}(k|k-1,\theta) + B(\theta)u(k) + K(\theta)\epsilon(k), \quad (8.4)$$
$$y(k) = C(\theta)\widehat{x}(k|k-1,\theta) + D(\theta)u(k) + \epsilon(k). \quad (8.5)$$

If we denote this innovation model by means of transfer functions, then, in analogy with the signal-generating system (8.1), we get the following parameterizations of the deterministic and stochastic part:

$$G(q,\theta) = D(\theta) + C(\theta)\Big(qI - A(\theta)\Big)^{-1} B(\theta),$$
$$H(q,\theta) = I + C(\theta)\Big(qI - A(\theta)\Big)^{-1} K(\theta).$$

Note that the matrix A appears both in $G(q)$ and in $H(q)$, therefore it characterizes the dynamics both of the deterministic and of the stochastic part of (8.1).

The four problems involved in estimating the parameters of a model defined in Section 7.2 will be addressed in the following subsections for the prediction-error problem. The prediction-error approach is illustrated in Figure 8.1. In this figure, $\widehat{y}(k,\theta)$ is derived from (8.5) as $C(\theta)\widehat{x}(k|k-1,\theta) + D(\theta)u(k)$.

8.2.1 Parameterizing an innovation state-space model

Corresponding to the innovation state-space model (8.4)–(8.5), we could represent conceptually the following parameterization of the

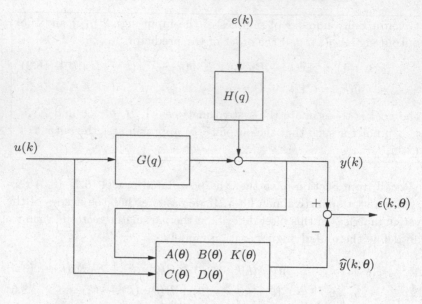

Fig. 8.1. The prediction-error model-estimation method.

one-step-ahead predictor:

$$\widehat{x}(k+1|k,\theta) = \Big(A(\theta) - K(\theta)C(\theta)\Big)\widehat{x}(k|k-1,\theta)$$

$$+ \Big(B(\theta) - K(\theta)D(\theta)\Big)u(k) + K(\theta)y(k), \quad (8.6)$$

$$\widehat{y}(k|k-1,\theta) = C(\theta)\widehat{x}(k|k-1,\theta) + D(\theta)u(k). \quad (8.7)$$

Various choices of parameterization for this predictor exist. The parameterization introduced in Section 7.3 for the output-error case can be used for the prediction-error case if the "A" matrix is taken as $A - KC$ and the "B" matrix as $[B - KD, K]$ and we use $[u(k), y(k)]^{\mathrm{T}}$ as the input to the system.

On making the evident assumption that the model derived from input–output data is reachable and observable, Theorem 5.4 may be used to impose on the system matrix $A - KC$ the additional constraint of asymptotic stability. This constraint then leads to the definition of the set Ω in the model structure $\mathcal{M}(\theta)$ (7.15) on page 217. Depending on the parameterization selected, the additional constraints in the parameter space on the one hand may be cumbersome to determine and on the other may complicate the numerical parameter search. In Example 7.3 it was illustrated how challenging it is to construct the constraints on the parameter set while restricting the parameterization to yield a stable model. Furthermore, extending the example to third- or fourth-order systems indicates that the analysis needs to be performed individually for each dedicated model parameterization. For such models of higher than second order, the parameter set Ω becomes *nonconvex*. This increases the complexity of the optimization problem involved in estimating the parameters. The advantage of the output normal form is that it inherently guarantees the asymptotic stability of the system matrix $A - KC$ of the one-step-ahead predictor as detailed in the following lemma.

Lemma 8.1 (Output normal form of the innovation model) *Let a predictor of the innovation model be given by*

$$\widehat{x}(k+1) = (A - KC)\,\widehat{x}(k) + (B - KD)u(k) + Ky(k), \qquad (8.8)$$

$$\widehat{y}(k) = C\widehat{x}(k) + Du(k), \qquad (8.9)$$

with the matrix $\overline{A} = A - KC$ asymptotically stable and the pair $(\overline{A}, C)$ observable, then a surjective parameterization is obtained by parameterizing the pair $(\overline{A}, C)$ in the output normal form given in Definition 7.1 on page 220 with the parameter vector $\theta_{\overline{A}C} \in \mathbb{R}^{n\ell}$ and parameterizing the triple of matrices $(\overline{B}, D, K)$ with the parameter vector $\theta_{\overline{B}DK} \in \mathbb{R}^{n(m+\ell)+m\ell}$ that contains all the entries of the matrices $\overline{B}$, D, and K, with $\overline{B} = B - KD$.

Proof The proof goes along the same lines as the proof of Lemma 7.2. □

To complete the parameter vector parameterizing (8.6)–(8.7) including the initial state conditions $\widehat{x}(0)$, we simply extend $\theta_{\overline{A}C}$ and $\theta_{\overline{B}DK}$ in the above lemma with these initial conditions to yield the parameter vector θ as

$$\theta = \begin{bmatrix} \widehat{x}(0) \\ \theta_{\overline{A}C} \\ \theta_{\overline{B}DK} \end{bmatrix}.$$

The total number of parameters in this case is $p = n(2\ell + m) + m\ell + n$.

8.2.2 The prediction-error cost function

The primary use of the innovation model structure (8.6)–(8.7) is to predict the output (or state) by making use of a particular value of the parameter vector θ and of the available input–output data sequences. To allow for on-line use of the predictor, the predictor needs to be causal. In off-line applications we may also operate with mixed causal, anti-causal predictors, such as the Wiener optimal filter (Hayes, 1996) and the Kalman-filter/smoothing combination discussed in Section 5.6. In this chapter we restrict the discussion to the causal multi-step-ahead prediction.

Definition 8.1 *For the innovation state-space model structure (8.4)–(8.5) on page 257, the N_p multi-step-ahead prediction of the output is a prediction of the output at a time instant $k + N_p$ making use of the input measurements $u(\ell)$, $\ell \leq k+N_p$ and the output measurements $y(\ell)$, $\ell \leq k$. This estimate is denoted by*

$$\widehat{y}(k + N_p|k, \theta).$$

The definition does not give a procedure for computing a multi-step-ahead prediction. The following lemma gives such a procedure based on the Kalman filter discussed in Chapter 5.

Lemma 8.2 *Given the model structure (8.4)–(8.5) and the quantities $\widehat{x}(k|k-1, \theta)$, $u(k)$, and $y(k)$ at time instant k, then the one-step-ahead prediction at time instant k is given as*

$$\widehat{x}(k + 1|k, \theta) = \Big(A(\theta) - K(\theta)C(\theta)\Big)\widehat{x}(k|k-1, \theta)$$

$$+\Big(B(\theta) - K(\theta)D(\theta)\Big)u(k) + K(\theta)y(k), \quad (8.10)$$

$$\widehat{y}(k + 1|k, \theta) = C(\theta)\widehat{x}(k + 1, k, \theta) + D(\theta)u(k), \quad (8.11)$$

and, on the basis of this one-step-ahead prediction, the multi-step-ahead prediction for $N_p > 1$ is given as

$$\widehat{x}(k + N_p|k, \theta) = A(\theta)^{N_p-1}\widehat{x}(k + 1|k, \theta)$$

$$+ \sum_{i=0}^{N_p-2} A(\theta)^{N_p-i-2}B(\theta)u(k + i + 1), \quad (8.12)$$

$$\widehat{y}(k + N_p|k, \theta) = C(\theta)\widehat{x}(k + N_p|k, \theta) + D(\theta)u(k + N_p). \quad (8.13)$$

The one-step-ahead prediction model (8.10)–(8.11) in this lemma directly follows from the parameterized innovation model (8.4)–(8.5) on page 257. On the basis of this estimate, the multi-step-ahead prediction can be found by computing the response to the system,

$$z(k + j, \theta) = A(\theta)z(k + j - 1, \theta) + B(\theta)u(k + j - 1),$$

for $j > 1$ with initial condition $z(k + 1, \theta) = \widehat{x}(k + 1|k, \theta)$. The multi-step-ahead prediction is then obtained by setting $\widehat{x}(k + N_p|k, \theta) = z(k + N_p, \theta)$. Thus, the multi-step-ahead prediction is obtained by iterating the system using the one-step-ahead predicted state as initial condition. It can be proven that the multi-step-ahead predictor in the lemma is the optimal predictor, in the sense that it solves the so-called Wiener problem. This proof is beyond the scope of this text. The interested reader is referred to the book of Hayes (1996, Chapter 7).

Given a finite number of measurements N of the input and output sequences of the data-generating system, we can estimate the parameters θ of the multi-step-ahead predictor (8.12)–(8.13) by minimizing a least-squares cost function

$$\min_{\theta} J_N(\theta, N_p) = \min_{\theta} \frac{1}{N} \sum_{k=0}^{N-1} \|y(k) - \widehat{y}(k|k - N_p, \theta)\|_2^2. \tag{8.14}$$

This least-squares criterion is inspired by the minimum-variance state-reconstruction property of the Kalman filter. To reveal this link, consider the data-generating system in innovation form (see Section 5.5.5) for the case $N_p = 1$,

$$\widehat{x}(k + 1, \theta_0) = A(\theta_0)\widehat{x}(k, \theta_0) + B(\theta_0)u(k) + K(\theta_0)e(k),$$
$$y(k) = C(\theta_0)\widehat{x}(k, \theta_0) + e(k),$$

with $\widehat{x}(0, \theta_0)$ given and with $K(\theta_0)$ derived from the solution of the DARE (5.73) via Equation (5.74) on page 162. From this innovation representation we can directly derive the Kalman filter as

$$\widehat{x}(k + 1, \theta_0) = A(\theta_0)\widehat{x}(k, \theta_0) + B(\theta_0)u(k) + K(\theta_0)$$
$$\times \Big(y(k) - C(\theta_0)\widehat{x}(k, \theta_0)\Big),$$
$$\widehat{y}(k, \theta_0) = C(\theta_0)\widehat{x}(k, \theta_0),$$

The minimum-variance property of the estimates obtained by use of the Kalman filter means that the variance of the prediction error $y(k) - \widehat{y}(k, \theta_0)$ is minimized. Therefore, if we denote $\widehat{y}(k, \theta)$ as the output of a Kalman filter as above but determined by the parameter vector θ instead of by θ_0, then the latter satisfies

$$\theta_0 = \arg\min \operatorname{tr} E\left[\Big(y(k) - \widehat{y}(k, \theta)\Big)\big(y(k) - \widehat{y}(k, \theta)\big)^{\mathrm{T}}\right].$$

Generally, it was shown that the Kalman filter is time-varying and, therefore, that the variance of the prediction error will change over time. However, if we make the assumption that the variance is constant and the prediction error is an ergodic sequence, an estimate of θ_0 may be obtained by means of the following optimization problem:

$$\widehat{\theta}_0 = \arg\min \lim_{N\to\infty} \frac{1}{N} \sum_{k=0}^{N-1} \|y(k) - \widehat{y}(k|k-1,\theta)\|_2^2.$$

The parameter-optimization problem (8.14) will be referred to as the *prediction-error estimation problem*. It forms a small part of the complete procedure of system identification, since it implicitly assumes the order of the state-space model (n) and the parameterization to be given. Finding the latter, structural information is generally far from a trivial problem, as will be outlined in the discussion of the identification cycle in Chapter 10.

Henceforth we will concentrate on the one-step-ahead prediction error, and thus consider the optimization problem

$$\min_{\theta} J_N(\theta) = \min_{\theta} \frac{1}{N} \sum_{k=0}^{N-1} \|y(k) - \widehat{y}(k|k-1,\theta)\|_2^2. \tag{8.15}$$

For innovation models, a more specific form of $J_N(\theta)$ is given in the following theorem (compare this with Theorem 7.1 on page 227).

Theorem 8.1 *For the innovation model (8.6)–(8.7) on page 258, the functional $J_N(\theta)$ can be written as*

$$J_N(\theta_{\bar{A}C}, \theta_{\bar{B}DK}) = \frac{1}{N} \sum_{k=0}^{N-1} \|y(k) - \phi(k,\theta_{\bar{A}C})\theta_{\bar{B}DK}\|_2^2, \tag{8.16}$$

with $\theta_{\bar{A}C}$ the parameters necessary to parameterize the pair $(\bar{A}, C)$ with $\bar{A} = A - KC$ and

$$\theta_{\bar{B}DK} = \begin{bmatrix} \widehat{x}(0) \\ \mathrm{vec}(\bar{B}) \\ \mathrm{vec}(K) \\ \mathrm{vec}(D) \end{bmatrix},$$

with $\bar{B} = B - KD$. The matrix $\phi(k,\theta_{\bar{A}C}) \in \mathbb{R}^{\ell\times(n+m(\ell+n)+n\ell)}$ is explicitly given as

$$\phi(k,\theta_{\bar{A}C}) = \Bigg[C(\theta_{\bar{A}C})\bar{A}(\theta_{\bar{A}C})^k \quad \sum_{\tau=0}^{k-1} u^{\mathrm{T}}(\tau) \otimes C(\theta_{\bar{A}C})\bar{A}(\theta_{\bar{A}C})^{k-1-\tau}$$

$$\sum_{\tau=0}^{k-1} y^{\mathrm{T}}(\tau) \otimes C(\theta_{\bar{A}C})\bar{A}(\theta_{\bar{A}C})^{k-1-\tau} \quad u^{\mathrm{T}}(k) \otimes I_\ell \Bigg].$$

Proof The one-step-ahead predictor related to the parameterized innovation model (8.6)–(8.7) on page 258 is

$$\widehat{x}(k+1, \theta_{\bar{A}C}, \theta_{\bar{B}DK}) = \overline{A}(\theta_{\bar{A}C})\widehat{x}(k, \theta_{\bar{A}C}, \theta_{\bar{B}DK}) + \overline{B}(\theta_{\bar{B}DK})u(k)$$
$$+ K(\theta_{\bar{B}DK})y(k),$$

$$\widehat{y}(k, \theta_{\bar{A}C}, \theta_{\bar{B}DK}) = C(\theta_{\bar{A}C})\widehat{x}(k, \theta_{\bar{A}C}, \theta_{\bar{B}DK}) + D(\theta_{\bar{B}DK})u(k),$$

with an initial state $\widehat{x}(0, \theta_{\bar{B}DK})$. The output of this state-space model
can explicitly be written in terms of the input, output, and initial state
$\widehat{x}(0, \theta_{\bar{B}DK})$ as (see Section 3.4.2)

$$\widehat{y}(k, \theta_{\bar{A}C}, \theta_{\bar{B}DK}) = C(\theta_{\bar{A}C})\overline{A}(\theta_{\bar{A}C})^k \widehat{x}(0, \theta_{\bar{B}DK})$$

$$+ \sum_{\tau=0}^{k-1} C(\theta_{\bar{A}C})\overline{A}(\theta_{\bar{A}C})^{k-1-\tau} B(\theta_{\bar{B}DK})u(\tau)$$

$$+ D(\theta_{\bar{B}DK})u(k)$$

$$+ \sum_{\tau=0}^{k-1} C(\theta_{\bar{A}C})\overline{A}(\theta_{\bar{A}C})^{k-1-\tau} K(\theta_{\bar{B}DK})y(\tau).$$

Application of the property that $\text{vec}(XYZ) = (Z^{\mathrm{T}} \otimes X)\text{vec}(Y)$ (see
Section 2.3) completes the proof.   □

The parameter vector θ in the original innovation model (8.6)–(8.7) on
page 258 could be constructed by simply stacking the vectors $\theta_{\bar{A}C}$ and
$\theta_{\bar{B}DK}$ of Theorem 8.1 as

$$\theta = \begin{bmatrix} \theta_{\bar{A}C} \\ \theta_{\bar{B}DK} \end{bmatrix}.$$

The output normal form presented in Lemma 8.1 can be used to
parameterize the formulation of the functional as expressed in
Theorem 8.1.

8.2.3 Numerical parameter estimation

To solve the prediction-error problem (8.14) on page 261, the iterative
methods described in Section 7.5 can be used. Of course, some minor
adjustments are necessary. For example, if the one-step-ahead prediction
is used, the cost function is computed by simulating the predictor given
by the system (8.6)–(8.7) on page 258, and the dynamic system (7.34)–
(7.35) on page 235 that needs to be simulated to obtain the Jacobian in

the Gauss–Newton method becomes

$$X_i(k+1,\theta) = \overline{A}(\theta)X_i(k,\theta) + \frac{\partial \overline{A}(\theta)}{\partial \theta(i)}\,\widehat{x}(k,\theta) + \frac{\partial \overline{B}(\theta)}{\partial \theta(i)}\,u(k)$$

$$+ \frac{\partial K(\theta)}{\partial \theta(i)}\,y(k),$$

$$\frac{\partial \widehat{y}(k,\theta)}{\partial \theta(i)} = C(\theta)X_i(k,\theta) + \frac{\partial C(\theta)}{\partial \theta(i)}\,\widehat{x}(k,\theta) + \frac{\partial D(\theta)}{\partial \theta(i)}\,u(k),$$

with

$$\overline{A}(\theta) = A(\theta) - K(\theta)C(\theta),$$
$$\overline{B}(\theta) = B(\theta) - K(\theta)D(\theta).$$

Similar straightforward adjustments are needed in the other numerical method discussed in Section 7.5.

8.2.4 Analyzing the accuracy of the estimates

To analyze the accuracy of the estimates obtained, the covariance matrix of the solution $\widehat{\theta}_N$ to the optimization problem (8.14) on page 261 can be used. The theory presented in Section 7.6 for the output-error methods applies also to the prediction-error methods. Using the covariance matrix to analyze the accuracy of the estimated model is done under the assumption that the system to be identified belongs to the assumed model set $\mathcal{M}(\theta)$ ((7.15) on page 217). Generally, in practice this assumption does not hold and the model parameters will be biased. For certain special model classes that will be introduced in Section 8.3 this bias is analyzed in Section 8.4.

Using an output-error or prediction-error method, the estimates of the model parameters are obtained from a finite number of input and output measurements as

$$\widehat{\theta}_N = \arg\min J_N(\theta).$$

The best possible model $\theta_\star$ within a given model structure is given by the minimizing parameter vector of the cost function $J_N(\theta)$ for $N \to \infty$:

$$\theta_\star = \arg\min \lim_{N\to\infty} J_N(\theta) = \arg\min \overline{J}(\theta).$$

The quality of an estimated model $\widehat{\theta}_N$ can now be measured using (Sjöberg *et al.*, 1995; Ljung, 1999)

$$E\overline{J}(\widehat{\theta}_N), \tag{8.17}$$

where the expectation E is with respect to the model $\widehat{\theta}_N$. The measure (8.17) describes the expected fit of the model to the true system, when the model is applied to a new set of input and output measurements that have the same properties (distributions) as the measurements used to determine $\widehat{\theta}_N$. This measure can be decomposed as follows (Sjöberg *et al.*, 1995; Ljung, 1999):

$$E\overline{J}(\widehat{\theta}_N) \approx \underbrace{E\|y(k) - y_0(k,\theta_0)\|_2^2}_{\text{noise}} + \underbrace{E\|y_0(k,\theta_0) - \widehat{y}(k,\theta_\star)\|_2^2}_{\text{bias}}$$
$$+ \underbrace{E\|\widehat{y}(k,\theta_\star) - \widehat{y}(k,\widehat{\theta}_N)\|_2^2}_{\text{variance}},$$

where $y_0(k,\theta_0)$ is the output of the predictor based on the true model, that is, $y(k) = y_0(k,\theta_0) + e(k)$, with $e(k)$ white-noise residuals. The three parts in this decomposition will now be discussed.

Noise part. The variance of the error between the measured output and a predictor based on the true model θ_0. This error is a white-noise sequence.

Bias part. The model structures of the true predictor $y_0(k,\theta_0)$ and of the model class adopted can be different. The bias error expresses the difference between the true predictor and the best possible approximation of the true predictor within the model class adopted.

Variance part. The use of a finite number of samples N to determine the model $\widehat{\theta}_N$ results in a difference from the best possible model (within the model class adopted) $\theta_\star$ based on an infinite number of samples.

8.3 Specific model parameterizations for SISO systems

For identification of SISO systems, various parameterizations of the innovation representation (8.6)–(8.7) on page 258 are in use (Ljung, 1999; Box and Jenkins, 1970; Johansson, 1993; Söderström and Stoica, 1989). It is shown in this section that these more-classical model parameterizations can be treated as *special cases* of the MIMO innovation model parameterization discussed in Section 8.2. We adopt the common practice of presenting these special SISO parameterizations in a transfer-function setting.

8.3.1 The ARMAX and ARX model structures

The ARMAX, standing for *Auto-Regressive Moving Average with eXoge-nous input,* model structure considers the following specific case of the general input–output description (8.1):

$$y(k) = \frac{b_1 q^{-1} + \cdots + b_n q^{-n}}{1 + a_1 q^{-1} + \cdots + a_n q^{-n}} u(k) + \frac{1 + c_1 q^{-1} + \cdots + c_n q^{-n}}{1 + a_1 q^{-1} + \cdots + a_n q^{-n}} e(k),$$

(8.18)

where $e(k) \in \mathbb{R}$ is again a zero-mean white-noise sequence that is inde-pendent from $u(k) \in \mathbb{R}$ and a_i, b_i, and c_i $(i = 1, 2, \ldots, n)$ are real-valued scalars. It is common practice to use negative powers of q in the descrip-tion of the ARMAX model, instead of positive powers as we did in Chapter 3.

A more general ARMAX representation exists, in which the order of the numerators and denominators may be different, and the transfer from $u(k)$ to $y(k)$ may contain an additional dead-time. To keep the nota-tion simple, these fine-tunings are not addressed in this book. When the order n is known, we can define an estimation problem to estimate the parameters a_i, b_i, and c_i $(i = 1, 2, \ldots, n)$ from a finite number of input–output measurements. The formulation and the solution of such an esti-mation problem is discussed next and is addressed by establishing a one-to-one correspondence between the ARMAX transfer-function descrip-tion (8.18) and a particular minimal parameterization of the state-space system (8.6)–(8.7) on page 258, as summarized in the following lemma.

Lemma 8.3 *There is a one-to-one correspondence between the ARMAX model given by Equation (8.18) and the following parameterization of a SISO state-space system in innovation form:*

$$x(k+1) = \begin{bmatrix} -a_1 & 1 & 0 & \cdots & 0 \\ -a_2 & 0 & 1 & \cdots & 0 \\ \vdots & & & \ddots & \vdots \\ -a_{n-1} & 0 & & \cdots & 1 \\ -a_n & 0 & & \cdots & 0 \end{bmatrix} x(k) + \begin{bmatrix} b_1 \\ b_2 \\ \vdots \\ b_{n-1} \\ b_n \end{bmatrix} u(k)$$

$$+ \begin{bmatrix} c_1 - a_1 \\ c_2 - a_2 \\ \vdots \\ c_{n-1} - a_{n-1} \\ c_n - a_n \end{bmatrix} e(k),$$

(8.19)

$$y(k) = \begin{bmatrix} 1 & 0 & 0 & \cdots & 0 \end{bmatrix} x(k) + e(k).$$

(8.20)

Proof The proof follows on showing that from the parameterization (8.19)–(8.20) we can obtain in a unique manner the difference equation (8.18). Let $x_i(k)$ denote the ith component of the vector $x(k)$, then Equation (8.19) is equivalent to the following set of equations:

$$x_1(k+1) = -a_1 x_1(k) + x_2(k) + b_1 u(k) + (c_1 - a_1)e(k),$$
$$x_2(k+1) = -a_2 x_1(k) + x_3(k) + b_2 u(k) + (c_2 - a_2)e(k),$$
$$\vdots \qquad\qquad \vdots$$
$$x_n(k+1) = -a_n x_1(k) + b_n u(k) + (c_n - a_n)e(k).$$

Making the substitution $y(k) = x_1(k) + e(k)$ yields

$$x_1(k+1) = -a_1 y(k) + x_2(k) + b_1 u(k) + c_1 e(k), \qquad\qquad \star$$
$$x_2(k+1) = -a_2 y(k) + x_3(k) + b_2 u(k) + c_2 e(k), \qquad\qquad \star$$
$$\vdots \qquad\qquad \vdots$$
$$x_{n-1}(k+1) = -a_{n-1} y(k) + x_n(k) + b_{n-1} u(k) + c_{n-1} e(k), \quad \star$$
$$x_n(k+1) = -a_n y(k) + b_n u(k) + c_n e(k).$$

Increasing the time index of all the equations indicated by a star ($\star$) and subsequently replacing $x_n(k+1)$ by the right-hand side of the last equation yields the following expressions for the indicated equations:

$$x_1(k+2) = -a_1 y(k+1) + x_2(k+1) + b_1 u(k+1) + c_1 e(k+1),$$
$$x_2(k+2) = -a_2 y(k+1) + x_3(k+1) + b_2 u(k+1) + c_2 e(k+1),$$
$$\vdots \qquad\qquad \vdots$$
$$x_{n-2}(k+2) = -a_{n-2} y(k+1) + x_{n-1}(k+1) + b_{n-2} u(k+1)$$
$$+ c_{n-2} e(k+1),$$
$$x_{n-1}(k+2) = -a_{n-1} y(k+1) - a_n y(k) + b_n u(k) + c_n e(k)$$
$$+ b_{n-1} u(k+1) + c_{n-1} e(k+1).$$

Implementing the above recipe $n-2$ times yields the single equation

$$x_1(k+n) = -a_1 y(k+n-1) - a_2 y(k+n-2) - \cdots - a_n y(k)$$
$$+ b_1 u(k+n-1) + b_2 u(k+n-2) + \cdots + b_n u(k)$$
$$+ c_1 e(k+n-1) + c_2 e(k+n-2) + \cdots + a_n e(k).$$

By making use of the output equation (8.20), we finally obtain

$$
\begin{aligned}
y(k+n) = &-a_1 y(k+n-1) - a_2 y(k+n-2) - \cdots - a_n y(k) \\
&+ b_1 u(k+n-1) + b_2 u(k+n-2) + \cdots + b_n u(k) \\
&+ e(k+n) + c_1 e(k+n-1) + c_2 e(k+n-2) + \cdots + a_n e(k).
\end{aligned}
$$

This is the difference equation (8.18).   □

The ARMAX model is closely related to the observer canonical form that we described in Sections 3.4.4 and 7.3. The ARMAX model can be converted into the observer canonical form and vice versa by turning the state-vector upside down.

The one-step-ahead predictor for the ARMAX model is summarized in the next lemma.

Lemma 8.4 *Let the differences $c_i - a_i$ be denoted by k_i for $i = 1, 2, \ldots, n$, then the one-step ahead predictor for the ARMAX model (8.18) is given by*

$$
\begin{aligned}
\widehat{y}(k|k-1) = &\frac{b_1 q^{-1} + \cdots + b_n q^{-n}}{1 + c_1 q^{-1} + \cdots + c_n q^{-n}} u(k) \\
&+ \frac{k_1 q^{-1} + \cdots + k_n q^{-n}}{1 + c_1 q^{-1} + \cdots + c_n q^{-n}} y(k).
\end{aligned} \tag{8.21}
$$

Proof Making use of the state-space parameterization of the ARMAX model given by (8.19) and (8.20), the one-step-ahead prediction based on Equations (8.10) and (8.11) on page 260 equals

$$
\widehat{x}(k+1|k) = \left(
\begin{bmatrix}
-a_1 & 1 & 0 & \cdots & 0 \\
-a_2 & 0 & 1 & \cdots & 0 \\
\vdots & & & \ddots & \vdots \\
-a_{n-1} & 0 & & \cdots & 1 \\
-a_n & 0 & & \cdots & 0
\end{bmatrix}
-
\begin{bmatrix}
k_1 \\
k_2 \\
\vdots \\
k_{n-1} \\
k_n
\end{bmatrix}
\begin{bmatrix} 1 & 0 & \cdots & 0 & 0 \end{bmatrix}
\right)
$$

$$
\times \widehat{x}(k|k-1) +
\begin{bmatrix}
b_1 \\
b_2 \\
\vdots \\
b_{n-1} \\
b_n
\end{bmatrix}
u(k) +
\begin{bmatrix}
k_1 \\
k_2 \\
\vdots \\
k_{n-1} \\
k_n
\end{bmatrix}
y(k),
$$

$$
\widehat{y}(k|k-1) = \begin{bmatrix} 1 & 0 & 0 & \cdots & 0 \end{bmatrix} \widehat{x}(k|k-1);
$$

with $c_i = k_i + a_i$, this equals

$$\widehat{x}(k+1|k) = \begin{bmatrix} -c_1 & 1 & 0 & \cdots & 0 \\ -c_2 & 0 & 1 & \cdots & 0 \\ \vdots & & & \ddots & \vdots \\ -c_{n-1} & 0 & & \cdots & 1 \\ -c_n & 0 & & \cdots & 0 \end{bmatrix} \widehat{x}(k|k-1) + \begin{bmatrix} b_1 \\ b_2 \\ \vdots \\ b_{n-1} \\ b_n \end{bmatrix} u(k)$$

$$+ \begin{bmatrix} k_1 \\ k_2 \\ \vdots \\ k_{n-1} \\ k_n \end{bmatrix} y(k),$$

$$\widehat{y}(k|k-1) = \begin{bmatrix} 1 & 0 & 0 & \cdots & 0 \end{bmatrix} \widehat{x}(k|k-1).$$

Following the proof of Lemma 8.3 on page 266, the transfer-function representation of this state-space model equals (8.21).   □

On introducing the following polynomials in the shift operator q,

$$A(q) = 1 + a_1 q^{-1} + \cdots + a_n q^{-n},$$
$$B(q) = b_1 q^{-1} + \cdots + b_n q^{-n},$$
$$C(q) = 1 + c_1 q^{-1} + \cdots + c_n q^{-n},$$

the ARMAX model can be denoted by

$$y(k) = \frac{B(q)}{A(q)} u(k) + \frac{C(q)}{A(q)} e(k). \tag{8.22}$$

The one-step-ahead predictor is denoted by

$$\widehat{y}(k|k-1) = \frac{B(q)}{C(q)} u(k) + \frac{C(q) - A(q)}{C(q)} y(k). \tag{8.23}$$

This is a stable predictor, provided that the polynomial $C(q)$ has all its roots within the unit circle.

The Auto-Regressive with eXogeneous input (ARX) model is a special case of the ARMAX model structure constraining the parameters $c_i = 0$ for $i = 1, 2, \ldots, n$, and thus $C(q) = 1$. Therefore, the ARX model is given by

$$y(k) = \frac{B(q)}{A(q)} u(k) + \frac{1}{A(q)} e(k),$$

and the associated predictor equals

$$\widehat{y}(k|k-1) = B(q)u(k) + [1 - A(q)]y(k). \qquad (8.24)$$

To identify a model in the ARMAX or ARX structure, we minimize the prediction-error cost function $J_N(\theta)$ described in Section 8.2.2. The methods for minimizing this cost function were described in Sections 7.5 and 8.2.3. They require the evaluation of the cost function and its Jacobian. This evaluation depends on the particular parameterization of the state-space innovation model. As pointed out in Section 8.2.3, the choice of a specific parameterization changes only the following matrices in the evaluation of the Jacobian:

$$\frac{\partial \overline{A}}{\partial \theta_i}, \quad \frac{\partial B}{\partial \theta_i}, \quad \frac{\partial C}{\partial \theta_i}, \quad \frac{\partial D}{\partial \theta_i}, \quad \frac{\partial K}{\partial \theta_i},$$

for $i = 1, 2, \ldots, p$. The following example shows that these quantities are easy to compute.

Example 8.1 (Jacobian calculations for ARMAX model structure) Given an ARMAX model, with matrices

$$A = \begin{bmatrix} -\theta_1 & 1 \\ -\theta_2 & 0 \end{bmatrix}, \quad B = \begin{bmatrix} \theta_3 \\ \theta_4 \end{bmatrix},$$

$$C = \begin{bmatrix} 1 & 0 \end{bmatrix}, \quad K = \begin{bmatrix} \theta_5 \\ \theta_6 \end{bmatrix},$$

it is easy to see that

$$\overline{A} = A - KC = \begin{bmatrix} -\theta_1 - \theta_5 & 1 \\ -\theta_2 - \theta_6 & 0 \end{bmatrix},$$

and therefore

$$\frac{\partial \overline{A}}{\partial \theta_i} = \begin{bmatrix} -1 & 0 \\ 0 & 0 \end{bmatrix}, \quad i = 1, 5,$$

$$\frac{\partial \overline{A}}{\partial \theta_i} = \begin{bmatrix} 0 & 0 \\ -1 & 0 \end{bmatrix}, \quad i = 2, 6,$$

$$\frac{\partial \overline{A}}{\partial \theta_i} = \begin{bmatrix} 0 & 0 \\ 0 & 0 \end{bmatrix}, \quad i = 3, 4.$$

The following example illustrates that, for an ARX model, minimization of the prediction-error cost function $J_N(\theta)$ described in Section 8.2.2 leads to a linear least-squares problem.

Example 8.2 (Prediction error for an ARX model) The ARX predictor is given by Equation (8.24). Taking

$$A(q) = 1 + a_1 q^{-1} + \cdots + a_n q^{-n},$$
$$B(q) = b_1 q^{-1} + \cdots + b_n q^{-n},$$

we can write

$$\widehat{y}(k|k-1) = \phi(k)^{\mathrm{T}}\theta,$$

with

$$\theta = \begin{bmatrix} -a_1 & -a_2 & \cdots & -a_n | b_1 & b_2 & \cdots & b_n \end{bmatrix}^{\mathrm{T}},$$
$$\phi(k) = \begin{bmatrix} y(k-1) & \cdots & y(k-n) | u(k-1) & \cdots & u(k-n) \end{bmatrix}^{\mathrm{T}}.$$

Thus, the prediction-error cost function is given by

$$J_N(\theta) = \frac{1}{N} \sum_{k=0}^{N-1} (y(k) - \phi(k)^{\mathrm{T}}\theta)^2.$$

Example 7.7 on page 235 shows that this form of the cost function leads to a linear least-squares problem.

8.3.2 The Box–Jenkins and output-error model structures

The Box–Jenkins (BJ) (Box and Jenkins, 1970) model structure parameterizes the input–output relationship (8.1) on page 256 as

$$y(k) = \frac{b_1 q^{-1} + \cdots + b_n q^{-n}}{1 + a_1 q^{-1} + \cdots + a_n q^{-n}} u(k) + \frac{1 + c_1 q^{-1} + \cdots + c_n q^{-n}}{1 + d_1 q^{-1} + \cdots + d_n q^{-n}} e(k). \tag{8.25}$$

On introducing the polynomials

$$A(q) = 1 + a_1 q^{-1} + \cdots + a_n q^{-n}, \tag{8.26}$$
$$B(q) = b_1 q^{-1} + \cdots + b_n q^{-n}, \tag{8.27}$$
$$C(q) = 1 + c_1 q^{-1} + \cdots + c_n q^{-n}, \tag{8.28}$$
$$D(q) = 1 + d_1 q^{-1} + \cdots + d_n q^{-n}, \tag{8.29}$$

the BJ model can be denoted by

$$y(k) = \frac{B(q)}{A(q)} u(k) + \frac{C(q)}{D(q)} e(k). \tag{8.30}$$

A similar result to that in Lemma 8.3, but now for the BJ model, is given next.

Lemma 8.5 *There is a one-to-one correspondence between the BJ model given by Equation (8.25) and the following parameterization of a SISO state-space system in innovation form:*

$$
x(k+1) = \left[
\begin{array}{ccccc|ccccc}
-a_1 & 1 & 0 & \cdots & 0 & 0 & 0 & 0 & \cdots & 0 \\
a_2 & 0 & 1 & \cdots & 0 & 0 & 0 & 0 & \cdots & 0 \\
\vdots & & & \ddots & \vdots & \vdots & & & \ddots & \vdots \\
-a_{n-1} & 0 & 0 & \cdots & 1 & 0 & 0 & 0 & \cdots & 0 \\
-a_n & 0 & 0 & \cdots & 0 & 0 & 0 & 0 & \cdots & 0 \\
\hline
0 & 0 & 0 & \cdots & 0 & -d_1 & 1 & 0 & \cdots & 0 \\
0 & 0 & 0 & \cdots & 0 & -d_2 & 0 & 1 & \cdots & 0 \\
\vdots & & & \ddots & \vdots & \vdots & & & \ddots & \vdots \\
0 & 0 & 0 & \cdots & 0 & -d_{n-1} & 0 & 0 & \cdots & 1 \\
0 & 0 & 0 & \cdots & 0 & -d_n & 0 & 0 & \cdots & 0
\end{array}
\right] x(k)
$$

$$
+ \left[
\begin{array}{c}
b_1 \\
b_2 \\
\vdots \\
b_{n-1} \\
b_n \\
\hline
0 \\
0 \\
\vdots \\
0 \\
0
\end{array}
\right] u(k) + \left[
\begin{array}{c}
0 \\
0 \\
\vdots \\
0 \\
0 \\
\hline
c_1 - d_1 \\
c_2 - d_2 \\
\vdots \\
c_{n-1} - d_{n-1} \\
c_n - d_n
\end{array}
\right] e(k), \tag{8.31}
$$

$$
y(k) = \begin{bmatrix} 1 & 0 & 0 & \cdots & 0 & | & 1 & 0 & 0 & \cdots & 0 \end{bmatrix} x(k) + e(k). \tag{8.32}
$$

Proof The proof is similar to the one given for Lemma 8.3 on page 266. □

On embedding the specific BJ model into the general state-space model considered in Theorem 5.4, we draw the conclusion that the asymptotic stability of the one-step-ahead predictor requires the roots of the deterministic polynomial $A(q)$ to be within the unit circle. This condition is necessary in order to make the pair $(A, Q^{1/2})$ of the BJ model (8.32) corresponding to the state-space model in Theorem 5.4 stabilizable.

The following lemma shows that the one-step-ahead predictor of the BJ model equals

$$\hat{y}(k|k-1) = \frac{D(q)}{C(q)}\frac{B(q)}{A(q)}u(k) + \frac{C(q)-D(q)}{C(q)}y(k).$$

Lemma 8.6 *The one-step-ahead predictor for the BJ model (8.25) is given by*

$$\hat{y}(k|k-1) = \frac{D(q)}{C(q)}\frac{B(q)}{A(q)}u(k) + \frac{C(q)-D(q)}{C(q)}y(k), \qquad (8.33)$$

where the polynomials $A(q), B(q), C(q),$ and $D(q)$ are given by Equations (8.26)–(8.29).

Proof Making use of the state-space parameterization of the BJ model given by Equations (8.30) and (8.31) and the definition $k_i = c_i - d_i$, the one-step-ahead prediction based on Equations (8.10)–(8.11) on page 260 equals

$$\hat{x}(k+1|k) = \begin{bmatrix} -a_1 & 1 & 0 & \cdots & 0 & 0 & 0 & 0 & \cdots & 0 \\ -a_2 & 0 & 1 & \cdots & 0 & 0 & 0 & 0 & \cdots & 0 \\ \vdots & & & \ddots & \vdots & \vdots & & & \ddots & \vdots \\ -a_{n-1} & 0 & 0 & \cdots & 1 & 0 & 0 & 0 & \cdots & 0 \\ -a_n & 0 & 0 & \cdots & 0 & 0 & 0 & 0 & \cdots & 0 \\ -k_1 & 0 & 0 & \cdots & 0 & -d_1-k_1 & 1 & 0 & \cdots & 0 \\ -k_2 & 0 & 0 & \cdots & 0 & -d_2-k_2 & 0 & 1 & \cdots & 0 \\ \vdots & & & \ddots & \vdots & \vdots & & & \ddots & \vdots \\ -k_{n-1} & 0 & 0 & \cdots & 0 & -d_{n-1}-k_{n-1} & 0 & 0 & \cdots & 1 \\ -k_n & 0 & 0 & \cdots & 0 & -d_n-k_n & 0 & 0 & \cdots & 0 \end{bmatrix}$$

$$\times \hat{x}(k|k-1) + \begin{bmatrix} b_1 \\ b_2 \\ \vdots \\ b_{n-1} \\ b_n \\ 0 \\ 0 \\ \vdots \\ 0 \\ 0 \end{bmatrix} u(k) + \begin{bmatrix} 0 \\ 0 \\ \vdots \\ 0 \\ 0 \\ k_1 \\ k_2 \\ \vdots \\ k_{n-1} \\ k_n \end{bmatrix} y(k),$$

$$\hat{y}(k|k-1) = \begin{bmatrix} 1 & 0 & 0 & \cdots & 0 & 1 & 0 & 0 & \cdots & 0 \end{bmatrix}\hat{x}(k|k-1).$$

This system is denoted briefly by

$$\widehat{x}(k+1|k) = \left[\begin{array}{c|c} A_{11} & 0 \\ \hline A_{21} & A_{22} \end{array}\right] \widehat{x}(k|k-1) + \left[\begin{array}{c} B \\ \hline 0 \end{array}\right] u(k) + \left[\begin{array}{c} 0 \\ \hline K \end{array}\right] y(k),$$

$$\widehat{y}(k|k-1) = \left[\begin{array}{c|c} C_1 & C_2 \end{array}\right] \widehat{x}(k|k-1).$$

Since $A_{21} = -KC_1$, we can write the one-step-ahead prediction of the output as

$$\widehat{y}(k|k-1) = \left[\begin{array}{c|c} C_1 & C_2 \end{array}\right] \left(qI - \left[\begin{array}{c|c} A_{11} & 0 \\ \hline -KC_1 & A_{22} \end{array}\right] \right)^{-1}$$

$$\times \left(\left[\begin{array}{c} B \\ \hline 0 \end{array}\right] u(k) + \left[\begin{array}{c} 0 \\ \hline K \end{array}\right] y(k) \right)$$

$$= \left[\begin{array}{c|c} C_1 & C_2 \end{array}\right] \left[\begin{array}{c|c} qI - A_{11} & 0 \\ \hline KC_1 & qI - A_{22} \end{array}\right]^{-1}$$

$$\times \left(\left[\begin{array}{c} B \\ \hline 0 \end{array}\right] u(k) + \left[\begin{array}{c} 0 \\ \hline K \end{array}\right] y(k) \right)$$

$$= \Big(C_1 (qI - A_{11})^{-1} B$$

$$- C_2 (qI - A_{22})^{-1} K C_1 (qI - A_{11})^{-1} B \Big) u(k)$$

$$+ C_2 (qI - A_{22})^{-1} K y(k)$$

$$= \Big(I - C_2 (qI - A_{22})^{-1} K \Big) \Big(C_1 (qI - A_{11})^{-1} B \Big) u(k)$$

$$+ C_2 (qI - A_{22})^{-1} K y(k). \tag{8.34}$$

Since

$$A_{22} + KC_2 = \left[\begin{array}{ccccc} -d_1 & 1 & 0 & \cdots & 0 \\ -d_2 & 0 & 1 & \cdots & 0 \\ \vdots & & & \ddots & \vdots \\ -d_{n-1} & 0 & 0 & \cdots & 1 \\ -d_n & 0 & 0 & \cdots & 0 \end{array}\right],$$

and $k_i = c_i - d_i$, it follows from Lemma 8.3 on page 266 and Exercise 8.4 on page 289 that

$$I - C_2 (qI - A_{22})^{-1} K = \left[\frac{C(q)}{D(q)}\right]^{-1}.$$

Therefore Equation (8.34) can be written in terms of the defined polynomials as Equation (8.33). $\qquad\square$

On putting the parameters c_i and d_i for $i = 1, 2, \ldots, n$ into the BJ model structure, we obtain a model and predictor that fit within the output-error model set discussed in Chapter 7. The resulting specific transfer-function parameterization has classically been referred to as the output-error (OE) model. In polynomial form, it reads as

$$y(k) = \frac{B(q)}{A(q)} u(k) + e(k),$$

and the associated predictor is given by

$$\widehat{y}(k|k-1) = \frac{B(q)}{A(q)} u(k). \tag{8.35}$$

Thus, if the OE model is stable, then also its predictor is stable.

8.4 Qualitative analysis of the model bias for SISO systems

The asymptotic variance analyzed in Sections 7.6 and 8.2.4 can be used as an indication of the accuracy of the estimated parameters if the system that generated the input–output data set belongs to the model set $\mathcal{M}(\theta)$. The latter hypothesis generally does not hold. Examples are when the underlying system has a very large state dimension, whereas for designing a controller one is interested in a low-dimensionality model.

Therefore, in addition to the variance, also the bias in the estimated parameters needs to be considered. In this section we will analyze the bias for some specific SISO systems. We first introduce some notation. Let $\theta_\star$ be the minimizing parameter vector of the cost function $J_N(\theta)$ for $N \to \infty$,

$$\theta_\star = \arg\min \lim_{N \to \infty} J_N(\theta) = \arg\min \overline{J}(\theta),$$

and let the system by which the input–output data were generated be described as

$$y(k) = \frac{B_0(q)}{A_0(q)} u(k) + v(k)$$
$$= \frac{b_1^0 q^{-1} + b_2^0 q^{-2} + \cdots + b_n^0 q^{-n}}{1 + a_1^0 q^{-1} + + a_2^0 q^{-2} + \cdots + a_n^0 q^{-n}} u(k) + v(k), \tag{8.36}$$

with n the order of the system and with $v(k)$ a stochastic perturbation that is independent from $u(k)$. Under these notions the *bias* is the difference between comparable quantities derived from the estimated model and from the true system that persists on taking the limit for $N \to \infty$. One such comparable quantity is the transfer function, which can, for example, be presented as a Bode plot.

To quantify the variance in the estimate $\widehat{\theta}_N$ given by

$$\widehat{\theta}_N = \arg\min J_N(\theta),$$

we should then analyze

$$E\left[[\widehat{\theta}_N - \theta_\star][\widehat{\theta}_N - \theta_\star]^{\mathrm{T}}\right],$$

instead of $E[[\widehat{\theta}_N - \theta_0][\widehat{\theta}_N - \theta_0]^{\mathrm{T}}]$ as was done in Section 7.6.

The bias of the estimated model is analyzed under the assumption that the time sequences are ergodic (see Section 4.3.4). In that case the following limit holds:

$$\lim_{N\to\infty} \frac{1}{N} \sum_{k=0}^{N-1} \left(y(k) - \widehat{y}(k|k-1)\right)^2 = E\left[\left(y(k) - \widehat{y}(k|k-1)\right)^2\right].$$

When the prediction of the output depends on the parameter vector θ, the above equation can be written

$$\lim_{N\to\infty} \frac{1}{N} \sum_{k=0}^{N-1} \left(y(k) - \widehat{y}(k|k-1,\theta)\right)^2 = \overline{J}(\theta), \qquad (8.37)$$

establishing the link with the cost function $\overline{J}(\theta)$. This cost function is now analyzed for the ARMAX and BJ model structures that were introduced in the previous section.

Lemma 8.7 *(Ljung, 1999) Let the LTI system that generates the output $y(k)$ for a given input sequence $u(k)$, $k = 0, 1, 2, \ldots, N-1$, with spectrum $\Phi^u(\omega)$ be denoted by*

$$y(k) = G_0(q)u(k) + v(k),$$

where $v(k)$ is a stochastic perturbation independent from $u(k)$ with spectrum $\Phi^v(\omega)$, and let the time sequences $v(k)$, $u(k)$, and $y(k)$ be ergodic and let the parameters a_i, b_i, and c_i of an ARMAX model be stored in the parameter vector θ, then the parameter vector $\theta_\star$ minimizing the cost function

$$\overline{J}(\theta) = \lim_{N\to\infty} \frac{1}{N} \sum_{k=0}^{N-1} \left(y(k) - \widehat{y}(k|k-1,\theta)\right)^2$$

satisfies

$$
\theta_\star = \arg\min \frac{1}{2\pi} \int_{-\pi}^{\pi} \left| G_0(e^{j\omega}) - \frac{B(e^{j\omega},\theta)}{A(e^{j\omega},\theta)} \right|^2 \left| \frac{A(e^{j\omega},\theta)}{C(e^{j\omega},\theta)} \right|^2 \Phi^u(\omega)
$$

$$
+ \left| \frac{A(e^{j\omega},\theta)}{C(e^{j\omega},\theta)} \right|^2 \Phi^v(\omega) d\omega. \tag{8.38}
$$

Proof The one-step-ahead predictor related to the ARMAX model structure is given by Equation (8.23) on page 269. Hence, the one-step-ahead prediction error $\epsilon(k|k-1) = y(k) - \hat{y}(k|k-1)$ is given by

$$
\epsilon(k|k-1) = \frac{A(q,\theta)}{C(q,\theta)} y(k) - \frac{B(q,\theta)}{C(q,\theta)} u(k).
$$

To express $\epsilon(k|k-1)$ as the sum of two statistically independent time sequences, simplifying the calculation of the spectrum of $\epsilon(k|k-1)$, we substitute into the above expression the model of the system that generated the sequence $y(k)$. This yields

$$
\epsilon(k|k-1) = \frac{A(q,\theta)}{C(q,\theta)} \left(G_0(q) - \frac{B(q,\theta)}{A(q,\theta)} \right) u(k) + \frac{A(q,\theta)}{C(q,\theta)} v(k).
$$

By virtue of the ergodic assumption,

$$
\overline{J}(\theta) = E[\epsilon(k|k-1)^2].
$$

Using Parseval's identity (4.10) on page 106 (assuming a sample time $T = 1$), this can be written as

$$
E[\epsilon(k|k-1)^2] = \frac{1}{2\pi} \int_{-\pi}^{\pi} \Phi^\epsilon(\omega) d\omega. \tag{8.39}
$$

An expression for $\Phi^\epsilon(\omega)$ can be derived by using Lemma 4.3 on page 106 and exploiting the independence between $u(k)$ and $v(k)$:

$$
\Phi^\epsilon(\omega) = \left| G_0(e^{j\omega}) - \frac{B(e^{j\omega},\theta)}{A(e^{j\omega},\theta)} \right|^2 \left| \frac{A(e^{j\omega},\theta)}{C(e^{j\omega},\theta)} \right|^2 \Phi^u(\omega) + \left| \frac{A(e^{j\omega},\theta)}{C(e^{j\omega},\theta)} \right|^2 \Phi^v(\omega).
$$

Substitution into (8.39) results in Equation (8.38). □

Since the ARX model structure is a special case of the ARMAX model structure, we can, with a redefinition of the parameter vector θ, immediately derive the expression for the parameter vector $\theta_\star$ minimizing $\overline{J}(\theta)$

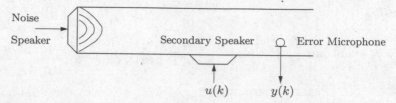

Fig. 8.2. A schematic representation of an acoustical duct to cancel out an unwanted sound field at the position of the microphone using the speaker driven by $u(k)$.

in Equation (8.37) as

$$\theta_\star = \arg\min \frac{1}{2\pi} \int_{-\pi}^{\pi} \left| G_0(e^{j\omega}) - \frac{B(e^{j\omega}, \theta)}{A(e^{j\omega}, \theta)} \right|^2 \left| A(e^{j\omega}, \theta) \right|^2 \Phi^u(\omega)$$
$$+ \left| A(e^{j\omega}, \theta) \right|^2 \Phi^v(\omega) d\omega. \tag{8.40}$$

The use of Lemma 8.7 in qualitatively analyzing the bias in the estimate obtained with the ARX model structure is highlighted in the following example.

Example 8.3 (Qualitative analysis of the bias in estimating low-order ARX models) The system to be modeled is an acoustical duct, depicted in Figure 8.2, which is used for active-noise-control experiments. At the left-hand end of the duct is mounted a loudspeaker that produces undesired noise. The goal is to drive the secondary loudspeaker mounted just before the other end of the duct such that at the far-right end of the duct a region of silence is created. Most control algorithms used in active noise control need a model of the transfer from the secondary loudspeaker to the error microphone.

A high-order approximation of the acoustical relationship between the speaker activated with the signal u, and the microphone producing the measurements y, is given by the following transfer function:

$$G(q) = \frac{\displaystyle\sum_{j=0}^{19} b_j q^{-j}}{\displaystyle\sum_{j=0}^{19} a_j q^{-j}},$$

with a_j and b_j listed in Table 8.1.

Table 8.1. *Coefficients of the transfer function between u and y in the model of the acoustical duct*

θ	Value	θ	Value
a_0	1	b_0	0
a_1	$-1.8937219532483\text{E } 0$	b_1	$-5.6534330123106\text{E} -6$
a_2	$9.2020408176247\text{E} -1$	b_2	$5.6870704280702\text{E} -6$
a_3	$8.4317527635808\text{E} -13$	b_3	$7.7870811926239\text{E} -3$
a_4	$-6.9870644340972\text{E} -13$	b_4	$1.3389477125431\text{E} -3$
a_5	$3.2703011891141\text{E} -13$	b_5	$-9.1260667240191\text{E} -3$
a_6	$-2.8053825784320\text{E} -14$	b_6	$1.4435759589218\text{E} -8$
a_7	$-4.8518619047975\text{E} -13$	b_7	$-1.2021568096247\text{E} -8$
a_8	$9.0515016323085\text{E} -13$	b_8	$-2.2746529807395\text{E} -9$
a_9	$-8.9573340462955\text{E} -13$	b_9	$6.3067990166664\text{E} -9$
a_{10}	$6.2104932381850\text{E} -13$	b_{10}	$9.1305924779895\text{E} -10$
a_{11}	$-4.0655443037130\text{E} -13$	b_{11}	$-7.5200613526843\text{E} -9$
a_{12}	$3.8448359402553\text{E} -13$	b_{12}	$1.9549739577695\text{E} -9$
a_{13}	$-4.9321540807220\text{E} -13$	b_{13}	$1.3891832078608\text{E} -8$
a_{14}	$5.3571245452629\text{E} -13$	b_{14}	$-1.6372496840947\text{E} -8$
a_{15}	$-6.7043859898372\text{E} -13$	b_{15}	$9.0003511972213\text{E} -3$
a_{16}	$6.5050860651120\text{E} -13$	b_{16}	$-1.9333235975678\text{E} -3$
a_{17}	$6.6499999999978\text{E} -1$	b_{17}	$-7.0669966879457\text{E} -3$
a_{18}	$-1.2593250989101\text{E } 0$	b_{18}	$-3.7850561971775\text{E} -6$
a_{19}	$6.1193571437226\text{E} -1$	b_{19}	$3.7590122810601\text{E} -6$

In the above values, E 0 means $\times 10^0$, E -6 means $\times 10^{-6}$, etc.

The magnitude of the Bode plot of the transfer function $G(e^{j\omega})$ is depicted by the thick line in the top part of Figure 8.3. The input sequence $u(k)$ is taken to be a zero-mean unit-variance white-noise sequence of length 10 000. With this input sequence, an output sequence $y(k)$ is generated using the high-order transfer function $G(q)$. These input and output sequences are then used to estimate a sixth-order ARX model via the use of a QR factorization to solve the related linear least-squares problem (see Example 7.7 on page 235). The estimated transfer function $\widehat{G}(e^{j\omega})$ is depicted by the thin line in the top part of Figure 8.3. We observe that, according to Equation (8.40) with $\Phi^v(\omega) = 0$, the estimated low-order model accurately matches the high-order model for those frequency values for which $|A(e^{j\omega})|$ is large. From the graph of $|A(e^{j\omega})|$ in the lower part of Figure 8.3, we observe that this holds in the high-frequency region above 100 Hz.

The following lemma gives a result similar to Lemma 8.7, but for the BJ model.

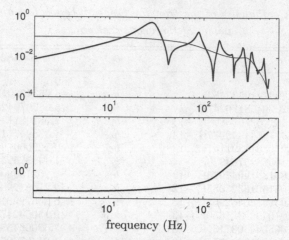

Fig. 8.3. Top: a magnitude plot of the transfer function between $u(k)$ and $y(k)$ of the true 19th-order situation (thick line) and the estimated sixth-order ARX model (thin line). Bottom: the weighting function $|A(e^{j\omega})|$ in the criterion (8.40) for the ARX estimation.

Lemma 8.8 *(Ljung, 1999) Let the LTI system that generates the output $y(k)$ for a given input sequence $u(k)$, $k = 0, 1, 2, \ldots, N-1$, with spectrum $\Phi^u(\omega)$ be denoted by*

$$y(k) = G_0(q)u(k) + v(k),$$

where $v(k)$ is a stochastic perturbation independent from $u(k)$ with spectrum $\Phi^v(\omega)$, let the time sequences $v(k)$, $u(k)$, and $y(k)$ be ergodic, and let the parameters a_i, b_i, c_i, and d_i of a BJ model be stored in the parameter vector θ, then the parameter vector $\theta_\star$ minimizing the cost function

$$\overline{J}(\theta) = \lim_{N\to\infty} \frac{1}{N} \sum_{k=0}^{N-1} \left(y(k) - \widehat{y}(k|k-1, \theta) \right)^2$$

satisfies

$$\theta_\star = \arg\min \frac{1}{2\pi} \int_{-\pi}^{\pi} \left| G_0(e^{j\omega}) - \frac{B(e^{j\omega}, \theta)}{A(e^{j\omega}, \theta)} \right|^2 \left| \frac{D(e^{j\omega}, \theta)}{C(e^{j\omega}, \theta)} \right|^2 \Phi^u(\omega)$$

$$+ \left| \frac{D(e^{j\omega}, \theta)}{C(e^{j\omega}, \theta)} \right|^2 \Phi^v(\omega) d\omega. \tag{8.41}$$

Proof The proof is similar to the proof of Lemma 8.7 on page 276 using the predictor related to the BJ model structure as given by Equation (8.33) on page 273.   □

Since the OE model structure is a special case of the BJ model structure, we can with a redefinition of the parameter vector θ immediately derive an expression for the parameter vector $\theta_\star$ of an OE model minimizing the cost function $\overline{J}(\theta)$:

$$\theta_\star = \arg\min \frac{1}{2\pi} \int_{-\pi}^{\pi} \left| G_0(e^{j\omega}) - \frac{B(e^{j\omega}, \theta)}{A(e^{j\omega}, \theta)} \right|^2 \Phi^u(\omega) + \Phi^v(\omega) d\omega. \quad (8.42)$$

The use of Lemma 1.8 in qualitatively analyzing the bias in the estimate obtained with the OE model structure is highlighted with a continuation of Example 8.3.

Example 8.4 (Qualitative analysis of the bias in estimating low-order OE models) Making use of the same acoustical model of the duct as analyzed in Example 8.3, we now attempt to estimate a sixth-order output-error model. By generating several realizations of the input and output data sequences with the same statistical properties as outlined in Example 8.3 on page 278, a series of sixth-order output-error models was estimated using the tools from the Matlab System Identification toolbox (MathWorks, 2000a). Because of the nonquadratic nature of the cost function to be optimized by the output-error method, the numerical search discussed in Section 8.2.3 "got stuck" in a local minimum a number of times. The best result obtained out of 30 trials is presented below.

A Bode plot of the transfer function $G(e^{j\omega})$ is depicted by the thick line in Figure 8.4. The transfer function $\widehat{G}(e^{j\omega})$ of one estimated sixth-order OE model is also depicted in Figure 8.4. Clearly the most dominant peak around 25 Hz is completely captured. According to the theoretical qualitative analysis summarized by Equation (8.42) for $\Phi^v(\omega) = 0$, it would be expected that the second most dominant peak around 90 Hz would be matched. However, this conclusion assumes that the *global* minimum of the cost function $\overline{J}(\theta)$ optimized by the output-error method has been found. The fact that the peak around 200 Hz is matched subsequently instead of the one around 90 Hz indicates that the global optimum still is not being found. In Chapter 10 it will be shown that the convergence of the prediction-error method can be greatly improved when it makes use

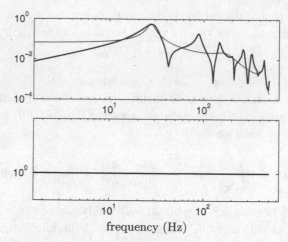

Fig. 8.4. Top: a magnitude plot of the transfer function between $u(k)$ and $y(k)$ of the true 19th-order situation (thick line) and the estimated sixth-order OE model (thin line). Bottom: the weighting function of the error on the transfer function estimate in the criterion (8.42).

of the initial estimates provided by the subspace-identification methods to be discussed in Chapter 9.

The BJ model structure allows us to estimate the parameters a_i and b_i for $i = 1, 2, \ldots, n$ *unbiasedly*, irrespective of the values of the parameters c_i, d_i, for $i = 1, 2, \ldots, n$, provided that they generate a stable predictor, and provided that n corresponds to the true order of the data-generating system. Let the data-generating system be represented as

$$y(k) = \frac{B_0(q)}{A_0(q)} u(k) + v(k),$$

with $v(k)$ a stochastic zero-mean perturbation that is independent from $u(k)$. The BJ model structure has the ability to estimate the deterministic part,

$$\frac{B(q)}{A(q)} u(k),$$

correctly even if the noise part,

$$\frac{C(q)}{D(q)} e(k),$$

does not correspond to that in the underlying signal-generating system. To see this, let θ_{ab} denote the vector containing the quantities a_i, b_i, $i = 1, 2, \ldots, n$, and let θ_{cd} denote the vector containing the quantities c_i, d_i,

$i = 1, 2, \ldots, n$. Consider the noise part of the BJ model to be fixed at some value $\overline{\theta}_{cd}$, then we can denote the criterion $J_N(\theta)$ as

$$J_N(\theta_{ab}, \overline{\theta}_{cd}) = \frac{1}{N} \sum_{k=0}^{N-1} \left(y(k) - \widehat{y}(k|k-1) \right)^2$$

$$= \frac{1}{N} \sum_{k=0}^{N-1} \left[\frac{D(q, \overline{\theta}_{cd})}{C(q, \overline{\theta}_{cd})} \left(\frac{B_0(q)}{A_0(q)} u(k) + v(k) - \frac{B(q, \theta_{ab})}{A(q, \theta_{ab})} u(k) \right) \right]^2.$$

When we take the limit $N \to \infty$ and assume ergodicity of the time sequences, then, by Parseval's identity (4.10) on page 106, the prediction-error methods will perform the following minimization:

$$\min_{\theta_{ab}} \frac{1}{2\pi} \int_{-\pi}^{\pi} \left| \frac{D(e^{j\omega}, \overline{\theta}_{cd})}{C(e^{j\omega}, \overline{\theta}_{cd})} \right|^2 \underbrace{\left| \frac{B_0(e^{j\omega})}{A_0(e^{j\omega})} - \frac{B(e^{j\omega}, \theta_{ab})}{A(e^{j\omega}, \theta_{ab})} \right|^2} \Phi^u(\omega)$$

$$+ \left| \frac{D(e^{j\omega}, \overline{\theta}_{cd})}{C(e^{j\omega}, \overline{\theta}_{cd})} \right|^2 \Phi^v(\omega) d\omega.$$

When n is correctly specified, or, more generally, when the orders of the polynomials $A_0(q)$ and $B_0(q)$ correspond exactly to the orders of the polynomials $A(q)$ and $B(q)$, respectively, the minimum that corresponds to the underbraced term is zero. Therefore, if the global optimum of the above criterion $\overline{J}(\theta_{ab})$ is found, the true values of the polynomials $A_0(q)$ and $B_0(q)$ are estimated.

8.5 Estimation problems in closed-loop systems

This section briefly highlights some of the complications that arise on using the prediction-error method with input and output samples recorded during a closed-loop experiment. We consider the closed-loop configuration of an LTI system P and an LTI controller C as depicted in Figure 8.5. In general, system identification is much more difficult in closed-loop identification experiments. This will be illustrated by means of a few examples to highlight that, when identifying innovation models, it is necessary to parameterize *both* the deterministic and the stochastic part of the model *exactly equal* to the corresponding parts of the signal-generating system. The first example assumes only a correct parameterization of the deterministic part, whereas in the second example both the stochastic and the deterministic part are correctly parameterized.

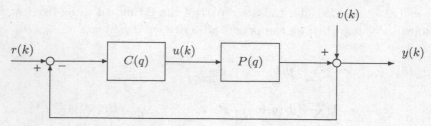

Fig. 8.5. A block scheme of an LTI system P in a closed-loop configuration with a controller C.

Example 8.5 (Biased estimation with closed-loop data) Consider the feedback configuration in Figure 8.5 driven by the external reference signal $r(k)$, with the system P given as

$$y(k) = b_1^0 u(k-1) + b_2^0 u(k-2) + v(k), \qquad (8.43)$$

where $v(k)$ is a zero-mean stochastic sequence that is independent from the external reference $r(k)$. The controller C is a simple proportional controller (Dorf and Bishop, 1998), of the form

$$u(k) = K\Big(r(k) - y(k)\Big). \qquad (8.44)$$

If we were to use an OE model structure with a correctly parameterized deterministic part corresponding to that of the system P, the one-step-ahead prediction error would be

$$\epsilon(k|k-1) = y(k) - \begin{bmatrix} u(k-1) & u(k-2) \end{bmatrix}\begin{bmatrix} b_1 \\ b_2 \end{bmatrix},$$

and with a prediction-error method we would solve the following least-squares problem:

$$\min_{b_1,b_2} \frac{1}{N} \sum_{k=0}^{N-1} \left(y(k) - \begin{bmatrix} u(k-1) & u(k-2) \end{bmatrix}\begin{bmatrix} b_1 \\ b_2 \end{bmatrix} \right)^2.$$

If we substitute for $y(k)$ the expression given in Equation (8.43), this problem can be written as

$$\min_{\beta} \frac{1}{N} \sum_{k=0}^{N-1} \left(\begin{bmatrix} u(k-1) & u(k-2) \end{bmatrix}\beta + v(k) \right)^2,$$

subject to

$$\beta = \begin{bmatrix} b_1^0 - b_1 \\ b_2^0 - b_2 \end{bmatrix}.$$

We assume the recorded time sequences to be ergodic. If the above least-squares problem has a unique solution in the limit of $N \to \infty$, this solution is zero ($\beta = 0$), *provided that* the following conditions are satisfied:

$$E[u(k-1)v(k)] = 0, \qquad E[u(k-2)v(k)] = 0. \qquad (8.45)$$

However, substituting Equation (8.43) into (8.44) yields

$$u(k) = \frac{K}{1 + Kb_1^0 q^{-1} + Kb_2^0 q^{-2}} r(k) - \frac{K}{1 + Kb_1^0 q^{-1} + Kb_2^0 q^{-2}} v(k),$$

which clearly shows that, for $K \neq 0$, the input $u(k)$ is not independent from the noise $v(k)$. For $K \neq 0$, the conditions (8.45) are satisfied only if $v(k)$ is a white-noise sequence. This corresponds to the correct parameterization of the stochastic part of the output-error model. If $v(k)$ were colored noise, biased estimates would result. This is in contrast to the open-loop case, for which the assumption that $u(k)$ and $v(k)$ are independent is sufficient to obtain unbiased estimates.

The final example in this chapter illustrates the necessity that the model set $\mathcal{M}(\theta)$ (Equation (7.15) on page 217) encompasses *both* the deterministic and the stochastic part of the signal-generating system.

Example 8.6 (Unbiased estimation with closed-loop data) Consider the feedback configuration in Figure 8.5 driven by the external reference signal $r(k)$, with the system P given as

$$y(k) = a^0 y(k-1) + b^0 u(k-1) + e(k), \qquad (8.46)$$

where $e(k)$ is a zero-mean white-noise sequence. The controller C has the following dynamic form:

$$u(k) = fu(k-1) + g\Big(r(k) - y(k)\Big), \qquad (8.47)$$

with $f, g \in \mathbb{R}$. If we were to use an ARX model structure with correctly parameterized deterministic and stochastic parts for the system P, the one-step-ahead prediction error would be

$$\epsilon(k|k-1) = y(k) - \begin{bmatrix} y(k-1) & u(k-1) \end{bmatrix} \begin{bmatrix} a \\ b \end{bmatrix}.$$

Following Example 8.5, the conditions for consistency become

$$E[y(k-1)e(k)] = 0, \qquad E[u(k-1)e(k)] = 0. \qquad (8.48)$$

These conditions hold since

$$u(k) = \frac{g(1 - a^0 q^{-1})}{1 - (f + a^0 - g b^0) q^{-1} + f a^0 q^{-2}} r(k)$$

$$- \frac{g}{1 - (f + a^0 - g b^0) q^{-1} + f a^0 q^{-2}} e(k),$$

$$y(k) = \frac{g b^0 q^{-1}}{1 - (f + a^0 - g b^0) q^{-1} + f a^0 q^{-2}} r(k)$$

$$+ \frac{1 - f q^{-1}}{1 - (f + a^0 - g b^0) q^{-1} + f a^0 q^{-2}} e(k),$$

and $e(k)$ is a white-noise sequence.

The consistency that is obtained in Example 8.6 with a correctly parameterized ARX model of a system operating in closed-loop mode can be generalized for the class of MIMO innovation model structures (8.6)–(8.7) on page 258 when the signal-generating system belongs to the model set.

Modifications based on the so-called instrumental-variable method, to be treated in Chapter 9, have been developed that can relax the stringent requirement of correctly parameterizing the stochastic part of the system to be identified. An example in a subspace identification context is given by Chou and Verhaegen (1997).

8.6 Summary

In this chapter an introduction to the estimation of unknown parameters with the so-called linear prediction-error method (PEM) is given.

First we considered the parameterization of a state-space model in the innovation form. It has been shown that many different choices can be made here. The output normal form and the tridiagonal parameterizations that were introduced in Chapter 7 can also be used to parameterize innovation-form state-space models. The prediction-error methods are based on minimizing the one-step-ahead prediction error. The parameters in the selected model structure are determined by minimizing the mean squared prediction error. The optimization methods discussed in Section 7.5 can be used for this purpose. For SISO systems, specific parameterizations have been presented. These include the classically known ARX, ARMAX, output-error, and Box–Jenkins models (Ljung, 1999). From the definition of the prediction for the general MIMO state-space innovation model, the predictions for the specific ARX, ARMAX, output-error, and Box–Jenkins models were derived. It has been shown that the selection of the particular model structure has great influence on the nature or difficulty of the numerical optimization problem. For most parameterizations considered in this chapter, the problem is *highly nonlinear*. The numerical, iterative procedures for solving such problems are guaranteed only to find local minima.

The quality of the estimates obtained can be evaluated in various ways. First, it was assumed that the model parameterization defined by the chosen model structure includes the signal-generating system. For this situation the asymptotic covariance of the estimated parameters was given. Second, for the more realistic assumption that the model parameterization does not include the signal-generating system, the bias was studied via Parseval's identity for the special classes of SISO models considered. The latter bias formulas have had and still do have a great impact on understanding and applying system-identification methods. They are also used for analyzing some of the consequences when performing identification experiments in closed-loop mode.

Exercises

8.1   Consider the transfer function

$$\mathcal{M}(z) = \begin{bmatrix} D & 0 \end{bmatrix} + C(zI - (A - KC))^{-1}\begin{bmatrix} B & K \end{bmatrix},$$

with arbitrary system matrices $A \in \mathbb{R}^{n \times n}$, $B \in \mathbb{R}^{n \times m}$, $C \in \mathbb{R}^{\ell \times n}$, $D \in \mathbb{R}^{\ell \times m}$, and $K \in \mathbb{R}^{n \times \ell}$.

(a) Let $a(z)$ be a scalar polynomial of order n given by

$$a(z) = z^n + a_1 z^{n-1} + \cdots + a_n.$$

Let $\mathcal{B}(z)$ and $\mathcal{K}(z)$ be polynomial matrices with polynomial entries of order $n - 1$ given as

$$\mathcal{B}(z) = \begin{bmatrix} b_{11}(z) & \cdots & b_{1m} \\ \vdots & & \vdots \\ b_{\ell 1}(z) & \cdots & b_{\ell m}(z) \end{bmatrix},$$

$$\mathcal{K}(z) = \begin{bmatrix} k_{11}(z) & \cdots & k_{1\ell} \\ \vdots & & \vdots \\ k_{\ell 1}(z) & \cdots & k_{\ell \ell}(z) \end{bmatrix}.$$

Show that the transfer function $\mathcal{M}(z)$ can be expressed as

$$\begin{bmatrix} D & 0 \end{bmatrix} + \frac{\begin{bmatrix} \mathcal{B}(z) & \mathcal{K}(z) \end{bmatrix}}{a(z)}.$$

(b) For the special case $\ell = 1$ show that the observable canonical form (7.13)–(7.14) on page 216 is a surjective parameterization of the transfer function $\mathcal{M}(z)$.

8.2   Consider the one-step-ahead predictor for a second-order ($n = 2$) ARMAX model as given in Lemma 8.4 on page 268. Let $c_i = a_i + k_i$ ($i = 1, 2$). The parameters in the one-step-ahead prediction will be estimated using N measurements of the input $u(k)$ and the output $y(k)$ of the system:

$$y(k) = \frac{q^{-1} + 0.5q^{-2}}{1 - 1.5q^{-1} + 0.7q^{-2}} u(k) + v(k),$$

with $u(k)$ and $v(k)$ zero-mean, statistically independent white-noise sequences of unit variance.

(a) Determine an expression for the matrix $\Phi(c_1, c_2)$ such that the prediction-error criterion $J_N(c_1, c_2, \theta_{bk})$ can be written as

$$J_N(c_1, c_2, \theta_{bk}) = \frac{1}{N} \|Y - \Phi(c_1, c_2)\theta_{bk}\|_2^2,$$

with

$$\theta_{bk} = \begin{bmatrix} \widehat{x}(0)^{\mathrm{T}} & k_1 & k_2 & b_1 & b_2 \end{bmatrix}^{\mathrm{T}},$$
$$Y = \begin{bmatrix} y(0) & y(1) & \cdots & y(N-1) \end{bmatrix}^{\mathrm{T}}.$$

(b) If the coefficient c_2 is fixed to its true value 0.7, derive the condition on c_1 such that the ARMAX predictor is asymptotically stable.

(c) Write a Matlab program that calculates the matrix $\Phi(c_1, c_2)$, and takes as input arguments the vector $c = \begin{bmatrix} c_1 & c_2 \end{bmatrix}^{\mathrm{T}}$, the output sequence Y, and the input sequence stored in the vector $U = \begin{bmatrix} u(1) & u(2) & \cdots & u(N) \end{bmatrix}^{\mathrm{T}}$.

(d) Let δ_S denote the interval on the real axis for which the ARMAX predictor with $c_2 = 0.7$ is asymptotically stable. Plot the prediction-error criterion $J_N(c_1, 0.7, \theta_{bk})$ as a function of $c_1 \in \delta_S$. Does the minimal value of this criterion indicate the correct value of c_1?

8.3 Consider the ARX predictor given by Equation (8.24) on page 270. Using the measurements $u(k)$ and $y(k)$ acquired in the closed-loop configuration with an LTI controller with transfer function $C(e^{j\omega})$ as depicted in Figure 8.5 on page 284, the task is to estimate an ARX model for the unknown plant P. Show that, in the limit of $N \to \infty$, the prediction-error method attempts to find the following estimate:

$$
\theta_\star = \arg\min \frac{1}{2\pi} \int_{-\pi}^{\pi} \left| P(e^{j\omega}) - \frac{B(e^{j\omega}, \theta)}{A(e^{j\omega}, \theta)} \right|^2
$$

$$
\times \left| \frac{A(e^{j\omega}, \theta) C(e^{j\omega})}{1 + P(e^{j\omega}) C(e^{j\omega})} \right|^2 \Phi^r(\omega)
$$

$$
+ \left| \frac{1 + \dfrac{B(e^{j\omega}, \theta)}{A(e^{j\omega}, \theta)} C(e^{j\omega})}{1 + P(e^{j\omega}) C(e^{j\omega})} \right|^2 \left| A(e^{j\omega}, \theta) \right|^2 \Phi^v(\omega) d\omega.
$$

8.4 Let the following state-space model be given:

$$
x(k+1) = Ax(k) + Bu(k),
$$
$$
y(k) = Cx(k) + u(k).
$$

(a) Show that the transfer function describing the transfer from $u(k)$ to $y(k)$ is given as

$$
y(k) = \left(I + C(qI - A)^{-1} B \right) u(k).
$$

(b) Show that the transfer function describing the transfer from $y(k)$ to $u(k)$ is given as

$$
u(k) = \left(I - C \Big(qI - (A - BC) \Big)^{-1} B \right) y(k).
$$

8.5 Consider the OE predictor given by Equation (8.35) on page 275. Using the measurements $u(k)$ and $y(k)$ acquired in the closed-loop configuration with the LTI controller with transfer function $C(e^{j\omega})$ as depicted in Figure 8.5 on page 284, the task is to estimate an OE model for the unknown plant P.

(a) Show that, in the limit of $N \to \infty$, the prediction-error method attempts to find the following estimate:

$$\theta_\star = \arg\min \frac{1}{2\pi} \int_{-\pi}^{\pi} \left| P(e^{j\omega}) - \frac{B(e^{j\omega}, \theta)}{A(e^{j\omega}, \theta)} \right|^2$$

$$\times \left| \frac{C(e^{j\omega})}{1 + P(e^{j\omega})C(e^{j\omega})} \right|^2 \Phi^r(\omega)$$

$$+ \left| \frac{1 + \dfrac{B(e^{j\omega}, \theta)}{A(e^{j\omega}, \theta)} C(e^{j\omega})}{1 + P(e^{j\omega})C(e^{j\omega})} \right|^2 \Phi^v(\omega) d\omega.$$

(b) Show that, for $v(k) = 0$, the model given by

$$\frac{B(e^{j\omega}, \theta)}{A(e^{j\omega}, \theta)}$$

approximates the system $P(e^{j\omega})$ accurately in the so-called *cross-over*-frequency region, that is, the frequency region in which the loop gain $P(e^{j\omega})C(e^{j\omega}) \approx -1$.

8.6   Adapted from Example 8.1 of Ljung (1999). We are given the system described by

$$y(k) = \frac{b_0 q^{-1}}{1 + a_0 q^{-1}} u(k) + \frac{1 + c_0 q^{-1}}{1 + a_0 q^{-1}} e(k),$$

with $u(k)$ and $e(k)$ ergodic, zero-mean and statistically independent white-noise sequences with variances σ_u^2 and σ_e^2, respectively. Using N measurements of the input and the output of this system, we attempt to estimate the two unknown coefficients a and b in a first-order ARX model.

(a) Show that, in the limit of $N \to \infty$,

$$E[y^2(k)] = \frac{b_0^2 \sigma_u^2 + (c_0(c_0 - a_0) - a_0 c_0 + 1)\sigma_e^2}{1 - a_0^2}.$$

(b) Show that, in the limit of $N \to \infty$, the prediction-error criterion $\overline{J}(a, b)$ that is minimized by the ARX method is given as

$$\overline{J}(a, b) = E[y^2(k)](1 + a^2 - 2aa_0) + (b^2 - 2bb_0)\sigma_u^2 + 2ac_0\sigma_e^2.$$

(c) Show that, in the limit of $N \to \infty$, the optimal parameter values for a and b that minimize the above criterion are

$$\widehat{a} = a_0 - \frac{c_0}{E[y^2(k)]} \sigma_e^2,$$

$$\widehat{b} = b_0.$$

(d) Show by explicitly evaluating the criterion values $\overline{J}(\widehat{a}, \widehat{b})$ and $\overline{J}(a_0, b_0)$ that, in the limit of $N \to \infty$, the following relationship holds:

$$\overline{J}(\widehat{a}, \widehat{b}) < \overline{J}(a_0, b_0).$$

9

Subspace model identification

After studying this chapter you will be able to

- derive the data equation that relates block Hankel matrices constructed from input–output data;
- exploit the special structure of the data equation for impulse input signals to identify a state-space model via subspace methods;
- use subspace identification for general input signals;
- use instrumental variables in subspace identification to deal with process and measurement noise;
- derive subspace identification schemes for various noise models;
- use the RQ factorization for a computationally efficient implementation of subspace identification schemes; and
- relate different subspace identification schemes via the solution of a least-squares problem.

9.1 Introduction

The problem of identifying an LTI state-space model from input and output measurements of a dynamic system, which we analyzed in the previous two chapters, is re-addressed in this chapter via a completely different approach. The approach we take is indicated in the literature (Verhaegen, 1994; Viberg, 1995; Van Overschee and De Moor, 1996b; Katayama, 2005) as the class of subspace identification methods. These methods are based on the fact that, by storing the input and output data in structured block Hankel matrices, it is possible to retrieve certain subspaces that are related to the system matrices of the signal-generating state-space model. Examples of such subspaces are the column space of the observability matrix, Equation (3.25) on page 67, and the row space of the state sequence of a Kalman filter. In this chapter we explain how subspace methods can be used to determine the system matrices of a linear time-invariant system up to a similarity transformation. Subspace methods have also been developed for the identification of linear parameter-varying systems (Verdult and Verhaegen, 2002) and certain classes of nonlinear systems. The interested reader should consult Verdult (2002) for an overview.

Unlike with the identification algorithms presented in Chapters 7 and 8, in subspace identification there is no need to parameterize the model. Furthermore, the system model is obtained in a noniterative way via the solution of a number of simple linear-algebra problems. The key linear-algebra steps are an RQ factorization, an SVD, and the solution of a linear least-squares problem. Thus, the problem of performing a nonlinear optimization is circumvented. These properties of subspace identification make it an attractive alternative to the output-error and prediction-error methods presented in the previous two chapters. However, the statistical analysis of the subspace methods is much more complicated than the statistical analysis of the prediction-error methods. This is because subspace identification methods do not explicitly minimize a cost function to obtain the system matrices. Although some results on the statistical analysis of subspace methods have been obtained, it remains a relevant research topic (Peternell *et al.*, 1996; Bauer, 1998; Jansson, 1997; Jansson and Wahlberg, 1998; Bauer and Jansson, 2000).

To explain subspace identification, we need some theory from linear algebra. Therefore, in this chapter we rely on the matrix results reviewed in Chapter 2. In Section 9.2 we describe the basics of subspace identification. In this section we consider only noise-free systems. We first describe subspace identification for impulse input signals and then switch to more general input sequences. Section 9.3 describes subspace identification in the presence of white measurement noise. To deal with more general additive noise disturbances, we introduce the concept of instrumental variables in Section 9.4. The instrumental-variables approach is used in Section 9.5 to deal with colored measurement noise and in Section 9.6

to deal with white process and white measurement noise simultaneously. In the latter section it is also shown that subspace identification for the case of white process and white measurement noise can be written as a least-squares problem. Finally, Section 9.7 illustrates the difficulties that arise when using the subspace methods presented here with data recorded in a closed-loop experiment.

9.2 Subspace model identification for deterministic systems

The subspace identification problem is formulated for deterministic LTI systems, that is, LTI systems that are not disturbed by noise. Let such a system be given by

$$x(k+1) = Ax(k) + Bu(k), \qquad (9.1)$$
$$y(k) = Cx(k) + Du(k), \qquad (9.2)$$

where $x(k) \in \mathbb{R}^n$, $u(k) \in \mathbb{R}^m$, and $y(k) \in \mathbb{R}^\ell$.

Given a finite number of samples of the input signal $u(k)$ and the output signal $y(k)$ of the minimal (reachable and observable) system (9.1)–(9.2), the goal is to determine the system matrices (A, B, C, D) and initial state vector up to a similarity transformation.

An important and critical step prior to the design (and use) of subspace identification algorithms is to find an appropriate relationship between the measured data sequences and the matrices that define the model. This relation will be derived in Section 9.2.1. We proceed by describing subspace identification for an autonomous system (Section 9.2.2) and for the special case when the input is an impulse sequence (Section 9.2.3). Finally, we describe subspace identification for more general input sequences (Section 9.2.4).

9.2.1 The data equation

In Section 3.4.2 we showed that the state of the system (9.1)–(9.2) with initial state $x(0)$ at time instant k is given by

$$x(k) = A^k x(0) + \sum_{i=0}^{k-1} A^{k-i-1} Bu(i). \qquad (9.3)$$

By invoking Equation (9.2), we can specify the following relationship between the input data batch $\{u(k)\}_{k=0}^{s-1}$ and the output data batch $\{y(k)\}_{k=0}^{s-1}$:

$$
\begin{bmatrix} y(0) \\ y(1) \\ y(2) \\ \vdots \\ y(s-1) \end{bmatrix} = \underbrace{\begin{bmatrix} C \\ CA \\ CA^2 \\ \vdots \\ CA^{s-1} \end{bmatrix}}_{\mathcal{O}_s} x(0)
$$

$$
+ \underbrace{\begin{bmatrix} D & 0 & 0 & \cdots & 0 \\ CB & D & 0 & \cdots & 0 \\ CAB & CB & D & & 0 \\ \vdots & & & \ddots & \ddots \\ CA^{s-2}B & CA^{s-3}B & \cdots & CB & D \end{bmatrix}}_{\mathcal{T}_s} \begin{bmatrix} u(0) \\ u(1) \\ u(2) \\ \vdots \\ u(s-1) \end{bmatrix},
$$

$$(9.4)$$

where s is some arbitrary positive integer. To use this relation in subspace identification, it is necessary to take $s > n$, as will be explained below. Henceforth the matrix $\mathcal{O}_s$ will be referred to as the *extended observability matrix*. Equation (9.4) relates vectors, derived from the input and output data sequences and the (unknown) initial condition $x(0)$, to the matrices $\mathcal{O}_s$ and $\mathcal{T}_s$, derived from the system matrices (A, B, C, D). Since the underlying system is time-invariant, we can relate time-shifted versions of the input, state, and output vectors in (9.4) to the same matrices $\mathcal{O}_s$ and $\mathcal{T}_s$. For example, consider a shift over k samples in Equation (9.4):

$$
\begin{bmatrix} y(k) \\ y(k+1) \\ \vdots \\ y(k+s-1) \end{bmatrix} = \mathcal{O}_s x(k) + \mathcal{T}_s \begin{bmatrix} u(k) \\ u(k+1) \\ \vdots \\ u(k+s-1) \end{bmatrix}. \tag{9.5}
$$

Now we can combine the relationships (9.4) and (9.5) for different time shifts, as permitted by the availability of input–output samples, to obtain

$$
\begin{bmatrix}
y(0) & y(1) & \cdots & y(N-1) \\
y(1) & y(2) & \cdots & y(N) \\
\vdots & \vdots & \ddots & \vdots \\
y(s-1) & y(s) & \cdots & y(N+s-2)
\end{bmatrix}
$$

$$
-\mathcal{O}_s X_{0,N} + \mathcal{T}_s
\begin{bmatrix}
u(0) & u(1) & \cdots & u(N-1) \\
u(1) & u(2) & \cdots & u(N) \\
\vdots & \vdots & \ddots & \vdots \\
u(s-1) & u(s) & & u(N+s-2)
\end{bmatrix}, \quad (9.6)
$$

where

$$
X_{i,N} = \begin{bmatrix} x(i) & x(i+1) & \cdots & x(i+N-1) \end{bmatrix},
$$

and in general we have $n < s \ll N$. The above equation is referred to as the *data equation*. The matrices constructed from the input and output data have (vector) entries that are constant along the block anti-diagonals. A matrix with this property is called a block *Hankel matrix*. For ease of notation we define the block Hankel matrix constructed from $y(k)$ as follows:

$$
Y_{i,s,N} =
\begin{bmatrix}
y(i) & y(i+1) & \cdots & y(i+N-1) \\
y(i+1) & y(i+2) & \cdots & y(i+N) \\
\vdots & \vdots & \ddots & \vdots \\
y(i+s-1) & y(i+s) & \cdots & y(i+N+s-2)
\end{bmatrix}.
$$

The first entry of the subscript of $Y_{i,s,N}$ refers to the time index of its top-left entry, the second refers to the number of block-rows, and the third refers to the number of columns. The block Hankel matrix constructed from $u(k)$ is defined in a similar way. These definitions allow us to denote the data equation (9.6) in a compact way:

$$
Y_{0,s,N} = \mathcal{O}_s X_{0,N} + \mathcal{T}_s U_{0,s,N}. \quad (9.7)
$$

The data equation (9.7) relates matrices constructed from the data to matrices constructed from the system matrices. We will explain that this representation allows us to derive information on the system matrices (A, B, C, D) from data matrices, such as $Y_{0,s,N}$ and $U_{0,s,N}$. This idea is first explored in the coming subsection for an autonomous system.

9.2.2 Identification for autonomous systems

The special case of the deterministic subspace identification problem for an autonomous system allows us to explain some of the basic operations in a number of subspace identification schemes. The crucial step is to use the data equation (9.7) to estimate the column space of the extended observability matrix $\mathcal{O}_s$. From this subspace we can then estimate the matrices A and C up to a similarity transformation. As we will see, the subspace identification method for autonomous systems is very similar to the Ho–Kalman realization algorithm (Ho and Kalman, 1966) based on Lemma 3.5 on page 75, which is described in Section 3.4.4.

For an autonomous system the B and D matrices equal zero, and thus the data equation (9.7) reduces to

$$Y_{0,s,N} = \mathcal{O}_s X_{0,N}. \tag{9.8}$$

This equation immediately shows that each column of the matrix $Y_{0,s,N}$ is a linear combination of the columns of the matrix $\mathcal{O}_s$. This means that the column space of the matrix $Y_{0,s,N}$ is contained in that of $\mathcal{O}_s$, that is, $\mathrm{range}(Y_{0,s,N}) \subseteq \mathrm{range}(\mathcal{O}_s)$. It is important to realize that we cannot conclude from Equation (9.8) that the column spaces of $Y_{0,s,N}$ and $\mathcal{O}_s$ are equal, because the linear combinations of the columns of $\mathcal{O}_s$ can be such that the rank of $Y_{0,s,N}$ is lower than the rank of $\mathcal{O}_s$.

However, if $s > n$ and $N \geq s$, it can be shown that under some mild conditions the column spaces of $Y_{0,s,N}$ and $\mathcal{O}_s$ are equal. To see this, observe that the matrix $X_{0,N}$ can be written as

$$X_{0,N} = \begin{bmatrix} x(0) & Ax(0) & A^2x(0) & \cdots & A^{N-1}x(0) \end{bmatrix}.$$

If the pair $(A, x(0))$ is reachable, the matrix $X_{0,N}$ has full row rank n (see Section 3.4.3). Since the system is assumed to be minimal (as stated at the beginning of Section 9.2), we have $\mathrm{rank}(\mathcal{O}_s) = n$. Application of Sylvester's inequality (Lemma 2.1 on page 16) to Equation (9.8) shows that $\mathrm{rank}(Y_{0,s,N}) = n$ and thus $\mathrm{range}(Y_{0,s,N}) = \mathrm{range}(\mathcal{O}_s)$.

Example 9.1 (Geometric interpretation of subspace identification) Consider an autonomous state-space system with $n = 2$. We take $s = 3$ and for a given and known value of $x(0)$ we plot for each of the vectors

$$\begin{bmatrix} y(k) \\ y(k+1) \\ y(k+2) \end{bmatrix}, \quad k = 0, 2, \ldots, N-1,$$

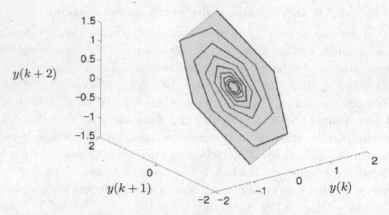

Fig. 9.1. Two state trajectories in a two-dimensional subspace of a three-dimensional ambient space.

a point in a three-dimensional space. Thus, each column of the matrix $Y_{0,s,N}$ corresponds to a point in a three-dimensional space. We connect these points by a line to obtain a curve. Figure 9.1 shows two such curves that correspond to two different values of $x(0)$. From the figure it becomes clear that the curves lie in a two-dimensional subspace (a plane). The plane is characteristic for the matrix pair (A, C) used to generate the data. A different pair (A, C) gives rise to a different plane.

This example illustrates that output data can be used to display information on the state dimension. A first-order autonomous system would have been displayed by state trajectories lying on a line in a three-dimensional plane.

An SVD of the matrix $Y_{0,s,N}$ allows us to determine the column space of $Y_{0,s,N}$ (see Section 2.5). Furthermore, because the column space of $Y_{0,s,N}$ equals that of $\mathcal{O}_s$, it can be used to determine the system matrices A and C up to an unknown similarity transformation T, in a similar way to that in the Ho–Kalman realization algorithm outlined in Section 3.4.4. Denote the SVD of $Y_{0,s,N}$ by

$$Y_{0,s,N} = U_n \Sigma_n V_n^{\mathrm{T}}, \qquad (9.9)$$

with $\Sigma_n \in \mathbb{R}^{n \times n}$ and $\mathrm{rank}(\Sigma_n) = n$, then U_n can be denoted by

$$U_n = \mathcal{O}_s T = \begin{bmatrix} CT \\ CT(T^{-1}AT) \\ \vdots \\ CT\left(T^{-1}AT\right)^{s-1} \end{bmatrix} = \begin{bmatrix} C_T \\ C_T A_T \\ \vdots \\ C_T A_T^{s-1} \end{bmatrix}.$$

Hence, the matrix C_T equals the first ℓ rows of U_n, that is, $C_T = U_n(1:\ell,:)$. The matrix A_T is computed by solving the following overdetermined equation, which due to the condition $s > n$ has a unique solution:

$$U_n(1:(s-1)\ell,:)A_T = U_n(\ell+1:s\ell,:). \tag{9.10}$$

Note that, from the SVD (9.9), a set of system matrices with a different similarity transformation can be obtained if we take for example $U_n\Sigma_n^{1/2}$ as the extended observability matrix $\mathcal{O}_s$ of the similarly equivalent state-space model.

9.2.3 Identification using impulse input sequences

Before treating the subspace identification problem for general inputs, we take a look at the special case of impulse input sequences. The first step is similar to the autonomous case: the column space of the extended observability matrix $\mathcal{O}_s$ is used to estimate the matrices A and C up to a similarity transformation. The second step is to determine the matrix B up to a similarity transformation and to determine the matrix D.

9.2.3.1 Deriving the column space of the observability matrix

We start the outline of the subspace method for impulse inputs for the system (9.1)–(9.2) with a single input, that is, $m = 1$. The impulse input signal equals

$$u(k) = \begin{cases} 1, & \text{for } k = 0, \\ 0, & \text{for } k \neq 0. \end{cases} \tag{9.11}$$

The multi-input case is dealt with in Exercise 9.1 on page 340. The data equation (9.7) for the impulse input takes the form

$$Y_{0,s,N+1} = \mathcal{O}_s X_{0,N+1} + \mathcal{T}_s \begin{bmatrix} 1 & 0 & \cdots & 0 \\ 0 & 0 & \cdots & 0 \\ \vdots & \vdots & \ddots & \vdots \\ 0 & 0 & \cdots & 0 \end{bmatrix}. \tag{9.12}$$

Therefore, we have

$$Y_{1,s,N} = \mathcal{O}_s X_{1,N}. \tag{9.13}$$

If $x(0) = 0$, $s > n$, and $N \geq s$, it can be shown that, under the conditions stipulated in the formulation of the deterministic subspace identification problem, the column spaces of $Y_{1,s,N}$ and $\mathcal{O}_s$ are equal. To see this, observe that, because $x(0) = 0$, the matrix $X_{1,N}$ can be written as

$$X_{1,N} = \begin{bmatrix} B & AB & A^2B & \cdots & A^{N-1}B \end{bmatrix} = \mathcal{C}_N,$$

and, therefore,

$$Y_{1,s,N} = \mathcal{O}_s \mathcal{C}_N. \tag{9.14}$$

Since $s > n$ and $N \geq s$, and since the system is minimal, we have $\mathrm{rank}(\mathcal{O}_s) = \mathrm{rank}(\mathcal{C}_N) = n$. Application of Sylvester's inequality (Lemma 2.1 on page 16) to Equation (9.14) shows that $\mathrm{rank}(Y_{1,s,N}) = n$ and thus $\mathrm{range}(Y_{1,s,N}) = \mathrm{range}(\mathcal{O}_s)$.

9.2.3.2 Computing the system matrices

Since the column space of $Y_{1,s,N}$ equals the column space of $\mathcal{O}_s$, an SVD of the matrix $Y_{1,s,N}$,

$$Y_{1,s,N} = U_n \Sigma_n V_n^{\mathrm{T}}, \tag{9.15}$$

can be used to compute the matrices A_T and C_T, in a similar fashion to in Section 9.2.2.

To determine the matrix B_T, observe that by virtue of the *choice* of U_n as the extended observability matrix of the similarly equivalent state-space model, we have that

$$\begin{aligned}
\Sigma_n V_n^{\mathrm{T}} &= T^{-1} \mathcal{C}_N \\
&= \begin{bmatrix} T^{-1}B & T^{-1}ATT^{-1}B & \cdots & (T^{-1}AT)^{N-1}T^{-1}B \end{bmatrix} \\
&= \begin{bmatrix} B_T & A_T B_T & \cdots & A_T^{N-1} B_T \end{bmatrix}.
\end{aligned}$$

So B_T equals the first column of the matrix $\Sigma_n V_n^{\mathrm{T}}$. The matrix $D_T = D$ equals $y(0)$, as can be seen from Equation (9.2), bearing in mind that $x(0) = 0$.

Example 9.2 (Impulse response subspace identification) Consider the LTI system (9.1)–(9.2) with system matrices given by

$$A = \begin{bmatrix} 1.69 & 1 \\ -0.96 & 0 \end{bmatrix}, \qquad B = \begin{bmatrix} 1 \\ 0.5 \end{bmatrix}, \qquad C = \begin{bmatrix} 1 & 0 \end{bmatrix}, \qquad D = 0.$$

The first 100 data points of the impulse response of this system are shown in Figure 9.2. These data points are used to construct the matrix $Y_{1,s,N}$ with $s = 3$. From the SVD of this matrix we determine the matrices A, B, and C up to a similarity transformation T. The computed singular

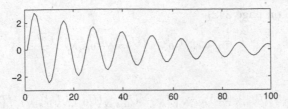

Fig. 9.2. The impulse response of the system used in Example 9.2.

values are, up to four digits, equal to 15.9425, 6.9597, and 0, and the system matrices that we obtain, up to four digits, are

$$A_T \approx \begin{bmatrix} 0.8529 & 1.0933 \\ -0.2250 & 0.8371 \end{bmatrix}, \quad B_T \approx \begin{bmatrix} 3.4155 \\ 1.2822 \end{bmatrix},$$
$$C_T \approx \begin{bmatrix} 0.5573 & -0.7046 \end{bmatrix}.$$

It is easy to verify that A_T has the same eigenvalues as the matrix A, and that the system (A_T, B_T, C_T) has the same impulse response as the original one.

9.2.4 Identification using general input sequences

In Section 9.2.3 we showed that, when an impulse input is applied to the system, we can exploit the special structure of the block Hankel matrix $U_{0,s,N}$ in Equation (9.7) on page 296, to get rid of the influence of the input and retrieve a matrix that has a column space equal to the column space of $\mathcal{O}_s$. In this section the retrieval of the column space of $\mathcal{O}_s$ is discussed for more general input sequences.

9.2.4.1 Deriving the column space of the observability matrix

Consider the multivariable system (9.1)–(9.2). We would like to find the column space of the extended observability matrix. If we know the matrix $\mathcal{T}_s$, we can obtain an estimate of this column space by subtracting $\mathcal{T}_s U_{0,s,N}$ from $Y_{0,s,N}$ followed by an SVD. However, since the system is unknown, $\mathcal{T}_s$ is also unknown and this trick is not appropriate; but we can instead apply this trick using an estimate of the matrix $\mathcal{T}_s$. A possible estimate of $\mathcal{T}_s$ can be obtained from the following least-squares problem (Viberg, 1995):

$$\min_{\mathcal{T}_s} ||Y_{0,s,N} - \mathcal{T}_s U_{0,s,N}||_\mathrm{F}^2.$$

When the input is such that the matrix $U_{0,s,N}$ has full (row) rank, the solution to the above least-squares problem is given by (see also

Exercise 9.5 on page 342)

$$\widehat{T}_s = Y_{0,s,N} U_{0,s,N}^{\mathrm{T}} (U_{0,s,N} U_{0,s,N}^{\mathrm{T}})^{-1}.$$

Now we get

$$Y_{0,s,N} - \widehat{T}_s U_{0,s,N} = Y_{0,s,N} \left(I_N - U_{0,s,N}^{\mathrm{T}} (U_{0,s,N} U_{0,s,N}^{\mathrm{T}})^{-1} U_{0,s,N} \right)$$

$$= Y_{0,s,N} \Pi_{U_{0,s,N}}^{\perp},$$

where the matrix

$$\Pi_{U_{0,s,N}}^{\perp} = I_N - U_{0,s,N}^{\mathrm{T}} (U_{0,s,N} U_{0,s,N}^{\mathrm{T}})^{-1} U_{0,s,N} \qquad (9.16)$$

is a projection matrix referred to as the orthogonal projection onto the column space of $U_{0,s,N}$, because it has the property $U_{0,s,N} \Pi_{U_{0,s,N}}^{\perp} = 0$ (see Section 2.6.1 for a review of projection matrices and their properties). The condition on the rank of $U_{0,s,N}$ restricts the type of input sequences that we can use to identify the system. This means that not every input can be used to identify a system. In Section 10.2.4 we will introduce the notion of persistency of excitation to characterize conditions the input has to satisfy to guarantee that subspace identification can be carried out. For now, we assume that the input is such that the matrix $U_{0,s,N} U_{0,s,N}^{\mathrm{T}}$ is of full rank.

Since $U_{0,s,N} \Pi_{U_{0,s,N}}^{\perp} = 0$, we can derive from Equation (9.7) that

$$Y_{0,s,N} \Pi_{U_{0,s,N}}^{\perp} = \mathcal{O}_s X_{0,N} \Pi_{U_{0,s,N}}^{\perp}. \qquad (9.17)$$

We have, in fact, removed the influence of the input on the output. What remains is the response of the system due to the state. Equation (9.17) shows that the column space of the matrix $Y_{0,s,N} \Pi_{U_{0,s,N}}^{\perp}$ is contained in the column space of the extended observability matrix. The next thing needed is, of course, to show that these column spaces are equal. This is equivalent to showing that $Y_{0,s,N} \Pi_{U_{0,s,N}}^{\perp}$ is of rank n. We have the following result.

Lemma 9.1 *Given the minimal state-space system (9.1)–(9.2), if the input $u(k)$ is such that*

$$\mathrm{rank}\left(\begin{bmatrix} X_{0,N} \\ U_{0,s,N} \end{bmatrix} \right) = n + sm, \qquad (9.18)$$

then

$$\mathrm{rank}\left(Y_{0,s,N} \Pi_{U_{0,s,N}}^{\perp} \right) = n$$

and

$$\text{range}\left(Y_{0,s,N}\Pi^{\perp}_{U_{0,s,N}}\right) = \text{range}\,(\mathcal{O}_s). \qquad (9.19)$$

Proof Equation (9.18) implies

$$\begin{bmatrix} X_{0,N}X^{\mathrm{T}}_{0,N} & X_{0,N}U^{\mathrm{T}}_{0,s,N} \\ U_{0,s,N}X^{\mathrm{T}}_{0,N} & U_{0,s,N}U^{\mathrm{T}}_{0,s,N} \end{bmatrix} > 0.$$

With the Schur complement (Lemma 2.3 on page 19) it follows that

$$\text{rank}(X_{0,N}X^{\mathrm{T}}_{0,N} - X_{0,N}U^{\mathrm{T}}_{0,s,N}(U_{0,s,N}U^{\mathrm{T}}_{0,s,N})^{-1}U_{0,s,N}X^{\mathrm{T}}_{0,N}) = n. \qquad (9.20)$$

Using the fact that $\Pi^{\perp}_{U_{0,s,N}}(\Pi^{\perp}_{U_{0,s,N}})^{\mathrm{T}} = \Pi^{\perp}_{U_{0,s,N}}$, we can write

$$Y_{0,s,N}\Pi^{\perp}_{U_{0,s,N}}(\Pi^{\perp}_{U_{0,s,N}})^{\mathrm{T}}Y^{\mathrm{T}}_{0,s,N} = Y_{0,s,N}\Pi^{\perp}_{U_{0,s,N}}Y^{\mathrm{T}}_{0,s,N},$$

and, with Equation (9.17), also

$$\begin{aligned} Y_{0,s,N}\Pi^{\perp}_{U_{0,s,N}}Y^{\mathrm{T}}_{0,s,N} &= \mathcal{O}_s X_{0,N}\Pi^{\perp}_{U_{0,s,N}}X^{\mathrm{T}}_{0,N}\mathcal{O}^{\mathrm{T}}_s \\ &= \mathcal{O}_s X_{0,N}X^{\mathrm{T}}_{0,N}\mathcal{O}^{\mathrm{T}}_s \\ &\quad -\mathcal{O}_s X_{0,N}U^{\mathrm{T}}_{0,s,N}(U_{0,s,N}U^{\mathrm{T}}_{0,s,N})^{-1}U_{0,s,N}X^{\mathrm{T}}_{0,N}\mathcal{O}^{\mathrm{T}}_s. \end{aligned}$$

Because $\text{rank}(\mathcal{O}_s) = n$, an application of Sylvester's inequality (Lemma 2.1 on page 16) shows that

$$\text{rank}\left(Y_{0,s,N}\Pi^{\perp}_{U_{0,s,N}}Y^{\mathrm{T}}_{0,s,N}\right) = n.$$

This completes the proof. $\qquad\qquad\qquad\qquad\qquad\qquad\qquad\square$

Lemma 9.1 provides a condition on the input sequence that allows us to recover the column space of the extended observability matrix. In Section 10.2.4.2 we examine this rank condition in more detail.

Example 9.3 (Input and state rank condition) Consider the state equation

$$x(k+1) = \frac{1}{2}x(k) + u(k).$$

If we take $x(0) = 0$, $u(0) = u(2) = 1$, and $u(1) = u(3) = 0$, then it is easy to see that

$$\begin{bmatrix} X_{0,3} \\ \hline U_{0,2,3} \end{bmatrix} = \begin{bmatrix} x(0) & x(1) & x(2) \\ u(0) & u(1) & u(2) \\ u(1) & u(2) & u(3) \end{bmatrix} = \begin{bmatrix} 0 & 1 & 1/2 \\ 1 & 0 & 1 \\ 0 & 1 & 0 \end{bmatrix}.$$

Because this matrix has full row rank, adding columns to it will not change its rank. Therefore, with $x(0) = 0$, the rank condition

$$\text{rank}\left(\begin{bmatrix} X_{0,N} \\ U_{0,2,N} \end{bmatrix}\right) = 3$$

is satisfied for any finite sequence $u(k)$, $0 \leq k \leq N$, for which $N \geq 3$, $u(0) = u(2) = 1$, and $u(1) = u(3) = 0$.

9.2.4.2 Improving the numerical efficiency by using the RQ factorization

We have seen above that, for a proper choice of the input sequence, the column space of the matrix $Y_{0,s,N}\Pi_{U_{0,s,N}}^{\perp}$ equals the column space of the extended observability matrix. Therefore, an SVD of this matrix (and a least-squares problem) can be used to determine the matrices A_T and C_T (similarly to in Section 9.2.2). However, this is not attractive from a computational point of view, because the matrix $Y_{0,s,N}\Pi_{U_{0,s,N}}^{\perp}$ has N columns and typically N is large. Furthermore, it requires the construction of the matrix $\Pi_{U_{0,s,N}}^{\perp}$, which is also of size N and involves the computation of a matrix inverse, as shown by Equation (9.16) on page 302. For a more efficient implementation with respect both to the number of flops and to the required memory storage, the explicit calculation of the product $Y_{0,s,N}\Pi_{U_{0,s,N}}^{\perp}$ can be avoided when using the following RQ factorization (Verhaegen and Dewilde, 1992a):

$$\begin{bmatrix} U_{0,s,N} \\ Y_{0,s,N} \end{bmatrix} = \begin{bmatrix} R_{11} & 0 & 0 \\ R_{21} & R_{22} & 0 \end{bmatrix} \begin{bmatrix} Q_1 \\ Q_2 \\ Q_3 \end{bmatrix}, \tag{9.21}$$

where $R_{11} \in \mathbb{R}^{sm \times sm}$ and $R_{22} \in \mathbb{R}^{s\ell \times s\ell}$. The relation between this RQ factorization and the matrix $Y_{0,s,N}\Pi_{U_{0,s,N}}^{\perp}$ is given in the following lemma.

Lemma 9.2 *Given the RQ factorization (9.21), we have* $Y_{0,s,N}\Pi_{U_{0,s,N}}^{\perp} = R_{22}Q_2$.

Proof From the RQ factorization (9.21) we can express $Y_{0,s,N}$ as

$$Y_{0,s,N} = R_{21}Q_1 + R_{22}Q_2.$$

Furthermore, it follows from the orthogonality of the matrix

$$\begin{bmatrix} Q_1 \\ Q_2 \\ Q_3 \end{bmatrix}$$

that

$$\begin{bmatrix} Q_1^T & Q_2^T & Q_3^T \end{bmatrix} \begin{bmatrix} Q_1 \\ Q_2 \\ Q_3 \end{bmatrix} = I_N \quad \Rightarrow \quad Q_1^T Q_1 + Q_2^T Q_2 + Q_3^T Q_3 = I_N$$

and

$$\begin{bmatrix} Q_1 \\ Q_2 \\ Q_3 \end{bmatrix} \begin{bmatrix} Q_1^T & Q_2^T & Q_3^T \end{bmatrix} = I_N \quad \Rightarrow \quad \begin{cases} Q_i Q_j^T = 0, & i \neq j, \\ Q_i Q_i^T = I. \end{cases}$$

With $U_{0,s,N} = R_{11} Q_1$ and Equation (9.16) on page 302, we can derive

$$\begin{aligned} \Pi_{U_{0,s,N}}^{\perp} &= I_N - Q_1^T R_{11}^T (R_{11} Q_1 Q_1^T R_{11}^T)^{-1} R_{11} Q_1 \\ &= I_N - Q_1^T Q_1 \\ &= Q_2^T Q_2 + Q_3^T Q_3, \end{aligned}$$

and, therefore,

$$Y_{0,s,N} \Pi_{U_{0,s,N}}^{\perp} = R_{21} Q_1 Q_2^T Q_2 + R_{22} Q_2 Q_2^T Q_2 = R_{22} Q_2,$$

which completes the proof. $\qquad \square$

From this lemma it follows that the column space of the matrix $Y_{0,s,N} \Pi_{U_{0,s,N}}^{\perp}$ equals the column space of the matrix $R_{22} Q_2$. Furthermore, we have the following.

Theorem 9.1 *Given the minimal system (9.1)–(9.2) and the RQ factorization (9.21), if $u(k)$ is such that*

$$\text{rank}\left(\begin{bmatrix} X_{0,N} \\ U_{0,s,N} \end{bmatrix} \right) = n + sm,$$

then

$$\text{range}(\mathcal{O}_s) = \text{range}(R_{22}). \tag{9.22}$$

Proof From Lemma 9.1 on page 302 we derived Equation (6.19). Combining this with the result of Lemma 9.2 yields

$$\text{range}(Y_{0,s,N}\Pi^{\perp}_{U_{0,s,N}}) = \text{range}(\mathcal{O}_s) = \text{range}(R_{22}Q_2).$$

Therefore the rank of $R_{22}Q_2$ is n. This, in combination with the fact that Q_2 has full row rank and the use of Sylvester's inequality (Lemma 2.1 on page 16), shows that R_{22} has rank n. This completes the proof. $\square$

This theorem shows that, to compute the column space of $\mathcal{O}_s$, we do not need to store the matrix Q_2, which is much larger than the matrix R_{22}: $Q_2 \in \mathbb{R}^{s\ell \times N}$, $R_{22} \in \mathbb{R}^{s\ell \times s\ell}$, with typically $N \gg s\ell$. Furthermore, to compute the matrices A_T and C_T, we can, instead of using an SVD of the matrix $Y_{0,s,N}\Pi^{\perp}_{U_{0,s,N}}$ having N columns, compute an SVD of the matrix R_{22} which has only $s\ell$ columns.

9.2.4.3 Computing the system matrices

Theorem 9.1 shows that an SVD of the matrix R_{22} allows us to determine the column space of $\mathcal{O}_s$, provided that the input satisfies the rank condition (9.18) on page 302. Hence, given the SVD

$$R_{22} = U_n \Sigma_n V_n^{\mathrm{T}}$$

with $\Sigma_n \in \mathbb{R}^{n\times n}$ and $\text{rank}(\Sigma_n) = n$, we can compute A_T and C_T as outlined in Section 9.2.2.

The matrices B_T and D_T, together with the initial state $x_T(0) = T^{-1}x(0)$, can be computed by solving a least-squares problem, as in the proof of Theorem 7.1. Given the matrices A_T and C_T, the output of the system (9.1)–(9.2) on page 294 can be expressed linearly in the matrices B_T and D_T:

$$y(k) = C_T A_T^k x_T(0) + \left(\sum_{\tau=0}^{k-1} u(\tau)^{\mathrm{T}} \otimes C_T A_T^{k-\tau-1}\right)\text{vec}(B_T)$$
$$+ (u(k)^{\mathrm{T}} \otimes I_\ell)\text{vec}(D_T).$$

Let $\widehat{A}_T$ and $\widehat{C}_T$ denote the estimates of A_T and C_T computed in the previous step. Now, taking

$$\phi(k)^{\mathrm{T}} = \left[\widehat{C}_T\widehat{A}_T^k \quad \left(\sum_{\tau=0}^{k-1} u(\tau)^{\mathrm{T}} \otimes \widehat{C}_T\widehat{A}_T^{k-\tau-1}\right) \quad (u(k)^{\mathrm{T}} \otimes I_\ell)\right] \quad (9.23)$$

and

$$\theta = \begin{bmatrix} x_T(0) \\ \text{vec}(B_T) \\ \text{vec}(D_T) \end{bmatrix}, \quad (9.24)$$

we can solve for θ in a least-squares setting

$$\min_\theta \frac{1}{N} \sum_{k=0}^{N-1} ||y(k) - \phi(k)^{\mathrm{T}}\theta||_2^2, \qquad (9.25)$$

as described in Example 7.7 on page 235.

Alternatively, the matrices B_T and D_T can be computed from the matrices R_{21} and R_{11} of the RQ factorization (9.21). This has been described by Verhaegen and Dewilde (1992a) and is based on exploiting the structure of the matrix T_s given in Equation (9.4) on page 295.

9.3 Subspace identification with white measurement noise

In the previous discussion the system was assumed to be noise-free. In practice this of course rarely happens. To discuss the treatment of more realistic circumstances, we now take the output-error estimation problem stated at the beginning of Chapter 7 in which the output is perturbed by an additive white-noise sequence. In the subsequent sections of this chapter it will be assumed that the noise sequences are ergodic (see Section 4.3.4).

Let the additive noise be denoted by $v(k)$, then the signal-generating system that will be considered can be written as

$$x(k+1) = Ax(k) + Bu(k), \qquad (9.26)$$
$$y(k) = Cx(k) + Du(k) + v(k). \qquad (9.27)$$

The data equation for this system is similar to (9.7) on page 296 and reads

$$Y_{i,s,N} = \mathcal{O}_s X_{i,N} + T_s U_{i,s,N} + V_{i,s,N}, \qquad (9.28)$$

where $V_{i,s,N}$ is a block Hankel matrix constructed from the sequence $v(k)$.

The next lemma shows that, in the limit of $N \to \infty$, the result of Lemma 9.1 can be extended to the case in which the additive noise at the output is white.

Lemma 9.3 *Given the minimal system (9.26)–(9.27) with $u(k)$, $x(k)$, and $v(k)$ ergodic stochastic processes, with the input $u(k)$ satisfying*

$$\mathrm{rank}\left(\lim_{N\to\infty} \frac{1}{N}\left(\begin{bmatrix} X_{i,N} \\ U_{i,s,N} \end{bmatrix} \begin{bmatrix} X_{i,N}^{\mathrm{T}} & U_{i,s,N}^{\mathrm{T}} \end{bmatrix} \right) \right) = n + sm \qquad (9.29)$$

and with $v(k)$ a white-noise sequence that is uncorrelated with $u(k)$ and satisfies

$$\lim_{N\to\infty} \frac{1}{N} V_{i,s,N} V_{i,s,N}^{\mathrm{T}} = \sigma^2 I_{s\ell},$$

the SVD of the matrix

$$\lim_{N\to\infty} \frac{1}{N} Y_{i,s,N} \Pi^{\perp}_{U_{i,s,N}} Y_{i,s,N}^{\mathrm{T}}$$

is given by

$$\lim_{N\to\infty} \frac{1}{N} Y_{i,s,N} \Pi^{\perp}_{U_{i,s,N}} Y_{i,s,N}^{\mathrm{T}} = \begin{bmatrix} U_1 & U_2 \end{bmatrix} \begin{bmatrix} \Sigma_n^2 + \sigma^2 I_n & 0 \\ 0 & \sigma^2 I_{s\ell-n} \end{bmatrix} \begin{bmatrix} U_1^{\mathrm{T}} \\ U_2^{\mathrm{T}} \end{bmatrix},$$

(9.30)

where the $n \times n$ diagonal matrix Σ_n^2 contains the nonzero singular values of the matrix $\mathcal{O}_s M_X \mathcal{O}_s^{\mathrm{T}}$ and

$$M_X = \lim_{N\to\infty} \frac{1}{N} X_{i,N} \Pi^{\perp}_{U_{i,s,N}} X_{i,N}^{\mathrm{T}}.$$

(9.31)

The matrix U_1 in this SVD satisfies

$$\mathrm{range}(U_1) = \mathrm{range}(\mathcal{O}_s).$$

(9.32)

Proof From the data equation it follows that

$$Y_{i,s,N} \Pi^{\perp}_{U_{i,s,N}} = \mathcal{O}_s X_{i,N} \Pi^{\perp}_{U_{i,s,N}} + V_{i,s,N} \Pi^{\perp}_{U_{i,s,N}}.$$

Using the fact that $\Pi^{\perp}_{U_{i,s,N}} (\Pi^{\perp}_{U_{i,s,N}})^{\mathrm{T}} = \Pi^{\perp}_{U_{i,s,N}}$, we can write

$$Y_{i,s,N} \Pi^{\perp}_{U_{i,s,N}} (\Pi^{\perp}_{U_{i,s,N}})^{\mathrm{T}} Y_{i,s,N}^{\mathrm{T}} = Y_{i,s,N} \Pi^{\perp}_{U_{i,s,N}} Y_{i,s,N}^{\mathrm{T}}$$

and also

$$Y_{i,s,N} \Pi^{\perp}_{U_{i,s,N}} Y_{i,s,N}^{\mathrm{T}} = \mathcal{O}_s X_{i,N} \Pi^{\perp}_{U_{i,s,N}} X_{i,N}^{\mathrm{T}} \mathcal{O}_s^{\mathrm{T}} + \mathcal{O}_s X_{i,N} \Pi^{\perp}_{U_{i,s,N}} V_{i,s,N}^{\mathrm{T}}$$
$$+ V_{i,s,N} \Pi^{\perp}_{U_{i,s,N}} X_{i,N}^{\mathrm{T}} \mathcal{O}_s^{\mathrm{T}} + V_{i,s,N} \Pi^{\perp}_{U_{i,s,N}} V_{i,s,N}^{\mathrm{T}}.$$

Since $u(k)$ is uncorrelated with the white-noise sequence $v(k)$, we have

$$\lim_{N\to\infty} \frac{1}{N} U_{i,s,N} V_{i,s,N}^{\mathrm{T}} = 0, \qquad \lim_{N\to\infty} \frac{1}{N} X_{i,N} V_{i,s,N}^{\mathrm{T}} = 0.$$

These two limits, in combination with the expression for $\Pi^{\perp}_{U_{i,s,N}}$ as given in (9.16) on page 302, imply

$$\lim_{N\to\infty} \frac{1}{N} X_{i,N} \Pi^{\perp}_{U_{i,s,N}} V_{i,s,N}^{\mathrm{T}} = \lim_{N\to\infty} \frac{1}{N} X_{i,N} V_{i,s,N}^{\mathrm{T}} - \lim_{N\to\infty} \frac{1}{N} (X_{i,N} U_{i,s,N}^{\mathrm{T}})$$
$$\times \left(\frac{1}{N} U_{i,N} U_{i,s,N}^{\mathrm{T}} \right)^{-1} \left(\frac{1}{N} U_{i,N} V_{i,s,N}^{\mathrm{T}} \right)$$
$$= 0.$$

In the same way, we have

$$\lim_{N\to\infty} \frac{1}{N} V_{i,s,N} \Pi^{\perp}_{U_{i,s,N}} V^{\mathrm{T}}_{i,s,N} = \lim_{N\to\infty} \frac{1}{N} V_{i,s,N} V^{\mathrm{T}}_{i,s,N}.$$

Thus, the fact that $u(k)$ is uncorrelated with the white-noise sequence $v(k)$ yields

$$\lim_{N\to\infty} \frac{1}{N} Y_{i,s,N} \Pi^{\perp}_{U_{i,s,N}} Y^{\mathrm{T}}_{i,s,N} = \lim_{N\to\infty} \frac{1}{N} \Big(\mathcal{O}_s X_{i,N} \Pi^{\perp}_{U_{i,s,N}} X^{\mathrm{T}}_{i,N} \mathcal{O}^{\mathrm{T}}_s$$
$$+ V_{i,s,N} V^{\mathrm{T}}_{i,s,N} \Big). \tag{9.33}$$

With the white-noise property of $v(k)$ and the definition of the matrix M_X, Equation (9.33) becomes

$$\lim_{N\to\infty} \frac{1}{N} Y_{i,s,N} \Pi^{\perp}_{U_{i,s,N}} Y^{\mathrm{T}}_{i,s,N} = \mathcal{O}_s M_X \mathcal{O}^{\mathrm{T}}_s + \sigma^2 I_{s\ell}. \tag{9.34}$$

As in the proof of Lemma 9.1, using condition (9.29) it can be shown that

$$\mathrm{rank}\,(M_X) = \mathrm{rank}\Big(\lim_{N\to\infty} \frac{1}{N} \Big(X_{i,N} X^{\mathrm{T}}_{i,N}$$
$$- X_{i,N} U^{\mathrm{T}}_{i,s,N} (U_{i,s,N} U^{\mathrm{T}}_{i,s,N})^{-1} U_{i,s,N} X^{\mathrm{T}}_{i,N} \Big) \Big)$$
$$= n.$$

Therefore, by Sylvester's inequality (Lemma 2.1 on page 16) the SVD

$$\mathcal{O}_s M_X \mathcal{O}^{\mathrm{T}}_s = \begin{bmatrix} U_1 & U_2 \end{bmatrix} \begin{bmatrix} \Sigma^2_n & 0 \\ 0 & 0 \end{bmatrix} \begin{bmatrix} U^{\mathrm{T}}_1 \\ U^{\mathrm{T}}_2 \end{bmatrix} \tag{9.35}$$

holds for $\Sigma^2_n > 0$ and $\mathrm{range}(U_1) = \mathrm{range}(\mathcal{O}_s)$. Since the matrix $\begin{bmatrix} U_1 & U_2 \end{bmatrix}$ is orthogonal, we can write Equation (9.34) as

$$\lim_{N\to\infty} \frac{1}{N} Y_{i,s,N} \Pi^{\perp}_{U_{i,s,N}} Y^{\mathrm{T}}_{i,s,N} = \begin{bmatrix} U_1 & U_2 \end{bmatrix} \begin{bmatrix} \Sigma^2_n & 0 \\ 0 & 0 \end{bmatrix} \begin{bmatrix} U^{\mathrm{T}}_1 \\ U^{\mathrm{T}}_2 \end{bmatrix}$$
$$+ \sigma^2 \begin{bmatrix} U_1 & U_2 \end{bmatrix} \begin{bmatrix} U^{\mathrm{T}}_1 \\ U^{\mathrm{T}}_2 \end{bmatrix}$$
$$= \begin{bmatrix} U_1 & U_2 \end{bmatrix} \begin{bmatrix} \Sigma^2_n + \sigma^2 I_n & 0 \\ 0 & \sigma^2 I_{s\ell-n} \end{bmatrix} \begin{bmatrix} U^{\mathrm{T}}_1 \\ U^{\mathrm{T}}_2 \end{bmatrix}.$$

This is a valid SVD. □

From this lemma we conclude that the computation of the column space $\mathcal{O}_s$ does not change in the presence of an additive white noise in the output. Therefore, we can still obtain estimates of the system matrices from an SVD of the matrix $Y_{i,s,N} \Pi^{\perp}_{U_{i,s,N}}$. We can use the algorithm described in Section 9.2.4 to obtain estimates of the system matrices

(A_T, B_T, C_T, D_T). The requirement $N \to \infty$ means that these estimates are asymptotically unbiased. In practice the system matrices will be estimated on the basis of a finite number of data samples. The accuracy of these estimates will be improved by using more samples.

We can again use an RQ factorization for a more efficient implementation, because of the following theorem.

Theorem 9.2 *Given the minimal system (9.26)–(9.27) and the RQ factorization (9.21), if $v(k)$ is an ergodic white-noise sequence with variance σ^2 that is uncorrelated with $u(k)$, and $u(k)$ is an ergodic sequence such that Equation (9.29) holds, then,*

$$\left(\lim_{N \to \infty} \frac{1}{N} R_{22} R_{22}^T \right) = \mathcal{O}_s M_X \mathcal{O}_s^T + \sigma^2 I_{s\ell},$$

with the matrix M_x defined by Equation (9.31).

The proof follows easily on combining Lemma 9.3 on page 307 and Lemma 9.2 on page 304.

In the noise-free case, described in Section 9.2.4, the number of nonzero singular values equals the order of the state-space system. Equation (9.30) shows that this no longer holds if there is additive white noise present at the output. In this case the order can be determined from the singular values if we can distinguish the n disturbed singular values of the system from the $s\ell - n$ remaining singular values that are due to the noise. Hence, the order can be determined if the smallest singular value of $\Sigma_n^2 + \sigma^2 I_n$ is larger than σ^2. In other words, the ability to determine that n is the correct system order depends heavily on the gap between the nth and the $(n+1)$th singular value of the matrix R_{22}. The use of the singular values in detecting the order of the system is illustrated in the following example. The practical usefulness of this way of detecting the "system order" is discussed for more realistic circumstances in Example 10.12.

Example 9.4 (Noisy singular values) Consider the LTI system (9.26)–(9.27) with system matrices given in Example 9.2 on page 300. Let the input $u(k)$ be a unit-variance zero-mean Gaussian white-noise sequence and the noise $v(k)$ a zero-mean Gaussian white noise with standard deviation σ. The input and the corresponding noise-free output of the system are shown in Figure 9.3. Figure 9.4 plots the singular values of the matrix $(R_{22} R_{22}^T)/N$ from the RQ factorization (9.21) on page 304 for four values of the standard deviation σ of the noise

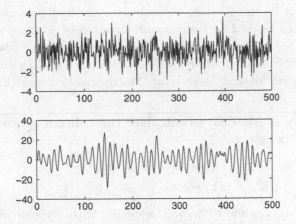

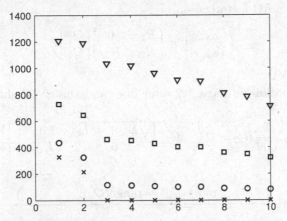

Fig. 9.3. Input and output data used in Example 9.4.

Fig. 9.4. Singular values of the matrix $(R_{22}R_{22}^{\mathrm{T}})/N$ corresponding to the system of Example 9.4 plotted for $\sigma = 1$ (crosses), $\sigma = 10$ (circles), $\sigma = 20$ (squares), and $\sigma = 30$ (triangles).

$v(k)$. From this we see that all the singular values differ from zero, and that they become larger with increasing standard deviation σ. Theorem 9.2 gives the relation between σ and the value of the singular values due to the noise. We clearly see that, when the signal-to-noise ratio decreases, the gap between the two dominant singular values from the system and the spurious singular values from the noise becomes smaller. This illustrates the fact that, for lower signal-to-noise ratio, it becomes more difficult to determine the order of the system from the singular values.

As shown in this section, the subspace method presented in Section 9.2.4 can be used for systems contaminated by additive white noise in the output. Because of this property, this subspace identification method is called the *MOESP* method (Verhaegen and Dewilde, 1992a; 1992b), where "MOESP" stands for "Multivariable Output-Error StatesPace."

Summary of MOESP to calculate the column space of $\mathcal{O}_s$

Consider the system

$$x(k+1) = Ax(k) + Bu(k),$$
$$y(k) = Cx(k) + Du(k) + v(k),$$

with $v(k)$ an ergodic white-noise sequence with variance $\sigma^2 I_\ell$ that is uncorrelated with $u(k)$, and $u(k)$ an ergodic sequence such that Equation (9.18) on page 302 is satisfied.

From the RQ factorization

$$\begin{bmatrix} U_{0,s,N} \\ Y_{0,s,N} \end{bmatrix} = \begin{bmatrix} R_{11} & 0 \\ R_{21} & R_{22} \end{bmatrix} \begin{bmatrix} Q_1 \\ Q_2 \end{bmatrix}$$

and the SVD

The equation on page 312 after line 'and the SVD' should look like:

$$\lim_{N \to \infty} \frac{1}{\sqrt{N}} R_{22} = \begin{bmatrix} U_n & U_2 \end{bmatrix} \begin{bmatrix} \sqrt{\Sigma_n^2 + \sigma^2 I_n} & 0 \\ 0 & \sigma I_{s\ell-n} \end{bmatrix} \begin{bmatrix} V_1^T \\ V_2^T \end{bmatrix}$$

we have

$$\text{range}(U_n) = \text{range}(\mathcal{O}_s).$$

9.4 The use of instrumental variables

When the noise v_k in Equation (9.27) is not a white-noise sequence but rather a colored noise, then the subspace method described in Section 9.2.4 will give biased estimates of the system matrices. This is because the column space of the matrix $Y_{i,s,N} \Pi_{U_{i,s,N}}^{\perp}$ no longer contains the column space of $\mathcal{O}_s$, as can be seen from the proof of Lemma 9.3 on page 307. This is illustrated in the following example.

Example 9.5 (MOESP with colored noise) Consider the system (9.26)–(9.27) on page 307 with system matrices

$$A = \begin{bmatrix} 1.5 & 1 \\ -0.7 & 0 \end{bmatrix}, \qquad B = \begin{bmatrix} 1 \\ 0.5 \end{bmatrix}, \qquad C = \begin{bmatrix} 1 & 0 \end{bmatrix}, \qquad D = 0.$$

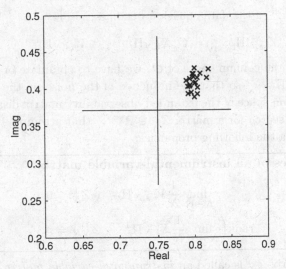

Fig. 9.5. One of the eigenvalues of the matrix A estimated by MOESP for 20 different realizations of colored measurement noise in Example 9.5. The big cross corresponds to the real value.

We take the input u_k equal to a unit-variance zero-mean white-noise sequence. The noise v_k is a colored sequence, generated as follows:

$$v_k = \frac{q^{-1} + 0.5q^{-2}}{1 - 1.69q^{-1} + 0.96q^{-2}} e_k,$$

where e_k is a zero-mean white-noise sequence with a variance equal to 0.2. We generate 1000 samples of the output signal, and use the MOESP method to estimate the matrices A and C up to a similarity transformation. To show that the MOESP method yields biased estimates, we look at the eigenvalues of the estimated A matrix. The real eigenvalues of this matrix are, up to four digits, equal to $0.75 \pm 0.3708j$. Figure 9.5 shows one of the eigenvalues of the estimated A matrix for 20 different realizations of the input–output sequences. We clearly see that these eigenvalues are biased.

It is possible to compute unbiased estimates of the system matrices by using so called *instrumental variables* (Söderström and Stoica, 1983), which is the topic of this section.

Recall that, after eliminating the influence of the input with the appropriate projection, that is multiplying the data equation (9.28) on the right by $\Pi^{\perp}_{U_{i,s,N}}$, the data equation becomes

$$Y_{i,s,N}\Pi^{\perp}_{U_{i,s,N}} = \mathcal{O}_s X_{i,N}\Pi^{\perp}_{U_{i,s,N}} + V_{i,s,N}\Pi^{\perp}_{U_{i,s,N}}. \tag{9.36}$$

To retrieve the column space of $\mathcal{O}_s$ we have to eliminate or modify the term $V_{i,s,N}\Pi^{\perp}_{U_{i,s,N}}$ so that the influence of the noise on the calculation of the column space of the extended observability matrix disappears. To do this, we search for a matrix $Z_N \in \mathbb{R}^{sz \times N}$ that attempts to eliminate the term via the following properties.

Properties of an instrumental-variable matrix:

$$\lim_{N \to \infty} \frac{1}{N} V_{i,s,N}\Pi^{\perp}_{U_{i,s,N}} Z_N^{\mathrm{T}} = 0, \tag{9.37}$$

$$\mathrm{rank}\left(\lim_{N \to \infty} \frac{1}{N} X_{i,N}\Pi^{\perp}_{U_{i,s,N}} Z_N^{\mathrm{T}} \right) = n. \tag{9.38}$$

Such a matrix Z_N is called an *instrumental-variable matrix*.

Because of property (9.37), we can indeed get rid of the term $V_{i,s,N}\Pi^{\perp}_{U_{i,s,N}}$ in Equation (9.36) by multiplying $Y_{i,s,N}\Pi^{\perp}_{U_{i,s,N}}$ on the right by Z_N and taking the limit for $N \to \infty$, that is,

$$\lim_{N \to \infty} \frac{1}{N} Y_{i,s,N}\Pi^{\perp}_{U_{i,s,N}} Z_N^{\mathrm{T}} = \lim_{N \to \infty} \frac{1}{N} \mathcal{O}_s X_{i,N}\Pi^{\perp}_{U_{i,s,N}} Z_N^{\mathrm{T}}.$$

Property (9.38) ensures that the multiplication by Z_N does not change the rank of the right-hand side of the last equation, and therefore we have

$$\mathrm{range}\left(\lim_{N \to \infty} \frac{1}{N} Y_{i,s,N}\Pi^{\perp}_{U_{i,s,N}} Z_N^{\mathrm{T}} \right) = \mathrm{range}(\mathcal{O}_s). \tag{9.39}$$

From this relation we immediately see that we can determine an asymptotically unbiased estimate of the column space of $\mathcal{O}_s$ from the SVD of the matrix $Y_{i,s,N}\Pi^{\perp}_{U_{i,s,N}} Z_N^{\mathrm{T}}$.

For an efficient implementation of the instrumental-variable method, we can again use an RQ factorization. The proposed RQ factorization is given as

$$\begin{bmatrix} U_{i,s,N} \\ Z_N \\ Y_{i,s,N} \end{bmatrix} = \begin{bmatrix} R_{11} & 0 & 0 & 0 \\ R_{21} & R_{22} & 0 & 0 \\ R_{31} & R_{32} & R_{33} & 0 \end{bmatrix} \begin{bmatrix} Q_1 \\ Q_2 \\ Q_3 \\ Q_4 \end{bmatrix}, \tag{9.40}$$

with $R_{11} \in \mathbb{R}^{sm \times sm}$, $R_{22} \in \mathbb{R}^{sz \times sz}$, and $R_{33} \in \mathbb{R}^{s\ell \times s\ell}$. The next lemma shows the relation between this RQ factorization and the matrix $Y_{i,s,N}\Pi^{\perp}_{U_{i,s,N}} Z_N^{\mathrm{T}}$.

Lemma 9.4 *Given the RQ factorization (9.40), we have*

$$Y_{i,s,N}\Pi^{\perp}_{U_{i,s,N}}Z_N^{\mathrm{T}} = R_{32}R_{22}^{\mathrm{T}}.$$

Proof The proof is similar to the proof of Lemma 9.2 on page 304. We can derive

$$\Pi^{\perp}_{U_{i,s,N}} = I_N - Q_1^{\mathrm{T}}Q_1 = Q_2^{\mathrm{T}}Q_2 + Q_3^{\mathrm{T}}Q_3 + Q_4^{\mathrm{T}}Q_4,$$

and therefore

$$\begin{aligned}
Y_{i,s,N}\Pi^{\perp}_{U_{i,s,N}}Z_N^{\mathrm{T}} &= (R_{31}Q_1 + R_{32}Q_2 + R_{33}Q_3)\\
&\quad \times (Q_2^{\mathrm{T}}Q_2 + Q_3^{\mathrm{T}}Q_3 + Q_4^{\mathrm{T}}Q_4)Z_N^{\mathrm{T}}\\
&= (R_{32}Q_2 + R_{33}Q_3)(Q_1^{\mathrm{T}}R_{21}^{\mathrm{T}} + Q_2^{\mathrm{T}}R_{22}^{\mathrm{T}})\\
&= R_{32}R_{22}^{\mathrm{T}},
\end{aligned}$$

which completes the proof. □

Lemma 9.4 highlights the fact that, if the properties (9.37) and (9.38) of the instrumental-variable matrix Z_N hold, the equivalence of the range spaces indicated in Equation (9.39) becomes

$$\mathrm{range}\left(\lim_{N\to\infty}\frac{1}{N}R_{32}R_{22}^{\mathrm{T}}\right) = \mathrm{range}(\mathcal{O}_s). \tag{9.41}$$

Hence, the matrix $R_{32}R_{22}^{\mathrm{T}}$ can be used to obtain asymptotically unbiased estimates of the matrices A_T and C_T. The question of how to choose Z_N remains to be answered. This will be dealt with in the subsequent sections.

9.5 Subspace identification with colored measurement noise

In this section we develop a subspace identification solution for the output-error estimation problem of Section 7.7, where the additive perturbation was considered to be a colored stochastic process. From Section 9.4 we know that, to deal with this case, we need to find an instrumental-variable matrix Z_N that satisfies both of the conditions (9.37) and (9.38). If we take for example $Z_N = U_{i,s,N}$, Equation (9.37) is satisfied, because u_k and v_k are uncorrelated, but Equation (9.38) is clearly violated for all possible input sequences, since $X_{i,N}\Pi^{\perp}_{U_{i,s,N}}U_{i,s,N}^{\mathrm{T}} = 0$. Hence, $Z_N = U_{i,s,N}$ is not an appropriate choice for this purpose. However, if we take a shifted version of the input to construct Z_N, like, for example, $Z_N = U_{0,s,N}$ and $i = s$, condition (9.37) holds, and, as explained below, there exist certain types of input sequences for which (9.38) also holds. Usually, to construct a suitable matrix Z_N, the data available for identification are split up into two overlapping parts. Among the many choices possible for splitting the data into two parts (Jansson, 1997), one that is often used is described below. The first part, from time instant 0 up to $N+s-2$, is used to construct the data matrix $U_{0,s,N}$; this can be thought of as the "past input." The second part, from time instant s up to $N+2s-2$, is used to construct the data matrices $U_{s,s,N}$ and $Y_{s,s,N}$, which can be thought of as the "future input" and "future output," respectively. With this terminology, we use the "future" input and output to identify the system, and the "past" input as the instrumental-variable matrix Z_N used to get rid of the influence of the noise. The next lemma shows that with this choice Equation (9.37) is satisfied.

Lemma 9.5 *Consider the system (9.26)–(9.27), with $x(k)$, $u(k)$, and $v(k)$ ergodic stochastic processes such that $v(k)$ is uncorrelated with $x(j)$ and $u(j)$ for all $k, j \in \mathbb{Z}$. Take as instrumental-variable matrix $Z_N = U_{0,s,N}$, then*

$$\lim_{N \to \infty} \frac{1}{N} V_{s,s,N} \Pi^{\perp}_{U_{s,s,N}} Z_N^{\mathrm{T}} = 0. \tag{9.42}$$

Proof Since $u(k)$ is uncorrelated with $v(k)$, we have

$$\lim_{N \to \infty} \frac{1}{N} V_{s,s,N} U_{0,s,N}^{\mathrm{T}} = 0, \qquad \lim_{N \to \infty} \frac{1}{N} V_{s,s,N} U_{s,s,N}^{\mathrm{T}} = 0.$$

This immediately implies that Equation (9.42) holds. □

The verification of the second condition, Equation (9.38) on page 314, is more difficult. An analysis can be performed for specific input sequences. This is done in our next lemma for the special case in which the input is a zero-mean white-noise sequence.

Lemma 9.6 *(Jansson and Wahlberg, 1998) Consider the minimal system (9.26)–(9.27), with $x(k)$, $u(k)$, and $v(k)$ ergodic stochastic processes. Let the input $u(k)$ be a zero-mean white-noise sequence and take as instrumental-variable matrix $Z_N = U_{0,s,N}$, $s \geq n$, then*

$$\mathrm{rank}\left(\lim_{N \to \infty} \frac{1}{N} X_{s,N} \Pi^{\perp}_{U_{s,s,N}} Z_N^{\mathrm{T}} \right) = n. \tag{9.43}$$

Proof Instead of verifying condition (9.43), we first rewrite this condition. With the specific choice of Z_N, Equation (9.43) is equivalent to

$$\text{rank}\left(\lim_{N\to\infty}\frac{1}{N}\left(X_{s,N}(I_N - U_{s,s,N}^T(U_{s,s,N}U_{s,s,N}^T)^{-1}U_{s,s,N})U_{0,s,N}^T\right)\right) = n.$$

Since the input $u(k)$ is white noise, the inverse of the matrix

$$\lim_{N\to\infty}\frac{1}{N}(U_{s,s,N}U_{s,s,N}^T)$$

exists. By virtue of the Schur complement (Lemma 2.3 on page 19), the condition (9.43) for the specific choice of Z_N is equivalent to the following condition:

$$\text{rank}\left(\lim_{N\to\infty}\frac{1}{N}\begin{bmatrix}X_{s,N}\\U_{s,s,N}\end{bmatrix}\begin{bmatrix}U_{0,s,N}^T & U_{s,s,N}^T\end{bmatrix}\right) = n + sm. \tag{9.44}$$

Because the input is a white-noise sequence, we have

$$\lim_{N\to\infty}\frac{1}{N}U_{s,s,N}U_{0,s,N}^T = 0, \qquad \lim_{N\to\infty}\frac{1}{N}U_{s,s,N}U_{s,s,N}^T = \Sigma_s,$$

where $\Sigma_s \in \mathbb{R}^{ms\times ms}$ is a diagonal matrix containing only positive entries. We can write the state sequence $X_{s,N}$ as

$$X_{s,N} = A^s X_{0,N} + \mathcal{C}_s^r U_{0,s,N},$$

where $\mathcal{C}_s^r$ denotes the reversed controllability matrix, that is,

$$\mathcal{C}_s^r = \begin{bmatrix}A^{s-1}B & A^{s-2}B & \cdots & B\end{bmatrix}.$$

By virtue of the white-noise property of the input $u(k)$, we have

$$\lim_{N\to\infty}\frac{1}{N}X_{0,N}U_{0,s,N}^T = 0.$$

Therefore,

$$\lim_{N\to\infty}\frac{1}{N}X_{s,N}U_{0,s,N}^T = \mathcal{C}_s^r\Sigma_s.$$

With the limits derived above, we can then write the matrix between round brackets in (9.44) as

$$\lim_{N\to\infty}\frac{1}{N}\begin{bmatrix}X_{s,N}\\U_{s,s,N}\end{bmatrix}\begin{bmatrix}U_{0,s,N}^T & U_{s,s,N}^T\end{bmatrix} = \begin{bmatrix}\mathcal{C}_s^r\Sigma_s & 0\\0 & \Sigma_s\end{bmatrix}.$$

We have $\Sigma_s > 0$. Since $s \geq n$, and since the system is minimal, it follows that $\text{rank}(\mathcal{C}_s^r\Sigma_s) = n$. This reasoning completes the proof. $\qquad\square$

In Section 10.2.4.2 we discuss more general conditions on the input signal that are related to the satisfaction of Equation (9.43). For now, we assume that the input signal is such that Equation (9.43) holds. With this assumption, we can use the RQ factorization (9.40) on page 314 with $Z_N = U_{0,s,N}$ and $i = s$ to compute unbiased estimates of the system matrices. We have the following important result.

Theorem 9.3 *Consider the minimal system (9.26)–(9.27), with $x(k)$, $u(k)$, and $v(k)$ ergodic stochastic processes such that $v(k)$ is uncorrelated with $x(j)$ and $u(j)$ for all $k, j \in \mathbb{Z}$. Let the input $u(k)$ be such that the matrix*

$$\lim_{N\to\infty} \frac{1}{N} U_{0,s,N} U_{0,s,N}^{\mathrm{T}} \tag{9.45}$$

has full rank and that Equation (9.44) is satisfied. Take $s \geq n$ in the RQ factorization

$$\begin{bmatrix} U_{s,s,N} \\ U_{0,s,N} \\ Y_{s,s,N} \end{bmatrix} = \begin{bmatrix} R_{11} & 0 & 0 & 0 \\ R_{21} & R_{22} & 0 & 0 \\ R_{31} & R_{32} & R_{33} & 0 \end{bmatrix} \begin{bmatrix} Q_1 \\ Q_2 \\ Q_3 \\ Q_4 \end{bmatrix}, \tag{9.46}$$

then

$$\mathrm{range}\left(\lim_{N\to\infty} \frac{1}{\sqrt{N}} R_{32}\right) = \mathrm{range}(\mathcal{O}_s).$$

Proof From Lemma 9.5 on page 316 it follows that the first condition, Equation (9.42), on the instrumental-variable matrix $Z_N = U_{0,s,N}$ is satisfied. By virtue of the choice of the instrumental-variable matrix, we have

$$\lim_{N\to\infty} \frac{1}{N} Y_{s,s,N} \Pi_{U_{s,s,N}}^{\perp} U_{0,s,N}^{\mathrm{T}} = \lim_{N\to\infty} \frac{1}{N} \mathcal{O}_s X_{s,N} \Pi_{U_{s,s,N}}^{\perp} U_{0,s,N}^{\mathrm{T}}. \tag{9.47}$$

Since $s \geq n$, and since condition (9.44) holds, the proof of Lemma 9.6 on page 316 implies

$$\mathrm{rank}\left(\lim_{N\to\infty} \frac{1}{N} X_{s,N} \Pi_{U_{s,s,N}}^{\perp} U_{0,s,N}^{\mathrm{T}}\right) = n.$$

Application of Sylvester's inequality (Lemma 2.1 on page 16) to Equation (9.47) shows that

$$\mathrm{rank}\left(\lim_{N\to\infty} \frac{1}{N} Y_{s,s,N} \Pi_{U_{s,s,N}}^{\perp} U_{0,s,N}^{\mathrm{T}}\right) = n.$$

Using the RQ factorization (9.46), we therefore have by virtue of Lemma 9.4 on page 315

$$\operatorname{rank}\left(\lim_{N\to\infty}\frac{1}{N}Y_{s,s,N}\Pi^{\perp}_{U_{s,s,N}}U^{\mathrm{T}}_{0,s,N}\right)=\operatorname{rank}\left(\lim_{N\to\infty}\frac{1}{N}R_{32}R^{\mathrm{T}}_{22}\right)=n.$$

Because of assumption (9.45), the matrix

$$\lim_{N\to\infty}\frac{1}{\sqrt{N}}R_{22}$$

is invertible. Application of Sylvester's inequality shows that

$$\operatorname{rank}\left(\lim_{N\to\infty}\frac{1}{\sqrt{N}}\mathcal{O}_{s}X_{s,N}\Pi^{\perp}_{U_{s,s,N}}U^{\mathrm{T}}_{0,s,N}\right)=\operatorname{rank}\left(\lim_{N\to\infty}\frac{1}{\sqrt{N}}R_{32}\right)=n.$$

This argumentation yields the desired result. $\qquad\square$

Theorem 9.3 shows that the matrices A_T and C_T can be estimated consistently from an SVD of the matrix R_{32} in a similar way to that described in Section 9.2.4.3. The matrices B_T and D_T and the initial state $x_T(0) = T^{-1}x(0)$ can be computed by solving a least-squares problem. Using Equations (9.23) and (9.24) on page 306, it is easy to see that

$$y(k) = \phi(k)^{\mathrm{T}}\theta + v(k).$$

Because $v(k)$ is not correlated with $\phi(k)$, an unbiased estimate of θ can be obtained by solving

$$\min_{\theta}\frac{1}{N}\sum_{k=0}^{N-1}||y(k)-\phi(k)^{\mathrm{T}}\theta||^2_2.$$

The subspace identification method presented in this section is called the *PI-MOESP* method (Verhaegen, 1993), where "PI" stands for "past inputs" and refers to the fact that the past-input data matrix is used as an instrumental-variable matrix.

Example 9.6 (PI-MOESP with colored noise) To show that the PI-MOESP method yields unbiased estimates, we perform the same experiment as in Example 9.5 on page 312. The eigenvalues of the estimated A matrix obtained by MOESP and by PI-MOESP are shown in Figure 9.6.

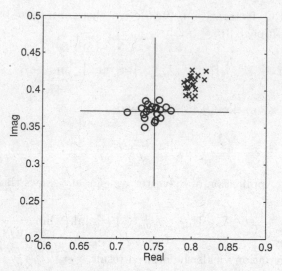

Fig. 9.6. One of the eigenvalues of the estimated A matrix for 20 different realizations of colored measurement noise in Example 9.6. The crosses are the eigenvalues obtained by MOESP; the circles are the eigenvalues obtained by PI-MOESP. The big cross corresponds to the real value.

We see that, whereas the eigenvalues obtained from MOESP are biased, those obtained from PI-MOESP are unbiased.

Summary of PI-MOESP to calculate the column space of $\mathcal{O}_s$

Consider the system

$$x(k+1) = Ax(k) + Bu(k),$$
$$y(k) = Cx(k) + Du(k) + v(k),$$

with $v(k)$ an ergodic noise sequence that is uncorrelated with $u(k)$, and $u(k)$ an ergodic sequence such that Equations (9.44), on page 317, and (9.45), on page 318, are satisfied.

From the RQ factorization

$$\begin{bmatrix} U_{s,s,N} \\ U_{0,s,N} \\ Y_{s,s,N} \end{bmatrix} = \begin{bmatrix} R_{11} & 0 & 0 \\ R_{21} & R_{22} & 0 \\ R_{31} & R_{32} & R_{33} \end{bmatrix} \begin{bmatrix} Q_1 \\ Q_2 \\ Q_3 \end{bmatrix},$$

we have

$$\mathrm{range}\left(\lim_{N \to \infty} \frac{1}{\sqrt{N}} R_{32} \right) = \mathrm{range}(\mathcal{O}_s).$$

9.6 Subspace identification with process and measurement noise

We now turn to the identification problem considered in Chapter 8. The refinement with respect to the previous section is that the noise $v(k)$ in (9.26) and (9.27) is obtained by filtering white-noise sequences as follows:

$$\widetilde{x}(k+1) = A\widetilde{x}(k) + Bu(k) + w(k), \tag{9.48}$$

$$y(k) = C\widetilde{x}(k) + Du(k) + v(k). \tag{9.49}$$

Throughout this section we assume that the process noise $w(k)$ and the measurement noise $v(k)$ are zero-mean white-noise sequences that are uncorrelated with the input $u(k)$. The relationship between the above signal-generating system and the system (9.26)–(9.27) can be made more explicit if we write the system (9.48)–(9.49) as

$$\overline{x}(k+1) = A\overline{x}(k) + Bu(k),$$

$$y(k) = C\overline{x}(k) + Du(k) + \overline{v}(k),$$

where $\overline{v}(k)$ is given by

$$\xi(k+1) = A\xi(k) + w(k),$$

$$\overline{v}(k) = C\xi(k) + v(k),$$

with $\xi(k) = \widetilde{x}(k) - \overline{x}(k)$. Hence,

$$\overline{v}(k) = CA^k\left(\widetilde{x}(0) - \overline{x}(0)\right) + \sum_{i=0}^{k-1} CA^{k-i-1}Bw(i) + v(k),$$

from which we clearly see that $\overline{v}(k)$ is a colored-noise sequence.

When we consider only the input–output transfer of the systems under investigation, we can formulate (9.48) and (9.49) in innovation form as in Section 8.2:

$$x(k+1) = Ax(k) + Bu(k) + Ke(k), \tag{9.50}$$

$$y(k) = Cx(k) + Du(k) + e(k), \tag{9.51}$$

where the innovation $e(k)$ is a white-noise sequence and K is the Kalman gain.

Using the system representation (9.50)–(9.51), we can relate the block Hankel matrices $U_{i,s,N}$ and $Y_{i,s,N}$ constructed from input–output data by the following data equation:

$$Y_{i,s,N} = \mathcal{O}_s X_{i,N} + \mathcal{T}_s U_{i,s,N} + \mathcal{S}_s E_{i,s,N}, \tag{9.52}$$

where $E_{i,s,N}$ is a block Hankel matrix constructed from the sequence $e(k)$, and

$$
S_s = \begin{bmatrix}
I_\ell & 0 & 0 & \cdots & 0 \\
CK & I_\ell & 0 & \cdots & 0 \\
CAK & CK & I_\ell & & 0 \\
\vdots & & & \ddots & \ddots \\
CA^{s-2}K & CA^{s-3}K & \cdots & CK & I_\ell
\end{bmatrix}
$$

describes the weighting matrix of the block Hankel matrix $E_{i,s,N}$.

In a subspace identification framework the solution to the identification problem considered in Section 8.2 starts with the estimation of the column space of the matrix $\mathcal{O}_s$. To achieve this, first the influence of the input is removed as follows:

$$
Y_{i,s,N}\Pi^{\perp}_{U_{i,s,N}} = \mathcal{O}_s X_{i,N}\Pi^{\perp}_{U_{i,s,N}} + S_s E_{i,s,N}\Pi^{\perp}_{U_{i,s,N}}.
$$

Next, we have to get rid of the term $S_s E_{i,s,N}\Pi^{\perp}_{U_{i,s,N}}$. As in Section 9.5, we will use the instrumental-variable approach from Section 9.4. We search for a matrix Z_N that has the following properties:

$$
\lim_{N\to\infty} \frac{1}{N} E_{i,s,N}\Pi^{\perp}_{U_{i,s,N}} Z_N^{T} = 0, \tag{9.53}
$$

$$
\mathrm{rank}\left(\lim_{N\to\infty} \frac{1}{N} X_{i,N}\Pi^{\perp}_{U_{i,s,N}} Z_N^{T} \right) = n. \tag{9.54}
$$

If we take $i = s$ as in Section 9.5, we can use $U_{0,s,N}$ as an instrumental variable. In addition, we can also use the past output $Y_{0,s,N}$ as an instrumental variable. Using both $U_{0,s,N}$ and $Y_{0,s,N}$ instead of only $U_{0,s,N}$ will result in better models when a finite number of data points is used. This is intuitively clear on looking at Equations (9.53) and (9.54) and keeping in mind that in practice we have only a finite number of data points available. This will be illustrated by an example.

Example 9.7 (Instrumental variables) Consider the system given by

$$
A = \begin{bmatrix} 1.5 & 1 \\ -0.7 & 0 \end{bmatrix}, \quad B = \begin{bmatrix} 1 \\ 0.5 \end{bmatrix}, \quad K = \begin{bmatrix} 2.5 \\ -0.5 \end{bmatrix}, \quad C = \begin{bmatrix} 1 & 0 \end{bmatrix}, \quad D = 0.
$$

This system is simulated for 2000 samples with an input signal $u(k)$ and a noise signal $e(k)$, both of which are white-noise zero-mean unit-variance

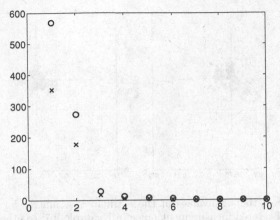

Fig. 9.7. Singular values of the matrix R_{32} from the RQ factorization (9.46) for Example 9.7. The crosses correspond to the case that $Z_N = U_{0,s,N}$ and the circles correspond to the case that Z_N contains both $U_{0,s,N}$ and $Y_{0,s,N}$.

sequences that are uncorrelated. Figure 9.7 compares the singular values of the matrix R_{32} from the RQ factorization (9.46) on page 318 with $i = s$, for the case that $Z_N = U_{0,s,N}$ and the case that Z_N contains both $U_{0,s,N}$ and $Y_{0,s,N}$. We see that in the latter case the first two singular values are larger than the ones obtained for $Z_N = U_{0,s,N}$, and that the singular values corresponding to the noise remain almost the same. This means that, for the case that Z_N contains both $U_{0,s,N}$ and $Y_{0,s,N}$, a better approximation of the column space of the extended observability matrix $\mathcal{O}_s$ is obtained than when using only $U_{0,s,N}$ as instrumental-variable matrix. This result can be proven using elementary numerical analysis of the SVD (Golub and Van Loan, 1996). That a more accurate model is obtained is illustrated in Figure 9.8 via the estimated eigenvalues of the A matrix for 20 different realizations of the input–output data. For the case in which Z_N contains both $U_{0,s,N}$ and $Y_{0,s,N}$, the variance of the estimated eigenvalues (indicated by the size of the cloud of the estimates) is smaller than for the case in which Z_N contains only $U_{0,s,N}$.

The previous example motivates the choice

$$Z_N = \begin{bmatrix} U_{0,s,N} \\ Y_{0,s,N} \end{bmatrix}. \tag{9.55}$$

The next lemma shows that with this choice Equation (9.53) is satisfied.

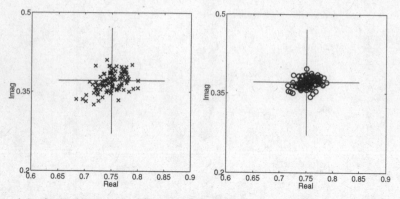

Fig. 9.8. One of the eigenvalues of the estimated A matrix for 20 different realizations of colored measurement noise in Example 9.7: (a) the crosses correspond to the case that $Z_N = U_{0,s,N}$ and (b) the circles correspond to the case that Z_N contains both $U_{0,s,N}$ and $Y_{0,s,N}$.

Lemma 9.7 *Consider the system (9.50)–(9.51), with $x(k)$, $u(k)$, and $e(k)$ ergodic stochastic processes such that the white-noise sequence $e(k)$ is uncorrelated with $x(j)$ and $u(j)$ for all $k, j \in \mathbb{Z}$. For*

$$Z_N = \begin{bmatrix} U_{0,s,N} \\ Y_{0,s,N} \end{bmatrix},$$

we have

$$\lim_{N \to \infty} \frac{1}{N} E_{s,s,N} \Pi^{\perp}_{U_{s,s,N}} Z_N^{\mathrm{T}} = 0. \qquad (9.56)$$

Proof Since $u(k)$ is uncorrelated with $e(k)$, we can first write the condition (9.56) as

$$\lim_{N \to \infty} \frac{1}{N} E_{s,s,N} \Pi^{\perp}_{U_{s,s,N}} Z_N^{\mathrm{T}} = \lim_{N \to \infty} \frac{1}{N} E_{s,s,N} Z_N^{\mathrm{T}},$$

then

$$\lim_{N \to \infty} \frac{1}{N} E_{s,s,N} U_{0,s,N}^{\mathrm{T}} = 0, \qquad \lim_{N \to \infty} \frac{1}{N} E_{s,s,N} U_{s,s,N}^{\mathrm{T}} = 0,$$

and then

$$\lim_{N \to \infty} \frac{1}{N} Y_{0,s,N} E_{s,s,N}^{\mathrm{T}} = \lim_{N \to \infty} \frac{1}{N} \left(\mathcal{O}_s X_{0,N} E_{s,s,N}^{\mathrm{T}} + \mathcal{T}_s U_{0,s,N} E_{s,s,N}^{\mathrm{T}} \right.$$

$$\left. + \mathcal{S}_s E_{0,s,N} E_{s,s,N}^{\mathrm{T}} \right)$$

$$= \lim_{N \to \infty} \frac{1}{N} \left(\mathcal{O}_s X_{0,N} E_{s,s,N}^{\mathrm{T}} + \mathcal{S}_s E_{0,s,N} E_{s,s,N}^{\mathrm{T}} \right).$$

Since $e(k)$ is a white-noise sequence, it follows that

$$\lim_{N\to\infty} \frac{1}{N} O_s X_{0,N} E_{s,s,N}^T = 0,$$

and

$$\lim_{N\to\infty} \frac{1}{N} E_{0,s,N} E_{s,s,N}^T = 0,$$

and therefore the proof is completed. $\qquad\square$

Now we turn our attention to the second condition, Equation (9.54) on page 322, for the instrumental variable matrix Z_N in Equation (9.55):

$$\text{rank}\left(\lim_{N\to\infty} \frac{1}{N} X_{s,N} \Pi_{U_{s,s,N}}^\perp \begin{bmatrix} U_{0,s,N}^T & Y_{0,s,N}^T \end{bmatrix}\right) = n. \tag{9.57}$$

For the projection matrix to exist, we need to assume that the matrix $\lim_{N\to\infty}(1/N)U_{s,s,N}U_{s,s,N}^T$ is invertible. Consider the following matrix partitioning:

$$\lim_{N\to\infty} \frac{1}{N} \begin{bmatrix} X_{s,N} \\ U_{s,s,N} \end{bmatrix} \begin{bmatrix} U_{0,s,N}^T & Y_{0,s,N}^T \,|\, U_{s,s,N}^T \end{bmatrix}$$

$$= \lim_{N\to\infty} \frac{1}{N} \begin{bmatrix} X_{s,N}[U_{0,s,N}^T \quad Y_{0,s,N}^T] & X_{s,N}U_{s,s,N}^T \\ U_{s,s,N}[U_{0,s,N}^T \quad Y_{0,s,N}^T] & U_{s,s,N}U_{s,s,N}^T \end{bmatrix}.$$

Application of Lemma 2.3 on page 19 shows that the Schur complement of the matrix $\lim_{N\to\infty}(1/N)U_{s,s,N}U_{s,s,N}^T$ equals

$$\lim_{N\to\infty} \frac{1}{N} X_{s,N}\left(I_N - U_{s,s,N}^T(U_{s,s,N}U_{s,s,N}^T)^{-1}U_{s,s,N}\right)\begin{bmatrix} U_{0,s,N}^T & Y_{0,s,N}^T \end{bmatrix}$$

$$= \lim_{N\to\infty} \frac{1}{N} X_{s,N} \Pi_{U_{s,s,N}}^\perp \begin{bmatrix} U_{0,s,N}^T & Y_{0,s,N}^T \end{bmatrix}.$$

Therefore, condition (9.57) is equivalent to

$$\text{rank}\left(\lim_{N\to\infty} \frac{1}{N} \begin{bmatrix} X_{s,N} \\ U_{s,s,N} \end{bmatrix} \begin{bmatrix} Y_{0,s,N}^T & U_{0,s,N}^T & U_{s,s,N}^T \end{bmatrix}\right) = n + sm. \tag{9.58}$$

Note that we have switched the positions of $Y_{0,s,N}$ and $U_{0,s,N}$, which, of course, does not change the rank condition. In general, Equation (9.58) is almost always satisfied. However, it is possible to construct special types of input signals $u(k)$ and noise signals $e(k)$ for which the rank condition fails (Jansson, 1997; Jansson and Wahlberg, 1998). An example input signal for which the rank condition always holds is a zero-mean white-noise sequence. For this signal we can state the following lemma, which you are requested to prove in Exercise 9.9 on page 344.

Lemma 9.8 *(Jansson and Wahlberg, 1998) Consider the minimal system (9.50)–(9.51), with $x(k)$, $u(k)$, and $e(k)$ ergodic stochastic processes such that the white-noise sequence $e(k)$ is uncorrelated with $x(j)$ and $u(j)$ for all $k, j \in \mathbb{Z}$. Let the input $u(k)$ be a zero-mean white-noise sequence and take*

$$Z_N = \begin{bmatrix} U_{0,s,N} \\ Y_{0,s,N} \end{bmatrix},$$

then

$$\mathrm{rank}\left(\lim_{N \to \infty} \frac{1}{N} X_{s,N} \Pi^{\perp}_{U_{s,s,N}} Z_N^{\mathrm{T}} \right) = n.$$

We refer to Section 10.2.4.2 for a more elaborate discussion on the relationship between conditions on the input signal and the rank condition (9.58).

9.6.1 The PO-MOESP method

The use of the instrumental-variable matrix Z_N in Equation (9.55) is the basis of the so-called *PO-MOESP* method (Verhaegen, 1994), where "PO" stands for "past outputs" and refers to the fact that the instrumental variables also contain the past output data. A key result on which the PO-MOESP method is based is presented in the following lemma.

Lemma 9.9 *Consider the minimal system (9.50)–(9.51), with $x(k)$, $u(k)$, and $e(k)$ ergodic stochastic processes such that the white-noise sequence $e(k)$ is uncorrelated with $x(j)$ and $u(j)$ for all $k, j \in \mathbb{Z}$. Let the state sequence $x(k)$ and the input sequence $u(k)$ be such that*

$$\mathrm{rank}\left(\lim_{N \to \infty} \frac{1}{N} \begin{bmatrix} X_{0,N} \\ U_{0,2s,N} \end{bmatrix} \begin{bmatrix} X_{0,N}^{\mathrm{T}} & U_{0,2s,N}^{\mathrm{T}} \end{bmatrix} \right) = n + 2sm, \qquad (9.59)$$

then

$$\mathrm{rank}\left(\lim_{N \to \infty} \frac{1}{N} \begin{bmatrix} Y_{0,s,N} \\ U_{0,2s,N} \end{bmatrix} \begin{bmatrix} Y_{0,s,N}^{\mathrm{T}} & U_{0,2s,N}^{\mathrm{T}} \end{bmatrix} \right) = s(\ell + 2m).$$

Proof We can write

$$\begin{bmatrix} Y_{0,s,N} \\ U_{0,2s,N} \end{bmatrix} = \underbrace{\begin{bmatrix} \mathcal{O}_s & [\mathcal{T}_s \ \ 0] & \mathcal{S}_s \\ 0 & I_{2sm} & 0 \end{bmatrix}}_{} \begin{bmatrix} X_{0,N} \\ U_{0,2s,N} \\ E_{0,s,N} \end{bmatrix}.$$

Since the matrix $\mathcal{S}_s$ is square and lower-triangular with diagonal entries equal to unity it has full rank. This explains the fact that the under-braced matrix above has full row rank. Application of Sylvester's inequality (Lemma 2.1 on page 16) shows that the lemma is proven if the matrix

$$\frac{1}{N}\begin{bmatrix} X_{0,N} \\ U_{0,2s,N} \\ E_{0,s,N} \end{bmatrix} \begin{bmatrix} X_{0,N}^{\mathrm{T}} & U_{0,2s,N}^{\mathrm{T}} & E_{0,s,N}^{\mathrm{T}} \end{bmatrix} \tag{9.60}$$

has full rank for $N \to \infty$. Observe that, because of the white-noise properties of $e(k)$, we have

$$\lim_{N\to\infty} \frac{1}{N}\begin{bmatrix} X_{0,N} \\ U_{0,2s,N} \\ E_{0,s,N} \end{bmatrix} \begin{bmatrix} X_{0,N}^{\mathrm{T}} & U_{0,2s,N}^{\mathrm{T}} & E_{0,s,N}^{\mathrm{T}} \end{bmatrix}$$

$$= \lim_{N\to\infty} \frac{1}{N}\begin{bmatrix} X_{0,N}X_{0,N}^{\mathrm{T}} & X_{0,N}U_{0,2s,N}^{\mathrm{T}} & 0 \\ U_{0,2s,N}X_{0,N}^{\mathrm{T}} & U_{0,2s,N}U_{0,2s,N}^{\mathrm{T}} & 0 \\ 0 & 0 & E_{0,s,N}E_{0,s,N}^{\mathrm{T}} \end{bmatrix},$$

and

$$\mathrm{rank}\left(\lim_{N\to\infty} \frac{1}{N}E_{0,s,N}E_{0,s,N}^{\mathrm{T}}\right) = \ell s.$$

Since

$$\mathrm{rank}\left(\lim_{N\to\infty} \frac{1}{N}\begin{bmatrix} X_{0,N}X_{0,N}^{\mathrm{T}} & X_{0,N}U_{0,2s,N}^{\mathrm{T}} \\ U_{0,2s,N}X_{0,N}^{\mathrm{T}} & U_{0,2s,N}U_{0,2s,N}^{\mathrm{T}} \end{bmatrix}\right) = n + 2ms,$$

it follows that the matrix (9.60) does indeed have full rank. $\qquad\square$

The foundation of the PO-MOESP method is presented in the following theorem.

Theorem 9.4 *Consider the minimal system (9.50)–(9.51), with $x(k)$, $u(k)$, and $e(k)$ ergodic stochastic processes such that the white-noise sequence $e(k)$ is uncorrelated with $x(j)$ and $u(j)$ for all $k, j \in \mathbb{Z}$. Let the state sequence $x(k)$ and the input sequence $u(k)$ be such that Equations (9.58) and (9.59) are satisfied. Take $s \geq n$ in the RQ factorization*

$$\begin{bmatrix} U_{s,s,N} \\ \begin{bmatrix} U_{0,s,N} \\ Y_{0,s,N} \end{bmatrix} \\ Y_{s,s,N} \end{bmatrix} = \begin{bmatrix} R_{11} & 0 & 0 & 0 \\ R_{21} & R_{22} & 0 & 0 \\ R_{31} & R_{32} & R_{33} & 0 \end{bmatrix} \begin{bmatrix} Q_1 \\ Q_2 \\ Q_3 \\ Q_4 \end{bmatrix}, \tag{9.61}$$

then

$$\text{range}\left(\lim_{N\to\infty}\frac{1}{\sqrt{N}}R_{32}\right) = \text{range}(\mathcal{O}_s).$$

Proof From Lemma 9.7 on page 324 it follows that the first condition (9.56) on the instrumental variable matrix

$$Z_N = \begin{bmatrix} U_{0,s,N} \\ Y_{0,s,N} \end{bmatrix}$$

is satisfied. By virtue of the choice of the instrumental-variable matrix, we have

$$\lim_{N\to\infty}\frac{1}{N}Y_{s,s,N}\Pi^{\perp}_{U_{s,s,N}}\begin{bmatrix} U_{0,s,N} \\ Y_{0,s,N} \end{bmatrix}^{\mathrm{T}} = \lim_{N\to\infty}\frac{1}{N}\mathcal{O}_s X_{s,N}\Pi^{\perp}_{U_{s,s,N}}\begin{bmatrix} U_{0,s,N} \\ Y_{0,s,N} \end{bmatrix}^{\mathrm{T}}.$$

$$(9.62)$$

Since $s \geq n$, and since condition (9.58) holds, the discussion just before Lemma 9.8 on page 326 implies

$$\text{rank}\left(\lim_{N\to\infty}\frac{1}{N}X_{s,N}\Pi^{\perp}_{U_{s,s,N}}\begin{bmatrix} U_{0,s,N} \\ Y_{0,s,N} \end{bmatrix}^{\mathrm{T}}\right) = n.$$

Application of Sylvester's inequality (Lemma 2.1 on page 16) to Equation (9.62) shows that

$$\text{rank}\left(\lim_{N\to\infty}\frac{1}{N}Y_{s,s,N}\Pi^{\perp}_{U_{s,s,N}}\begin{bmatrix} U_{0,s,N} \\ Y_{0,s,N} \end{bmatrix}^{\mathrm{T}}\right) = n.$$

Using the RQ factorization (9.61), we therefore have, by virtue of Lemma 9.4 on page 315,

$$\text{rank}\left(\lim_{N\to\infty}\frac{1}{N}Y_{s,s,N}\Pi^{\perp}_{U_{s,s,N}}\begin{bmatrix} U_{0,s,N} \\ Y_{0,s,N} \end{bmatrix}^{\mathrm{T}}\right) = \text{rank}\left(\lim_{N\to\infty}\frac{1}{N}R_{32}R_{22}^{\mathrm{T}}\right)$$
$$= n.$$

Because of Equation (9.59), the matrix

$$\lim_{N\to\infty}\frac{1}{\sqrt{N}}R_{22}$$

is invertible. Application of Sylvester's inequality shows that

$$\text{rank}\left(\lim_{N\to\infty}\frac{1}{\sqrt{N}}\mathcal{O}_s X_{s,N}\Pi^{\perp}_{U_{s,s,N}}\begin{bmatrix} U_{0,s,N} \\ Y_{0,s,N} \end{bmatrix}^{\mathrm{T}}\right) = \text{rank}\left(\lim_{N\to\infty}\frac{1}{\sqrt{N}}R_{32}\right)$$
$$= n.$$

This argumentation yields the desired result.   □

So the matrices A_T and C_T can be estimated consistently from an SVD of the matrix R_{32} in a similar way to that described in Section 9.2.4.3. The matrices B_T and D_T and the initial state $x_T(0) = T^{-1}x(0)$ can be computed by solving a least-squares problem. Using Equations (9.23) and (9.24) on page 306, it is easy to see that

$$y(k) = \phi(k)^{\mathrm{T}}\theta + \left(\sum_{\tau=0}^{k-1} C_T A_T^{k-\tau-1} K_T e(\tau)\right) + e(k).$$

Because $e(k)$ is not correlated with $\phi(k)$, an unbiased estimate of θ can be obtained by solving

$$\min_{\theta} \frac{1}{N} \sum_{k=0}^{N-1} ||y(k) - \phi(k)^{\mathrm{T}}\theta||_2^2.$$

Summary of PO-MOESP to calculate the column space of $\mathcal{O}_s$

Consider the system

$$x(k+1) = Ax(k) + Bu(k) + Ke(k),$$
$$y(k) = Cx(k) + Du(k) + e(k),$$

where $e(k)$ is an ergodic white-noise sequence that is uncorrelated with the ergodic sequence $u(k)$, and $u(k)$ and $e(k)$ are such that Equations (9.58) and (9.59) on page 326 are satisfied.
From the RQ factorization

$$\begin{bmatrix} U_{s,s,N} \\ \begin{bmatrix} U_{0,s,N} \\ Y_{0,s,N} \end{bmatrix} \\ Y_{s,s,N} \end{bmatrix} = \begin{bmatrix} R_{11} & 0 & 0 \\ R_{21} & R_{22} & 0 \\ R_{31} & R_{32} & R_{33} \end{bmatrix} \begin{bmatrix} Q_1 \\ Q_2 \\ Q_3 \end{bmatrix},$$

we have

$$\mathrm{range}\left(\lim_{N\to\infty} \frac{1}{\sqrt{N}} R_{32}\right) = \mathrm{range}(\mathcal{O}_s).$$

9.6.2 Subspace identification as a least-squares problem

In this section we reveal a close link between the RQ factorization (9.61) used in the PO-MOESP scheme and the solution to a least-squares problem. First, it will be shown, in the next theorem, that the extended observability matrix $\mathcal{O}_s$ can also be derived from the solution of a least-squares problem. Second, it will be shown that this solution enables the approximation of the state sequence of a Kalman filter. The latter idea was exploited in Van Overschee and De Moor (1994; 1996b) to derive another subspace identification method.

Theorem 9.5 *(Peternell et al., 1996) Consider the minimal system (9.50)–(9.51), with $x(k)$, $u(k)$, and $e(k)$ ergodic stochastic processes such that the white-noise sequence $e(k)$ is uncorrelated with $x(j)$ and $u(j)$ for all $k, j \in \mathbb{Z}$. Let the state sequence $x(k)$ and the input sequence $u(k)$ be such that Equation (9.59) is satisfied. Take the instrumental-variable matrix Z_N equal to*

$$\begin{bmatrix} U_{0,s,N} \\ Y_{0,s,N} \end{bmatrix}$$

and consider the following least-squares problem:

$$\begin{bmatrix} \widehat{L}_N^u & \widehat{L}_N^z \end{bmatrix} = \arg \min_{L^u, L^z} \left\| Y_{s,s,N} - \begin{bmatrix} L^u & L^z \end{bmatrix} \begin{bmatrix} U_{s,s,N} \\ Z_N \end{bmatrix} \right\|_F^2, \tag{9.63}$$

then

$$\lim_{N \to \infty} \widehat{L}_N^z = \mathcal{O}_s \mathcal{L}_s + \mathcal{O}_s (A - KC)^s \Delta_z, \tag{9.64}$$

with

$$\mathcal{L}_s = \begin{bmatrix} \mathcal{L}_s^u & \mathcal{L}_s^y \end{bmatrix},$$
$$\mathcal{L}_s^u = \begin{bmatrix} (A - KC)^{s-1}(B - KD) & (A - KC)^{s-2}(B - KD) & \cdots & (B - KD) \end{bmatrix},$$
$$\mathcal{L}_s^y = \begin{bmatrix} (A - KC)^{s-1}K & (A - KC)^{s-2}K & \cdots & K \end{bmatrix}, \tag{9.65}$$

and the bounded matrix $\Delta_z \in \mathbb{R}^{n \times s(\ell+m)}$ given by

$$\lim_{N \to \infty} \frac{1}{N} X_{0,N} \begin{bmatrix} U_{s,s,N}^T & Z_N^T \end{bmatrix} \left(\frac{1}{N} \begin{bmatrix} U_{s,s,N} \\ Z_N \end{bmatrix} \begin{bmatrix} U_{s,s,N}^T & Z_N^T \end{bmatrix} \right)^{-1} = \begin{bmatrix} \Delta_u & \Delta_z \end{bmatrix}.$$

Proof Substitution of Equation (9.51) on page 321 into Equation (9.50) yields

$$x(k+1) = (A - KC)x(k) + (B - KD)u(k) + Ky(k);$$

therefore,

$$X_{s,N} = (A - KC)^s X_{0,N} + \mathcal{L}_s \begin{bmatrix} U_{0,s,N} \\ Y_{0,s,N} \end{bmatrix}, \tag{9.66}$$

and the data equation for $i = s$ can be written as

$$Y_{s,s,N} = \mathcal{O}_s \mathcal{L}_s Z_N + \mathcal{T}_s U_{s,s,N} + \mathcal{S}_s E_{s,s,N} + \mathcal{O}_s (A - KC)^s X_{0,N}. \tag{9.67}$$

The normal equations corresponding to the least-squares problem (9.63) read

$$\begin{bmatrix} L^u & L^z \end{bmatrix} \begin{bmatrix} U_{s,s,N} \\ Z_N \end{bmatrix} \begin{bmatrix} U_{s,s,N}^T & Z_N^T \end{bmatrix}$$

$$= Y_{s,s,N} \begin{bmatrix} U_{s,s,N}^T & Z_N^T \end{bmatrix}$$

$$= \begin{bmatrix} \mathcal{T}_s & \mathcal{O}_s \mathcal{L}_s \end{bmatrix} \begin{bmatrix} U_{s,s,N} \\ Z_N \end{bmatrix} \begin{bmatrix} U_{s,s,N}^T & Z_N^T \end{bmatrix} + \mathcal{S}_s E_{s,s,N} \begin{bmatrix} U_{s,s,N}^T & Z_N^T \end{bmatrix}$$

$$+ \mathcal{O}_s (A - KC)^s X_{0,N} \begin{bmatrix} U_{s,s,N}^T & Z_N^T \end{bmatrix}.$$

Since $e(k)$ is white noise and is independent from $u(k)$, we get

$$\lim_{N \to \infty} \frac{1}{N} E_{s,s,N} \begin{bmatrix} U_{s,s,N}^T & Z_N^T \end{bmatrix} = 0.$$

By Lemma 9.9 we have that

$$\text{rank}\left(\lim_{N \to \infty} \frac{1}{N} \begin{bmatrix} U_{s,s,N} \\ Z_N \end{bmatrix} \begin{bmatrix} U_{s,s,N}^T & Z_N^T \end{bmatrix} \right) = s(\ell + 2m),$$

and thus

$$\lim_{N \to \infty} \begin{bmatrix} \widehat{L}_N^u & \widehat{L}_N^z \end{bmatrix} = \begin{bmatrix} \mathcal{T}_s & \mathcal{O}_s \mathcal{L}_s \end{bmatrix} + \mathcal{O}_s (A - KC)^s \begin{bmatrix} \Delta_u & \Delta_z \end{bmatrix},$$

and the proof is completed. $\qquad\square$

If we consider the instrumental-variable matrix Z_N as in the above theorem in the RQ factorization (9.61) on page 327, the part $\widehat{L}_N^z$ of the least-squares solution to the problem (9.63) can be written as

$$\widehat{L}_N^z = R_{32} R_{22}^{-1}. \qquad (9.68)$$

You are requested to verify this result in Exercise 9.10 on page 344. Because of this result we have that

$$\lim_{N \to \infty} R_{32} R_{22}^{-1} = \mathcal{O}_s \left(\mathcal{L}_s + (A - KC)^s \Delta_z \right).$$

The rank of the matrix R_{32} has already been investigated in Theorem 9.4 and, provided that (9.58) is satisfied, it is given by

$$\text{rank}\left(\lim_{N \to \infty} \frac{1}{\sqrt{N}} R_{32} \right) = \text{rank}\left(\lim_{N \to \infty} \frac{1}{N} R_{32} R_{22}^T \right)$$

$$= \text{rank}\left(\lim_{N \to \infty} \frac{1}{N} Y_{s,s,N} \Pi_{U_{s,s,N}}^{\perp} Z_N^T \right)$$

$$= \text{rank}\left(\lim_{N \to \infty} \frac{1}{N} X_{s,N} \Pi_{U_{s,s,N}}^{\perp} Z_N^T \right).$$

On the basis of the above theorem and representation of the least-squares solution in terms of the quantities computed in the RQ factorization of the PO-MOESP scheme, we have shown in another way that in the limit $N \to \infty$ the range spaces of the matrices R_{32} of extended observability matrix $\mathcal{O}_s$ are equal.

Using the result of Theorem 9.5, the definition of the matrix Z_N in this theorem and the expression (9.66) for $X_{s,N}$ given in the proof of the theorem, we have the following relationship:

$$
\begin{aligned}
\left(\lim_{N \to \infty} \widehat{L}_N^z \right) Z_N &= \left(\lim_{N \to \infty} R_{32} R_{22}^{-1} \right) Z_N \\
&= \mathcal{O}_s \mathcal{L}_s Z_N + \mathcal{O}_s (A - KC)^s \Delta_z Z_N \\
&= \mathcal{O}_s X_{s,N} + \underbrace{\mathcal{O}_s (A - KC)^s \left(\Delta_z Z_N - X_{0,N} \right)}.
\end{aligned}
$$

The matrix $X_{s,N}$ contains the state sequence of a Kalman filter. Since the system matrix $(A - KC)$ is asymptotically stable, it was argued in Van Overschee and De Moor (1994) that, for large enough s, the underbraced term in the above relationship is small. Therefore, the SVD

$$
R_{32} R_{22}^{-1} Z_N = U_n \Sigma_n V_n^{\mathrm{T}}
$$

can be used to approximate the column space of $\mathcal{O}_s$ by that of the matrix U_n, and to approximate the row space of the state sequence of a Kalman filter by

$$
\widehat{X}_{s,N} = \Sigma_n^{1/2} V_n^{\mathrm{T}}.
$$

The system matrices A_T, B_T, C_T, and D_T can now be estimated by solving the least-squares problem

$$
\min_{A_T, B_T, C_T, D_T} \left\| \begin{bmatrix} \widehat{X}_{s+1,N} \\ Y_{s,1,N-1} \end{bmatrix} - \begin{bmatrix} A_T & B_T \\ C_T & D_T \end{bmatrix} \begin{bmatrix} \widehat{X}_{s,N-1} \\ U_{s,1,N-1} \end{bmatrix} \right\|_{\mathrm{F}}^2 . \tag{9.69}
$$

The approximation of the state sequence as outlined above was originally proposed in the so-called *N4SID* subspace method (Van Overschee and De Moor, 1994; 1996b), where "N4SID" stands for "Numerical algorithm for Subspace IDentification." In Section 9.6.4 we show that the respective first steps of the N4SID and PO-MOESP methods, in which the SVD of a certain matrix is computed, differ only up to certain nonsingular weighting matrices.

Summary of N4SID to calculate the column space of $\mathcal{O}_s$

Consider the system

$$x(k+1) = Ax(k) + Bu(k) + Ke(k),$$
$$y(k) = Cx(k) + Du(k) + e(k),$$

with $e(k)$ a white-noise sequence that is uncorrelated with $u(k)$, and $u(k)$ and $e(k)$ such that Equations (9.58) and (9.59) on page 326 are satisfied.

From the RQ factorization

$$\begin{bmatrix} U_{s,s,N} \\ \begin{bmatrix} U_{0,s,N} \\ Y_{0,s,N} \end{bmatrix} \\ Y_{s,s,N} \end{bmatrix} = \begin{bmatrix} R_{11} & 0 & 0 \\ R_{21} & R_{22} & 0 \\ R_{31} & R_{32} & R_{33} \end{bmatrix} \begin{bmatrix} Q_1 \\ Q_2 \\ Q_3 \end{bmatrix},$$

we have for $N \to \infty$

$$\mathcal{O}_s X_{s,N} \approx R_{32} R_{22}^{-1} \begin{bmatrix} U_{0,s,N} \\ Y_{0,s,N} \end{bmatrix}$$

and

$$\mathrm{rank}\left(R_{32} R_{22}^{-1} \begin{bmatrix} U_{0,s,N} \\ Y_{0,s,N} \end{bmatrix} \right) = n.$$

9.6.3 Estimating the Kalman gain K_T

The estimated state sequence in the matrix $\widehat{X}_{s,N}$ is an approximation of the state sequence of the innovation model (9.50)–(9.51). Estimates of the covariance matrices that are related to this innovation model can be obtained from the state estimate, together with the estimated system matrices from the least-squares problem (9.69). Let the estimated system matrices be denoted by $\widehat{A}_T$, $\widehat{B}_T$, $\widehat{C}_T$, and $\widehat{D}_T$, then the residuals of the least-squares problem (9.69) are given by

$$\begin{bmatrix} \widehat{W}_{s,1,N-1} \\ \widehat{V}_{s,1,N-1} \end{bmatrix} = \begin{bmatrix} \widehat{X}_{s+1,N} \\ Y_{s,1,N-1} \end{bmatrix} - \begin{bmatrix} \widehat{A}_T & \widehat{B}_T \\ \widehat{C}_T & \widehat{D}_T \end{bmatrix} \begin{bmatrix} \widehat{X}_{s,N-1} \\ U_{s,1,N-1} \end{bmatrix}. \tag{9.70}$$

These residuals can be used to estimate the covariance matrices as follows:

$$\begin{bmatrix} \widehat{Q} & \widehat{S} \\ \widehat{S}^{\mathrm{T}} & \widehat{R} \end{bmatrix} = \lim_{N \to \infty} \frac{1}{N} \begin{bmatrix} \widehat{W}_{s,1,N} \\ \widehat{V}_{s,1,N} \end{bmatrix} \begin{bmatrix} \widehat{W}_{s,1,N}^{\mathrm{T}} & \widehat{V}_{s,1,N}^{\mathrm{T}} \end{bmatrix}. \tag{9.71}$$

The solution P of the following Riccati equation, that was derived in Section 5.7,

$$\widehat{P} = \widehat{A}_T \widehat{P} \widehat{A}_T^T + \widehat{Q} - (\widehat{S} + \widehat{A}_T \widehat{P} \widehat{C}_T^T)(\widehat{C}_T \widehat{P} \widehat{C}_T^T + \widehat{R})^{-1}(\widehat{S} + \widehat{A}_T \widehat{P} \widehat{C}_T^T)^T,$$

can be used to obtain an estimate of the Kalman gain $\widehat{K}_T$ for the system $(\widehat{A}_T, \widehat{B}_T, \widehat{C}_T, \widehat{D}_T)$:

$$\widehat{K}_T = (\widehat{S} + \widehat{A}_T \widehat{P} \widehat{C}_T^T)(\widehat{R} + \widehat{C}_T \widehat{P} \widehat{C}_T^T)^{-1}.$$

9.6.4 Relations among different subspace identification methods

The least-squares formulation in Theorem 9.5 on page 330 can be used to relate different subspace identification schemes for the estimation of the system matrices in (9.50)–(9.51). To show these relations, we present in the next theorem the solution to the least-squares problem (9.63) on page 330 in another, alternative, manner.

Theorem 9.6 *If Equation* (9.59) *is satisfied, then the solution* $\widehat{L}_N^z$ *to* (9.63) *can be formulated as*

$$\lim_{N\to\infty} \widehat{L}_N^z = \lim_{N\to\infty} \left(\frac{1}{N} Y_{s,s,N} \Pi_{U_{s,s,N}}^\perp Z_N^T \right) \left(\frac{1}{N} Z_N \Pi_{U_{s,s,N}}^\perp Z_N^T \right)^{-1}.$$

Proof Lemma 9.9 on page 326 shows that Equation (9.59) implies that the matrix

$$\lim_{N\to\infty} \frac{1}{N} \begin{bmatrix} Z_N \\ U_{s,s,N} \end{bmatrix} [Z_N^T \quad U_{s,s,N}^T]$$

is invertible. Application of the Schur complement (Lemma 2.3 on page 19) shows that this is equivalent to invertibility of the matrix

$$\lim_{N\to\infty} \frac{1}{N} Z_N \Pi_{U_{s,s,N}}^\perp Z_N^T$$

$$= \lim_{N\to\infty} \frac{1}{N} \left(Z_N Z_N^T - Z_N U_{s,s,N}^T (U_{s,s,N} U_{s,s,N}^T)^{-1} U_{s,s,N} Z_N^T \right).$$

The RQ factorization (9.61) on page 327 with

$$Z_N = \begin{bmatrix} U_{0,s,N} \\ Y_{0,s,N} \end{bmatrix}, \quad s \geq n,$$

allows us to derive

$$Z_N \Pi_{U_{s,s,N}}^\perp Z_N^T = (R_{21}Q_1 + R_{22}Q_2)(Q_2^T Q_2)(Q_1^T R_{21}^T + Q_2^T R_{22}^T)$$

$$= R_{22} R_{22}^T. \tag{9.72}$$

From Lemma 9.4 on page 315, we have

$$Y_{s,s,N}\Pi_{U_{s,s,N}}^{\perp}Z_N^{\mathrm{T}} = R_{32}R_{22}^{\mathrm{T}}.$$

Thus, we arrive at

$$\widehat{L}_N^z = R_{32}R_{22}(R_{22}R_{22}^{\mathrm{T}})^{-1} = R_{32}R_{22}^{-1},$$

which equals Equation (9.68) on page 331. This completes the proof.

□

Theorem 9.6 can be used to obtain an approximation of $\mathcal{O}_s\mathcal{L}_s$ and subsequently of the state sequence of the innovation model (9.50)–(9.51). Given the weighted SVD

$$W_1\Big((Y_{s,s,N}\Pi_{U_{s,s,N}}^{\perp}Z_N^{\mathrm{T}})(Z_N\Pi_{U_{s,s,N}}^{\perp}Z_N^{\mathrm{T}})^{-1}\Big)W_2 = U_n\Sigma_n V_n^{\mathrm{T}},$$

where W_1 and W_2 are nonsingular weighting matrices, we can estimate $\mathcal{O}_s$ as

$$\widehat{\mathcal{O}}_s = W_1^{-1}U_n\Sigma_n^{1/2} \tag{9.73}$$

and $\mathcal{L}_s$ as

$$\widehat{\mathcal{L}}_s = \Sigma_n^{1/2}V_n^{\mathrm{T}}W_2^{-1}. \tag{9.74}$$

We can use this estimate of $\mathcal{L}_s$ to reconstruct the state sequence, because

$$\widehat{X}_{s,N} = \widehat{\mathcal{L}}_s Z_N.$$

The variety of possible choices for the weighting matrices induces a whole set of subspace identification methods. The PO-MOESP method has weighting matrices

$$W_1 = I_{s\ell}, \qquad W_2 = (Z_N\Pi_{U_{s,s,N}}^{\perp}Z_N^{\mathrm{T}})^{1/2}.$$

To see this, note that the PO-MOESP scheme is based on computing the SVD of the matrix R_{32}, which, by Lemma 9.4 on page 315, is equal to

$$R_{32} = Y_{s,s,N}\Pi_{U_{s,s,N}}^{\perp}Z_N^{\mathrm{T}}(R_{22}^{\mathrm{T}})^{-1}.$$

The matrix R_{22}^{T} is the matrix square root of $Z_N\Pi_{U_{s,s,N}}^{\perp}Z_N^{\mathrm{T}}$, because of Equation (9.72).

The N4SID method has weighting matrices

$$W_1 = I_{s\ell}, \qquad W_2 = (Z_N Z_N^{\mathrm{T}})^{1/2},$$

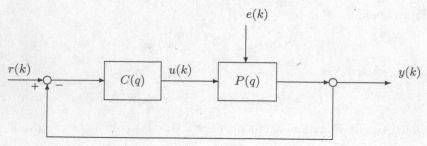

Fig. 9.9. A block scheme of an LTI innovation model P in a closed-loop configuration with a controller C.

because it is based on computing the SVD of the matrix $\mathcal{O}_s \mathcal{L}_s Z_N$. To see this, denote the SVD of the matrix $\mathcal{O}_s \mathcal{L}_s Z_N$ by

$$\mathcal{O}_s \mathcal{L}_s Z_N = \overline{U}_n \overline{\Sigma}_n \overline{V}_n^{\mathrm{T}}.$$

Taking $W_1 = I_{s\ell}$, Equation (9.73) yields $\overline{U}_n = U_n$ and $\overline{\Sigma}_n = \Sigma_n$. Now, because of Equation (9.74), the matrix W_2 must satisfy $V_n^{\mathrm{T}} W_2^{-1} Z_N = \overline{V}_n^{\mathrm{T}}$. Since $\overline{V}_n^{\mathrm{T}} \overline{V}_n = I_n$, we have $W_2^{-1} Z_N Z_N^{\mathrm{T}} W_2^{-\mathrm{T}} = I_{s(m+\ell)}$, which is satisfied for $W_2 = (Z_N Z_N^{\mathrm{T}})^{1/2}$.

We see that both the PO-MOESP and the N4SID method have a weighting matrix W_1 equal to the identity matrix. There exist methods in which the weighting matrix W_1 differs from the identity matrix. An example of such a method is the method of CVA (canonical variate analysis) described by Larimore (1990). This method involves taking the weighting matrices equal to

$$W_1 = (Y_{s,s,N} \Pi_{U_{s,s,N}}^{\perp} Y_{s,s,N}^{\mathrm{T}})^{-1/2}, \qquad W_2 = (Z_N \Pi_{U_{s,s,N}}^{\perp} Z_N^{\mathrm{T}})^{1/2}.$$

9.7 Using subspace identification with closed-loop data

To get a consistent estimate of the system matrices, all the subspace identification schemes presented in the previous sections require that the input of the system to be identified is *uncorrelated* with the additive perturbation $v(k)$ to the output. We refer, for example, to Theorems 9.3, 9.4, and 9.5.

This assumption on the input is easily violated when the data are acquired in a closed-loop configuration as illustrated in Figure 9.9. The problems caused by such a closed-loop experiment should be addressed for each identification method individually. To illustrate this, we consider the subspace identification method based on Theorem 9.5. The result is summarized in our final theorem of this chapter.

Theorem 9.7 *Consider the system P in Figure 9.9 given by (9.50) and (9.51) with $x(k)$, $u(k)$, and $e(k)$ ergodic stochastic processes and $e(k)$ a zero-mean white-noise sequence $e(k)$. The controller C is causal and the loop gain contains at least one sample delay. Take the instrumental-variable matrix Z_N equal to*

$$\begin{bmatrix} U_{0,s,N} \\ Y_{0,s,N} \end{bmatrix},$$

then

$$\lim_{N \to \infty} \frac{1}{N} E_{s,s,N} U_{s,s,N}^{\mathrm{T}} \neq 0, \qquad (9.75)$$

$$\lim_{N \to \infty} \frac{1}{N} E_{s,s,N} Z_N^{\mathrm{T}} = 0. \qquad (9.76)$$

Proof The data equation for the plant P in innovation form reads

$$Y_{0,s,N} = \mathcal{O}_s X_{0,N} + \mathcal{T}_s U_{0,s,N} + \mathcal{S}_s E_{0,s,N}.$$

Because of the white-noise property of $e(k)$, the closed-loop configuration, and the causality of the controller C, the state $x(k)$ and the input $u(k)$ satisfy

$$E[x(k)e(j)^{\mathrm{T}}] = 0, \qquad E[u(k)e(j)^{\mathrm{T}}] = 0, \quad \text{for all } j \geq k.$$

By virtue of the ergodicity of the sequences $x(k)$, $u(k)$, and $e(k)$, and, as a consequence, the ergodicity of $y(k)$, we have that

$$\lim_{N \to \infty} \frac{1}{N} E_{s,s,N} U_{0,s,N}^{\mathrm{T}} = 0, \qquad \lim_{N \to \infty} \frac{1}{N} E_{s,s,N} Y_{0,s,N}^{\mathrm{T}} = 0.$$

This proves Equation (9.76). Owing to the presence of the feedback loop, $u(k)$ contains a linear combination of the perturbations $e(j)$ for $j < k$, such that

$$E[u(k)e(j)^{\mathrm{T}}] \neq 0, \quad \text{for all } j < k.$$

This proves Equation (9.75). □

If we inspect the proof of Theorem 9.5 on page 330, then we see that the limit (9.75) prevents the possibility that Equation (9.64) on page 330 holds. As a consequence, an additive term results on the right-hand side of (9.64), causing the column space of the part $\widehat{\mathcal{L}}_N^z$ of the least-squares problem (9.63) to differ from the extended observability matrix $\mathcal{O}_s$. Therefore, it is not possible to retrieve as in the open-loop case a matrix that (in the limit $N \to \infty$) has $\mathcal{O}_s$ as its column space. This would result in biased estimates of the system matrices.

Several alternatives have been developed in the literature in order to remove such a bias. A summary of some existing strategies follows.

(i) Accept the bias on the system matrices calculated by one of the subspace identification schemes presented here and use the calculated model as an initial estimate of a prediction-error estimation problem. In Example 8.6 on page 285 it was shown that, when the innovation model is correctly parameterized, prediction-error methods are capable of providing consistent estimates.

(ii) Restrict the type of reference sequence $r(k)$ in Figure 9.9. For example, restrict it to being a white-noise sequence as in Chou and Verhaegen (1997) so that a number of calculations of the subspace identification schemes presented here still provide consistent estimates *without requiring knowledge of the controller C*.

(iii) Modify the subspace algorithms in such a way as to make them applicable to closed-loop data, while *not assuming knowledge of the controller C*. Examples of this approach are presented in Chou and Verhaegen (1997), Jansson (2003), Qin and Ljung (2003), and Chiuso and Picci (2005)

(iv) Use knowledge of the controller C that is assumed to be LTI to modify the subspace algorithms so that consistent estimates of the system matrices (and Kalman gain) can be obtained. An example of such an approach is given by Van Overschee and De Moor (1996a).

9.8 Summary

In this chapter we described several subspace identification methods. These methods are based on deriving a certain subspace that contains information about the system, from structured matrices constructed from the input and output data. To estimate this subspace, the SVD is used. The singular values obtained from this decomposition can be used to estimate the order of the system.

First, we described subspace identification methods for the special case when the input equals an impulse. In these cases, it is possible to exploit the special structure of the data matrices to get an estimate of the extended observability matrix. From this estimate it is then easy to derive the system matrices (A, B, C, D) up to a similarity transformation.

Next, we described how to deal with more general input sequences. Again we showed that it is possible to get an estimate of the extended observability matrix. From this estimate we computed the system matrices A and C up to a similarity transformation. The corresponding matrices B and D can then be found by solving a linear least-squares problem. We showed that the RQ factorization can be used for a computationally efficient implementation of this subspace identification method.

We continued by describing how to deal with noise. It was shown that, in the presence of white noise at the output, the subspace identification method for general inputs yields asymptotically unbiased estimates of the system matrices. This subspace method is therefore called the MOESP (Multivariable Output-Error State-sPace) method. To deal with colored noise at the output, the concept of instrumental variables was introduced. Different choices of instrumental variables lead to the PI-MOESP and PO-MOESP methods. The PI-MOESP method handles arbitrarily colored measurement noise, whereas the PO-MOESP method deals with white process and white measurement noise. Again we showed how to use the RQ factorization for an efficient implementation. It has been shown that various alternative subspace identification schemes can be derived from a least-squares problem formulation based on the structured data matrices treated in the PO-MOESP scheme. The least-squares approach enables the approximation of the state sequence of the innovation model, as was originally proposed in the N4SID subspace identification method. On the basis of this approximation of the state sequence we explained how to approximate the Kalman gain for the PO-MOESP and N4SID methods, in order to be able to construct an approximation of the one-step-ahead predictor. Though a theoretical foundation for the accuracy of this approximation is not given, experimental evidence has shown that it has proven its practical relevance. For an illustration we refer to Chapter 10 (see Example 10.13 on page 385). Finally, we showed that the subspace identification methods that were described cannot be used to obtain unbiased estimates of the system matrices if the input and output data are collected in a closed-loop manner.

Exercises

9.1   Consider the system

$$x(k+1) = Ax(k) + Bu(k),$$
$$y(k) = Cx(k) + Du(k),$$

where $u(k) \in \mathbb{R}^m$ and $y(k) \in \mathbb{R}^m$ with $m > 1$. We are given the sequences of input–output data pairs

$$\{u_i(k), y_i(k)\}_{k=1}^N \quad \text{for } i = 1, 2, \ldots, m$$

such that

$$u_i(k) = \begin{cases} e_i, & \text{for } k = 0, \\ 0, & \text{for } k \neq 0, \end{cases}$$

with e_i being the ith column of the $n \times n$ identity matrix. Assume that $N \gg s$, and show how the output data sequences $\{y_i(k)\}_{k=1}^{N_i}$ for $i = 1, 2, \ldots, m$ must be stored in the matrix $\mathcal{Y}$ such that

$$\mathcal{Y} = \begin{bmatrix} C \\ CA \\ \vdots \\ CA^{s-1} \end{bmatrix} \begin{bmatrix} B & AB & \cdots & A^{N-1}B \end{bmatrix}.$$

9.2   Consider the subspace identification method for impulse input signals, described in Section 9.2.3.

   (a) Write a Matlab program to determine the system matrices A, B, and C up to a similarity transformation.
   (b) Test this program using 20 data points obtained from the following system:

$$A = \begin{bmatrix} -0.5 & 1 \\ 0 & -0.5 \end{bmatrix}, \qquad B = \begin{bmatrix} 0 \\ 1 \end{bmatrix}, \qquad C = \begin{bmatrix} 1 & 0 \end{bmatrix}.$$

   Check the eigenvalues of the estimated A matrices, and compare the outputs from the models with the real output of the system.

9.3   Consider the minimal and asymptotically stable LTI system

$$x(k+1) = Ax(k) + Bu(k),$$
$$y(k) = Cx(k) + Du(k),$$

with $x(0) = 0$ and $u(k)$ equal to a step sequence given by

$$u(k) = \begin{cases} 1, & \text{for } k \geq 0, \\ 0, & \text{for } k < 0. \end{cases}$$

(a) Show that the data equation (9.7) on page 296 can be written as

$$Y_{0,s,N} = \mathcal{O}_s X_{0,N} + \mathcal{T}_s(\mathbb{E}_s \mathbb{E}_N^T),$$

where $\mathbb{E}_j \in \mathbb{R}^j$ denotes the vector with all entries equal to unity.

(b) Show that

$$\text{range}\left(Y_{0,s,N}\left(I_N - \frac{\mathbb{E}_N \mathbb{E}_N^T}{N} \right) \right) \subseteq \text{range}(\mathcal{O}_s).$$

(c) Show that

$$\lim_{N \to \infty} \frac{1}{N} \sum_{k=0}^{N-1} x(k) = (I_n - A)^{-1}B,$$

and use it to prove that

$$\lim_{N \to \infty} Y_{0,s,N}\left(I_N - \frac{\mathbb{E}_N \mathbb{E}_N^T}{N} \right) = -\mathcal{O}_s(I_n - A)^{-1}$$
$$\times \begin{bmatrix} B & AB & A^2B & \cdots \end{bmatrix}.$$

(d) Use the results derived above to prove that

$$\text{rank}\left(\lim_{N \to \infty} Y_{0,s,N}\left(I_N - \frac{\mathbb{E}_N \mathbb{E}_N^T}{N} \right) \right) = \text{rank}(\mathcal{O}_s).$$

9.4   When the input to the state-space system

$$x(k+1) = Ax(k) + Bu(k),$$
$$y(k) = Cx(k) + Du(k),$$

is periodic, the output will also be periodic. In this case there is no need to build a block Hankel matrix from the output, since one period of the output already contains all the information. Assume that the system is minimal and asymptotically stable and that $D = 0$. Let the input be periodic with period N_0, that is, $u(k) = u(k + N_0)$.

(a) Show that

$$\begin{aligned}&[y(0) \quad y(1) \quad \cdots \quad y(N_0 - 1)] \\ &= [\bar{g}(N_0) \quad \bar{g}(N_0 - 1) \quad \cdots \quad \bar{g}(1)] U_{0,N_0,N_0},\end{aligned}$$

where $\bar{g}(i) = CA^{i-1}(I_n - A^{N_0})^{-1}B$.

(b) What condition must the input $u(k)$ satisfy in order to determine the sequence $\bar{g}(i)$, $i = 1, 2, \ldots, N_0$, from this equation?

(c) Explain how the sequence $\bar{g}(i)$ can be used to determine the system matrices (A, B, C) up to a similarity transformation if $N_0 > 2n + 1$.

9.5   We are given the input sequence $u(k) \in \mathbb{R}^m$ and output sequence $y(k) \in \mathbb{R}^\ell$ of an observable LTI system for $k = 0, 1, 2, \ldots, N + s - 2$ stored in the block Hankel matrices $U_{0,s,N}$ and $Y_{0,s,N}$, with $N \gg s$. These Hankel matrices are related by

$$Y_{0,s,N} = \mathcal{O}_s X_{0,N} + \mathcal{T}_s U_{0,s,N},$$

for appropriately defined (but unknown) matrices $\mathcal{O}_s \in \mathbb{R}^{s\ell \times n}$, $\mathcal{T}_s \in \mathbb{R}^{s\ell \times sm}$, and $X_{0,N} \in \mathbb{R}^{n \times N}$, with $n < s$.

(a) Prove that the minimizer $\widehat{\mathcal{T}}_s$ of the least-squares problem

$$\min_{\mathcal{T}_s} \|Y_{0,s,N} - \mathcal{T}_s U_{0,s,N}\|_F^2$$

is

$$\widehat{\mathcal{T}}_s = Y_{0,s,N} U_{0,s,N}^{\mathrm{T}} (U_{0,s,N} U_{0,s,N}^{\mathrm{T}})^{-1}.$$

(b) Show that the column space of the matrix $(Y_{0,s,N} - \widehat{\mathcal{T}}_s U_{0,s,N})$ is contained in the column space of the matrix $\mathcal{O}_s$.

(c) Derive conditions on the matrices $X_{0,N}$ and $U_{0,s,N}$ for which the column spaces of the matrices $(Y_{0,s,N} - \widehat{\mathcal{T}}_s U_{0,s,N})$ and $\mathcal{O}_s$ coincide.

9.6   We are given a SISO LTI system $P(q)$ in innovation form,

$$\begin{aligned}x(k+1) &= Ax(k) + Bu(k) + Ke(k), \\ y(k) &= Cx(k) + e(k),\end{aligned}$$

with $x(k) \in \mathbb{R}^n$ and $e(k)$ zero-mean white noise with unit variance. This system is operating in closed-loop mode with the controller $C(q)$ as shown in Figure 9.10.

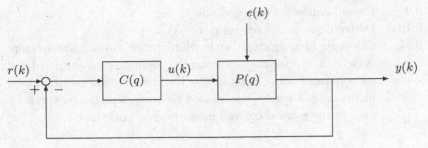

Fig. 9.10. System $P(q)$ in a closed-loop connection with controller $C(q)$.

(a) Derive a condition on the matrix $(A - KC)$ such that the input–output relationship between $u(k)$ and $y(k)$ can be written as an ARX model of order $s > n$:

$$y(k) = a_1 y(k-1) + a_2 y(k-2) + \cdots + a_s y(k-s)$$
$$+ b_1 u(k-1) + b_2 u(k-2) + \cdots + b_s u(k-s)$$
$$+ e(k). \tag{E9.1}$$

(b) Assuming that the derived condition on the matrix $(A - KC)$ holds, express the coefficients a_i and b_i for $i = 1, 2, \ldots, s$ in terms of the matrices A, B, C, and K.

(c) Prove that the coefficients a_i and b_i for $i = 1, 2, \ldots, s$ of the ARX model (E9.1) can be estimated unbiasedly in the closed-loop scenario.

9.7 Consider the MOESP subspace identification method summarized on page 312.

(a) Write a Matlab program to determine the matrices A and C based on the RQ factorization (9.21) on page 304 as described in Section 9.2.4.

(b) Test the program on the system used in Exercise 9.2 with a white-noise input sequence. Check the eigenvalues of the estimated A matrix.

9.8 Consider the system

$$x(k+1) = Ax(k) + Bu(k),$$
$$y(k) = Cx(k) + Du(k) + v(k),$$

with $v(k)$ a colored-noise sequence. Show that, when unbiased estimates $\widehat{A}_T$ and $\widehat{C}_T$ are used in Equation (9.25) on page 307 to estimate B_T and D_T, these estimates are also unbiased.

9.9   Prove Lemma 9.8 on page 326.

9.10   Derive Equation (9.68) on page 331.

9.11   Subspace identification with white process and measurement noise is a special case of subspace identification with colored measurement noise. Explain why the instrumental variables used in the case of white process and measurement noise cannot be used in the general colored-measurement-noise case.

10

The system-identification cycle

After studying this chapter you will be able to

- explain that the identification of an LTI model making use of real-life measurements is more then just estimating parameters in a user-defined model structure;
- identify an LTI model in a cyclic manner of iteratively refining data and models and progressively making use of more complex numerical optimization methods;
- explain that the identification cycle requires many choices to be made on the basis of cautious experiments, the user's expertise, and prior knowledge about the system to be identified or about systems bearing a close relationship with, or resemblance to, the target system;
- argue that a critical choice in system identification is the selection of the input sequence, both in terms of acquiring qualitative information for setting or refining experimental conditions and in terms of accurately estimating models;
- describe the role of the notion of persistency of excitation in system identification;
- use subspace identification methods to initialize prediction-error methods in identifying state-space models in the innovation form; and
- understand that the art of system identification is mastered by applying theoretical insights and methods to real-life experiments and working closely with an expert in the field.

345

10.1 Introduction

In the previous chapters, it was assumed that time sequences of input and output quantities of an unknown dynamical system were given. The task was to estimate parameters in a user-specified model structure on the basis of these time sequences. Although parameter identification is the nontrivial core part of system identification, many choices need to be made before arriving at an adequate data set and a suitable model parameterization. The choices at the start of a system-identification task, such as the selection of the digital data-acquisition infrastructure, the sampling rate, and the type of anti-aliasing filters, comprise the *experiment design.* Designing a system-identification experiment requires a balanced integration of engineering intuition, knowledge about systems and control theory, and domain-specific knowledge of the system to be studied. This combination makes it impossible to learn experiment design from this book, but the material therein can give you at least a starting point. In this chapter we discuss a number of relevant choices that may play a role in designing a system-identification experiment.

Since many of the choices in designing a system-identification experiment require (qualitative) knowledge of the underlying unknown system, the design often starts with preliminary experiments with simple standard test input sequences, such as a step input or impulse input. Qualitative information about the system to be identified is retrieved from these experiments, often by visual inspection or by making use of the DFT. The qualitative insights, such as the order of magnitude of the time constants, allow a refinement of the experimental conditions in subsequent experiments. This brief exposure already highlights the fact that system identification is an *iterative* process of gradual and cautious discovery and exploration. A flow chart of the iterative identification process or cycle is depicted in Figure 10.1 (Ljung, 1999). This chart highlights the fact that, prior to estimating the model (parameters), the data may need to be polished and pre-filtered to remove deficiencies (outliers, noise, trends) and to accentuate a certain frequency band of interest. This is indicated by the *data pre-processing* step.

Having acquired a polished data set, the next challenge to be tackled is the actual estimation of the model (parameters). The parameter-estimation step has to be preceded by the crucial *choice of model structure.* The model structure was formally defined in Section 7.3 as the mapping between the parameter space and the model to be identified. In Chapters 7 and 8, we listed a large number of possible parameter

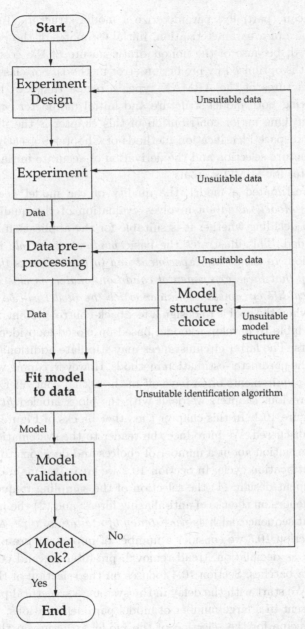

Fig. 10.1. A schematic view of the key elements in the system-identification cycle.

sets that can (partially) parameterize a model structure. In addition to the choice of a parameterization, initial estimates of the parameters are required, because of the nonquadratic nature of the cost function $J_N(\theta)$ that is optimized by prediction-error methods. For classical SISO model structures of the ARMAX type, a recipe that can be helpful in discovering both model structure and initial parameter estimates is highlighted. One major contribution of this chapter is the illustration that the subspace identification methods of Chapter 9 greatly simplify model-structure selection and the derivation of accurate initial parameter estimates for MIMO systems.

Having estimated a model, the quality of the model needs to be addressed. *Model validation* involves evaluation of the quality of the model and deciding whether it is suitable for the application for which it is intended. This illustrates the basic *guidance principle* in system identification: *to design the experiment and to (pre-)process the data in such a way that these experimental conditions match as closely as possible the real-life circumstances under which the model will be used.* For example, when the goal is to design a feedback-control system, it may be recommendable to develop a model based on closed-loop identification experiments. The latter circumstances may stipulate additional requirements on the parameter-estimation methods. However, coping with these additional requirements often pays off in the actual controller design.

In the previous chapters we dealt with the block labeled *fit model to data* in Figure 10.1. In this chapter the other blocks of Figure 10.1 will be briefly discussed. We introduce the reader to the systematic acquisition of information about a number of choices and decisions to be made in the identification cycle. In Section 10.2 we treat two main choices in the experiment design: (1) the selection of the sampling frequency and its consequences on the use of anti-aliasing filters; and (2) the properties of the input sequence, such as *persistency of excitation*, type, and duration. In Section 10.3 we consider a number of important pre-processing steps, such as decimation, trend removal, pre-filtering, and concatenation of data batches. Section 10.4 focuses on the selection of the model structure. We start with the delay in the system, a structural parameter that is present in a large number of model parameterizations. Next, we present a recipe for the selection of the model structure in the identification of SISO ARMAX models. For more general MIMO state-space models, the selection of the order of the model is treated in the context of subspace identification methods. Finally, in Section 10.5 some techniques to validate an identified model are presented.

10.2 Experiment design

The experiment-design considerations rely on the physical knowledge we have available about the system to be identified. A source of such knowledge is provided by physical models derived from the laws of Nature, such as Newton's law involving the dynamics of mechanical systems. These physical models are generally described by continuous-time (differential) equations. In this section we attempt to derive guidelines for generating sampled input and output data sequences from simple continuous-time models. We also discuss the duration of the experiment and discuss how the properties of the input signal influence the experiment design.

10.2.1 *Choice of sampling frequency*

Shannon's sampling theorem (Kamen, 1990) offers important guidelines in choosing the sampling frequency. Shannon's theorem states that, when a band-limited signal with frequency content in the band $[-\omega_B, \omega_B]$ (rad/s) is sampled with a sampling frequency $\omega_S = 2\omega_B$ (rad/s), it is possible to reconstruct the signal perfectly from the recorded samples. Shannon's theorem can be used to avoid *aliasing*. The effect of aliasing is illustrated in the following example.

Example 10.1 (Aliasing) Consider the harmonic signals $x(t)$ and $y(t)$:

$$x(t) = \cos(3t),$$
$$y(t) = \cos(5t).$$

Both signals are depicted in Figure 10.2 on the interval $t \in [0, 10]$ seconds. In this figure the thick line represents $x(t)$ and the thin line $y(t)$. On sampling both sequences with a sampling frequency of $\omega_S = 8$ rad/s, the two sampled data sequences coincide at the circles in Figure 10.2. Thus, the two sampled signals are indistinguishable. In the frequency domain, we observe that $|Y_N(\omega)|$ is mirrored (aliased) to the location of $|X_N(\omega)|$.

More generally, aliasing means that all frequency components of a signal with a frequency higher than half the sampling frequency ω_S are mirrored across the line $\omega = \omega_S/2$ in the frequency band $[-\omega_S/2, \omega_S/2]$. Therefore, when the frequency band of interest of a signal is $[-\omega_B, \omega_B]$, distortion of the signal in that band can be prevented if we pre-filter the signal by a band-pass filter. Such a filter is called an anti-aliasing filter. Ideally, a low-pass filter with frequency function as shown in Figure 10.3

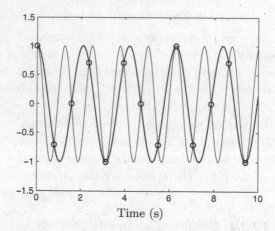

Fig. 10.2. Aliasing as a consequence of sampling the signals defined in Example 10.1 with a sampling time of $2\pi/8$ s. The circles indicate the sampling points.

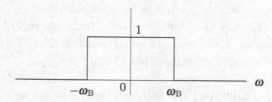

Fig. 10.3. The magnitude plot of an ideal low-pass filter with bandwidth ω_B.

should be taken as anti-aliasing filter. Realizable approximations of this frequency function, such as Chebyshev or Butterworth filters, are used in practice (MathWorks, 2000a).

In system identification, the frequency band of interest of the signals involved is dominantly characterized by the bandwidth of interest of the system to be identified. The *bandwidth* of a linear, time-invariant system is generally referred to as the frequency ω_B at which the magnitude of the frequency-response function has declined by 3 dB (equivalent to a factor of $10^{3/20} \approx \sqrt{2}$) from its value at frequency zero.

When the goal is to identify the system in the frequency band $[0,\ \omega_B]$ we have to excite the system in this frequency band and, subsequently, sampling has to take place with a frequency of at least $\omega_S = 2\omega_B$. In this sense, ω_B does not necessarily represent the bandwidth of the system, but rather, is the bandwidth of interest. A rule of thumb, based on a rough estimate of the bandwidth (of interest), is to select the sampling

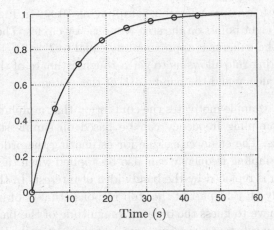

Fig. 10.4. Derivation of the order of magnitude of a system's bandwidth from the step response as explained in Example 10.2. Putting eight circles on the curve would result in an estimate of the required sampling frequency.

frequency $\omega_S = 10\omega_B$. Prior to sampling, an anti-aliasing filter with an appropriate band-pass frequency function should then be used.

At the start of an identification experiment we usually do not know the bandwidth of the system. Various characteristic quantities of the system allow us to get a rough estimate of its bandwidth. One such quantity is the rise time of the step response (Powell *et al.*, 1994). Its relationship with the selection of the sampling frequency is illustrated in Example 10.2.

Example 10.2 (Relationship between system bandwidth and rise time) Consider the first-order differential equation involving the scalar variables $y(t)$ and $u(t)$:

$$\tau \frac{dy(t)}{dt} + y(t) = u(t).$$

If we take as output signal $y(t)$ and as input signal $u(t)$, then the transfer function is given by

$$G(s) = \frac{1}{1 + \tau s},$$

with s a complex variable. If $\tau = 10\,s$, then the bandwidth ω_B of this system is $1/\tau = 0.1$ rad/s. The step response of the system is plotted in Figure 10.4. The rise time of this overdamped system (meaning that there is no overshoot) is approximately equal to the time period necessary to reach steady state. If we take about eight or nine samples during this period, we are approximately sampling at $10\omega_B$. In Figure 10.4 we have marked eight points on the step response by circles. The time interval between these marks is approximately equal to $T = 2\pi/10\omega_B = 2\pi$ s. This engineering rule allows us to get a rough estimate of the necessary sampling frequency and the bandwidth of the system.

The above example motivates the contention that a preliminary selection of the sampling frequency can be based on simple step-response measurements. The engineering rule for estimating the order of magnitude of the sampling frequency can also be applied when the bandwidth of the system is replaced by the bandwidth of interest. In that case, we will focus only on that part of the step response that is of interest.

When we have to guess the order of magnitude of the bandwidth (or dominant time constants, rise time, etc.) of the system to be identified, it is wise to sample at a fast sampling rate determined by the capabilities of the data-acquisition system and later on decimate (treated in Section 10.3.1) the sampled data sequences to a lower sampling rate. In this way, it is less likely that we will conclude at the data-processing stage that too low a sampling rate was used and that the experiment has to be repeated.

10.2.2 Transient-response analysis

In the previous subsection we discovered that simple step-response measurements can help in extracting qualitative information about the order of magnitude of system properties that are relevant for selecting the experimental conditions. Other types of elementary input–response measurements can supply additional or similar information. An example is a pulse of short duration, such as hitting a mechanical structure with a hammer. The measurement and analysis of the response with respect to elementary input sequences like pulses and steps is often called *transient-response analysis*.

In addition to helpful insights into selecting the sample frequency, other examples of the use of the transient-response analysis are

- determining which measurable (output) signals are affected by a certain input variable;
- testing the linearity of the system to be analyzed, for example, by using a number of step inputs of different amplitudes; and
- guessing the order of magnitude of the (dominant) time constants from a step-response experiment.

The latter practice is based on the analysis of the step response of a second-order system. Example 10.3 illustrates the rationale behind this common practice.

Example 10.3 (Transient analysis for determining the order of magnitude of the time constant of an LTI system) Consider the second-order continuous-time system with a transfer function given by

$$G(s) = \frac{\omega_n^2}{s^2 + 2\zeta\omega_n s + \omega_n^2}, \tag{10.1}$$

with ω_n the natural frequency and ζ the damping ratio. Since the Laplace transform of a step signal equals $1/s$, the response $y(t)$ to a step signal can be found by taking the inverse Laplace transform of

$$Y(s) = \frac{\omega_n^2}{s(s^2 + 2\zeta\omega_n s + \omega_n^2)},$$

which equals, for $t \geq 0$,

$$y(t) = 1 - \frac{1}{\sqrt{1-\zeta^2}}e^{-\zeta\omega_n t}\sin(\sqrt{1-\zeta^2}\,\omega_n t + \phi), \tag{10.2}$$

with $\tan\phi = (\sqrt{1-\zeta^2})/\zeta$. When $0.2 \leq \zeta \leq 0.6$, the second-order system shows an oscillatory response. The system is called underdamped. Such step responses are displayed in Figure 10.5 for $\omega_n = 1$. From the figure, we observe that the time interval of the first cycle (from zero to about 6.5 s, indicated by the vertical line in Figure 10.5), is approximately equal for all the step responses. Let the length of this interval be denoted by t_{cycle}, then, from Equation (10.2), we obtain the following expression for t_{cycle}:

$$2\pi = \sqrt{1-\zeta^2}\,\omega_n t_{\text{cycle}},$$

and thus

$$\omega_n \approx \frac{2\pi}{t_{\text{cycle}}}. \tag{10.3}$$

A rough estimate of the natural frequency is

$$\omega_n \approx \frac{2\pi}{6.5} \approx 0.97.$$

On the basis of Equation (10.2), an estimate of the time constant is $\tau = 1/(\zeta\omega_n)$. To get a rough estimate, we count in the step response the number of cycles before steady state is reached. We denote this number

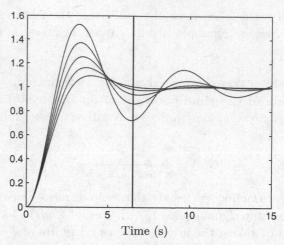

Fig. 10.5. Step responses of the second-order system (10.1) for $\omega_n = 1$ and various values of ζ. At time $t = 4$ s, from top to bottom the lines correspond to $\zeta = 0.2, 0.3, 0.4, 0.5$, and 0.6.

by κ. The time taken to reach steady state is approximately four time constants. Therefore,

$$\sqrt{1 - \zeta^2}\,\omega_n 4\tau \approx 2\pi\kappa.$$

Using Equation (10.3) yields

$$\frac{2\pi}{t_{\text{cycle}}}4\tau \approx 2\pi\kappa.$$

Therefore,

$$\tau \approx \frac{\kappa t_{\text{cycle}}}{4}.$$

For $\zeta = 0.4$, we have $\kappa \approx 1.5$, and we get an approximation of the time constant $\tau \approx 2.44$ s. which is indeed a rough estimate of the true time constant $\tau = 2.5$ s. This is a useful estimate from which to determine the sampling frequency and the system bandwidth.

In addition to retrieving a rough estimate of the (dominant) time constants, the step response may yield further qualitative system information. For example, an oscillatory step response indicates that the poles of the system have an imaginary part. Such qualitative information may be used as a constraint on the parameters of a parametric model to be estimated. In the field of system identification, integrating such qualitative insights into the process of parametric identification is generally referred to as *gray-box model identification.*

10.2.3 Experiment duration

When estimating the parameters in a parameterized model, such as the AR(MA)X model, the accuracy of the estimated parameters is inversely proportional to the number of samples, as indicated by the covariance expression (7.50) on page 245 derived in Section 7.6. Hence, the accuracy of the model is inversely proportional to the duration of the experiment. A direct conclusion of this fact is that one should make the experiment duration as long as possible. However, in a number of application domains constraining the duration of the identification experiment is an important issue. An example is the processing industry, in which, because of the slow time constants of distillation columns, identification experiments recording 1000 samples easily run over a period of a week. The product produced by the distillation column during the identification experiment does not in general satisfy the specifications and thus restricting the duration of the experiments is of great economic importance here. A rule of thumb for the duration of an identification experiment is that it should be about ten times the longest time constant of the system that is of interest. Some consequences of this rule of thumb are illustrated in the following example.

Example 10.4 (Multi-rate sampling) Consider a continuous-time system with a transfer function given by

$$G(s) = \frac{1}{(1 + \tau_1 s)(1 + \tau_2 s)},$$

with $\tau_1 = 1$ s and $\tau_2 = 0.01$ s. Following the rule of thumb outlined in Section 10.2.1 for selecting the sampling frequency, the smallest time constant indicates that we should use a sampling frequency of 1000 rad/s, or, equivalently, 160 Hz. The largest time constant, τ_1, indicates that we require an experiment duration of 10 s. This would result in a requirement to collect at least 1600 data points. However, when we are able to use two different sampling frequencies, this number of data points can be greatly reduced. The sampling frequency related to the smallest time constant should be used only for the first time interval of 0.1 s. This results in 16 data points. The second sampling rate is related to the largest time constant and is approximately equal to

1.6 Hz. Using this sampling frequency for 10 s results in another 16 data points. Using this pair of 16 data points, in comparison with the 1600 recorded when using a single sampling rate, gives a reduction by a factor of 50. The price to be paid for this data reduction is that the parameter-estimation method should have the capacity to treat multi-rate sampled data sequences. In the context of subspace identification this topic is treated in Haverkamp (2000).

10.2.4 Persistency of excitation of the input sequence

To be able to estimate a model from measured input and output data, the data should contain enough information. In the extreme case that the input sequence $u(k) = 0$ for all $k \in \mathbb{Z}$, no information about the transfer from $u(k)$ to $y(k)$ can be retrieved. Therefore, the input should be different from zero in some sense so that we will be able to identify particular transfer functions. This property of the input in relationship to system identification is generally indicated by the notion of *persistency of excitation*. Before defining this notion, an example that motivates the definition is given.

Example 10.5 (Persistency of excitation) Consider the FIR prediction model that relates $\widehat{y}(k)$ and $u(k)$ as

$$\widehat{y}(k) = \theta_1 u(k) + \theta_2 u(k-1) + \theta_3 u(k-2).$$

Let the input be a sinusoid with frequency ω, $u(k) = \sin(\omega k)$, $k = 0, 1, 2, \ldots, N$, then finding the parameters θ_i can be achieved by addressing the following optimization problem:

$$\min_{\theta_i} \sum_{k=2}^{N} \left(y(k) - \begin{bmatrix} \sin(\omega k) & \sin(\omega k - \omega) & \sin(\omega k - 2\omega) \end{bmatrix} \begin{bmatrix} \theta_1 \\ \theta_2 \\ \theta_3 \end{bmatrix} \right)^2,$$

$$(10.4)$$

with $y(k)$ given by

$$y(k) = \sin(\omega k) + \sin(\omega k - \omega) + \sin(\omega k - 2\omega).$$

The above functional is quadratic in the parameters θ_i ($i = 1, 2, 3$), a property we have also seen in Example 7.7 on page 235. However, because of the *combination* of the particular input and model structure, the optimization problem does not yield a unique estimate of the parameters

θ_i $(i = 1, 2, 3)$. To see this, we use the following goniometric relationship:

$$\sin(\omega k - \alpha\omega) = \begin{bmatrix} \cos(\omega k) & \sin(\omega k) \end{bmatrix} \begin{bmatrix} \cos(\alpha\omega) \\ -\sin(\alpha\omega) \end{bmatrix}.$$

Equation (10.4) can be written as

$$\min_{\theta_i} \sum_{k=2}^{N} \left(y(k) - \begin{bmatrix} \cos(\omega k) & \sin(\omega k) \end{bmatrix} \begin{bmatrix} 0 & \cos(\omega) & \cos(2\omega) \\ 1 & -\sin(\omega) & -\sin(2\omega) \end{bmatrix} \begin{bmatrix} \theta_1 \\ \theta_2 \\ \theta_3 \end{bmatrix} \right)^2.$$

By a redefinition of the unknown parameters, we can also consider

$$\min_{\theta_i} \sum_{k=2}^{N} \left(y(k) - \begin{bmatrix} \cos(\omega k) & \sin(\omega k) \end{bmatrix} \begin{bmatrix} \gamma_1 \\ \gamma_2 \end{bmatrix} \right)^2,$$

subject to

$$\begin{bmatrix} 0 & \cos(\omega) & \cos(2\omega) \\ 1 & -\sin(\omega) & -\sin(2\omega) \end{bmatrix} \begin{bmatrix} \theta_1 \\ \theta_2 \\ \theta_3 \end{bmatrix} = \begin{bmatrix} \gamma_1 \\ \gamma_2 \end{bmatrix}.$$

Using 1000 samples of the output and $\omega = \pi/10$, the minimizing set of parameters yielding a criterion value 0 satisfies

$$\begin{bmatrix} \hat{\gamma}_1 \\ \hat{\gamma}_2 \end{bmatrix} = \begin{bmatrix} -0.8968 \\ 2.7600 \end{bmatrix} = \begin{bmatrix} \theta_2 \cos(\omega) + \theta_3 \cos(2\omega) \\ \theta_1 - \theta_2 \sin(\omega) - \theta_3 \sin(2\omega) \end{bmatrix}.$$

These equations describe the intersection of two hyperplanes in $\mathbb{R}^3$. If the intersection is nonzero and the two hyperplanes do not coincide, the solution is a straight line in $\mathbb{R}^3$. All the points on this line are solutions to the parameter-optimization problem. Therefore, for the specific input signal $u(k) = \sin(\omega k)$, no unique estimate of the parameters θ_i $(i = 1, 2, 3)$ can be obtained. To obtain a unique solution, we may use the input sequence

$$u(k) = \sin(\omega k) + \sin(2\omega k).$$

The nonuniqueness of the solution to the parameter estimation problem in Example 10.5 is a direct consequence of the fact that the rows of the matrix

$$U_{0,3,N-1} = \begin{bmatrix} u(0) & u(1) & \cdots & u(N-2) \\ u(1) & u(2) & \cdots & u(N-1) \\ u(2) & u(3) & \cdots & u(N) \end{bmatrix}$$

are *not* independent and therefore the matrix

$$\frac{1}{N} U_{0,3,N-1} U_{0,3,N-1}^{\mathrm{T}}$$

that needs to be inverted in solving the least-squares problem (10.4) (see also Example 7.7 on page 235) becomes singular.

To avoid such singularity, the input sequence $u(k)$ should be "rich" enough. The mathematical notion used to express this richness property of the input is called *persistency of excitation*. The definition is as follows.

Definition 10.1 *The sequence $u(k)$, $k = 0, 1, 2, \ldots$ is persistently exciting of order n if and only if there exists an integer N such that the matrix*

$$U_{0,n,N} = \begin{bmatrix} u(0) & u(1) & \cdots & u(N-1) \\ u(1) & u(2) & \cdots & u(N) \\ \vdots & & \ddots & \vdots \\ u(n-1) & u(n) & \cdots & u(N+n-2) \end{bmatrix} \tag{10.5}$$

has full rank n.

When the sequence $u(k)$ is ergodic, or the sequence is deterministic, the condition in Definition 10.1 is equivalent to the nonsingularity of the auto-correlation matrix R^u:

$$\begin{aligned} R^u &= \lim_{N \to \infty} \frac{1}{N} U_{0,n,N} U_{0,n,N}^{\mathrm{T}} \\ &= \begin{bmatrix} R_u(0) & R_u(1) & \cdots & R_u(n-1) \\ R_u(1) & R_u(0) & \cdots & R_u(n-2) \\ \vdots & & \ddots & \\ R_u(n-1) & R_u(n-2) & \cdots & R_u(0) \end{bmatrix}, \end{aligned} \tag{10.6}$$

provided that this limit exists. An interpretation in the frequency domain is provided by the following lemma.

Lemma 10.1 *(Ljung, 1999) The ergodic input sequence with spectrum $\Phi^u(\omega)$ and sampling period 1 is persistently exciting of order n if and only if*

$$\frac{1}{2\pi} \int_{-\pi}^{\pi} |M_n(\mathrm{e}^{\mathrm{j}\omega})|^2 \Phi^u(\omega) \mathrm{d}\omega > 0 \tag{10.7}$$

for all filters $M_n(q) = m_1 q^{-1} + m_2 q^{-2} + \cdots + m_n q^{-n}$ with $m_i \neq 0$ for some $1 \leq i \leq n$.

Proof We have that, for all $m = \begin{bmatrix} m_n & m_{n-1} & \cdots & m_1 \end{bmatrix}^{\mathrm{T}}$ with $m_i \neq 0$ for some $1 \leq i \leq n$,

$$
m^{\mathrm{T}} R^u m = E\left[\begin{bmatrix} m_n & m_{n-1} & \cdots & m_1 \end{bmatrix} \begin{bmatrix} u(k-n) \\ u(k-n+1) \\ \vdots \\ u(k-1) \end{bmatrix} \right.
$$

$$
\left. \times \begin{bmatrix} u(k-n) & u(k-n+1) & \cdots & u(k-1) \end{bmatrix} \begin{bmatrix} m_n \\ m_{n-1} \\ \vdots \\ m_1 \end{bmatrix} \right]
$$

$$
= E\left[|(m_1 q^{-1} + m_2 q^{-2} + \cdots + m_n q^{-n})u(k)|^2 \right],
$$

where we have made use of the ergodicity of the input sequence $u(k)$. Let $v(k) = (m_1 q^{-1} + m_2 q^{-2} + \cdots + m_n q^{-n})u(k)$, then, according to Parseval's identity (4.10) on page 106, we have

$$
E[v^2(k+n-1)] = m^{\mathrm{T}} R^u m = \frac{1}{2\pi} \int_{-\pi}^{\pi} |M_n(\mathrm{e}^{\mathrm{j}\omega})|^2 \Phi^u(\omega) \mathrm{d}\omega.
$$

Since persistency of excitation requires $m^{\mathrm{T}} R^u m > 0$ for all nonzero vectors m, this completes the proof. $\qquad\qquad\qquad\qquad\qquad\square$

Lemma 10.1 indicates that a persistently exciting input sequence of order n cannot be filtered away (to zero) by an nth-order FIR filter of the form $m_1 q^{-1} + m_2 q^{-2} + \cdots + m_n q^{-n}$.

The above notion of persistency of excitation is not only relevant for estimating the parameters of a FIR model. It can also be used to express the conditions on the input sequence, to uniquely estimate the parameters in more general model structures with prediction-error methods and subspace identification methods.

10.2.4.1 Persistency of excitation related to identification of ARX models

We will first discuss persistency of excitation for the ARX model structure discussed in Section 8.3.1. The condition for persistency of excitation on the input is given in the next lemma.

Lemma 10.2 *Consider the signal-generating system*

$$
y(k) = \frac{B_{n_b}(q)}{A_{n_a}(q)} u(k) + v(k), \tag{10.8}
$$

with ergodic input and output sequences $u(k)$ and $y(k)$, with $u(k)$ independent from the perturbation $v(k)$, and with $v(k)$ persistently exciting of any order. The parameters a_i and b_i in the one-step-ahead prediction of a SISO ARX model defined by

$$\widehat{y}(k|k-1) = (b_1 q^{-1} + \cdots + b_{n_b} q^{-n_b}) u(k - n_k)$$
$$+ (a_1 q^{-1} + \cdots + a_{n_a} q^{-n_a}) y(k)$$

can be uniquely determined by minimizing the prediction-error criterion

$$\lim_{N \to \infty} \frac{1}{N} \sum_{k=1}^{N} \Big(y(k) - \widehat{y}(k|k-1) \Big)^2,$$

provided that the input $u(k)$ is persistently exciting of order n_b.

Proof Introduce the notation $n = \max(n_a, n_b + n_k)$. In minimizing the postulated prediction-error cost function, the matrix

$$\lim_{N \to \infty} \frac{1}{N} \begin{bmatrix} U_{n-n_k-n_b, n_b, N} \\ Y_{n-n_a, n_a, N} \end{bmatrix} \begin{bmatrix} U^{\mathrm{T}}_{n-n_k-n_b, n_b, N} & Y^{\mathrm{T}}_{n-n_a, n_a, N} \end{bmatrix} \qquad (10.9)$$

needs to be invertible. Now let m_u and m_y define the vectors

$$m_u = \begin{bmatrix} m_{u,1} & m_{u,2} & \cdots & m_{u,n_b} \end{bmatrix}^{\mathrm{T}},$$
$$m_y = \begin{bmatrix} m_{y,1} & m_{y,2} & \cdots & m_{y,n_a} \end{bmatrix}^{\mathrm{T}},$$

and related polynomials

$$M_{n_b}(q) = m_{u,1} q^{-1} + m_{u,2} q^{-2} + \cdots + m_{u,n_b} q^{-n_b},$$
$$M_{n_a}(q) = m_{y,1} q^{-1} + m_{y,2} q^{-2} + \cdots + m_{y,n_a} q^{-n_a}.$$

Using Parseval's identity (4.10) on page 106 (assuming a sample time $T = 1$), using Lemma 4.3 on page 106, and exploiting the mutual independence of $u(k)$ and $v(k)$, we can write

$$\lim_{N \to \infty} \frac{1}{N} \begin{bmatrix} m_u^{\mathrm{T}} & m_y^{\mathrm{T}} \end{bmatrix} \begin{bmatrix} U_{n-n_k-n_b, n_b, N} \\ Y_{n-n_a, n_a, N} \end{bmatrix} \begin{bmatrix} U^{\mathrm{T}}_{n-n_k-n_b, n_b, N} & Y^{\mathrm{T}}_{n-n_a, n_a, N} \end{bmatrix} \begin{bmatrix} m_u \\ m_y \end{bmatrix}$$

$$= \frac{1}{2\pi} \int_{-\pi}^{\pi} \left| M_{n_b}(e^{j\omega}) + M_{n_a}(e^{j\omega}) \frac{B_{n_b}(e^{j\omega})}{A_{n_a}(e^{j\omega})} \right|^2 \Phi^u(\omega)$$
$$+ |M_{n_a}(e^{j\omega})|^2 \Phi^v(\omega) d\omega \geq 0. \qquad (10.10)$$

This expression can become zero only if both terms of the integrand are identically zero. Since $v(k)$ is persistently exciting of any order, the second term of the integrand (10.10) can become zero only if $|M_{n_a}(e^{j\omega})| = 0$ for all ω. However, in that case the first part of the integrand becomes zero only if $M_{n_b}(e^{j\omega}) = 0$ for all ω, since $u(k)$ is persistently exciting of order n_b. Therefore, the integral in (10.10) can become zero only when all $m_{u,i}$, $i = 1, 2, \ldots, n_b$ and all $m_{y,i}$, $i = 1, 2, \ldots, n_a$ are zero. This establishes the lemma. $\qquad\square$

The proof of Lemma 10.2 indicates that the matrix (10.9) remains invertible irrespective of the parameter n_a. This is a direct consequence of the presence of the perturbation $v(k)$ as summarized in the following lemma.

Lemma 10.3 *Consider the estimation of an ARX model as in Lemma 10.2 with $v(k) = 0$ for all $k \in \mathbb{Z}$, then the parameters a_i and b_i in Lemma 10.2 can be uniquely determined provided that*

 (i) *the polynomials A_{n_a} and B_{n_b} are co-prime, that is, they have no nontrivial common factors; and*
 (ii) *the input is persistently exciting of order $n_a + n_b$.*

Proof For the case $v(k) = 0$, Equation (10.10) reduces to

$$\frac{1}{2\pi} \int_{-\pi}^{\pi} \left| M_{n_b}(e^{j\omega}) + M_{n_a}(e^{j\omega})\frac{B_{n_b}(e^{j\omega})}{A_{n_a}(e^{j\omega})} \right|^2 \Phi^u(\omega)\mathrm{d}\omega$$

$$= \frac{1}{2\pi} \int_{-\pi}^{\pi} \frac{\left| M_{n_b}(e^{j\omega})A_{n_a}(e^{j\omega}) + M_{n_a}(e^{j\omega})B_{n_b}(e^{j\omega}) \right|^2}{|A_{n_a}(e^{j\omega})|^2} \Phi^u(\omega)\mathrm{d}\omega.$$

To show that this quantity remains strictly positive for some filters $M_{n_b}(q)$ and $M_{n_a}(q)$ different from zero, we use the Bezout equality (Söderström and Stoica, 1983). This equality states that, when the polynomials A_{n_a} and B_{n_b} are co-prime,

$$M_{n_b}(q)A_{n_a}(q) + M_{n_a}(q)B_{n_b}(q) = 0$$

holds only provided that both $M_{n_b}(q) = 0$ and $M_{n_a}(q) = 0$. The condition of persistency of excitation on the input $u(k)$ guarantees that the above integral remains positive, since $M_{n_b}(q)A_{n_a}(q) + M_{n_a}(q)B_{n_b}(q)$ is a polynomial of order $n_a + n_b$. $\qquad\square$

Lemma 10.2 and Lemma 10.3 indicate that the solution to the parameter-estimation problem obtained by prediction-error methods is strongly influenced by (1) the presence of perturbations on the data and

(2) knowledge of the correct order of the polynomials that determine the dynamics of the data-generating system. If this information cannot be provided, the parameter values that minimize the cost function (8.14) on page 261 cannot be uniquely determined. An illustration is provided in the next example.

Example 10.6 (Nonuniqueness in minimizing the prediction-error cost function) Consider the data-generating model

$$y(k) = a_1 y(k-1) + \cdots + a_n y(k-n) + b_1 u(k-1) + \cdots + b_n u(k-n), \tag{10.11}$$

with $a_i \neq 0$ and $b_i \neq 0$ for $i = 1, 2, \ldots, n$. The matrix

$$\lim_{N \to \infty} \frac{1}{N} \begin{bmatrix} U_{k-1,n,N} \\ Y_{k,n+1,N} \end{bmatrix} \begin{bmatrix} U_{k-1,n,N} \\ Y_{k,n+1,N} \end{bmatrix}^{\mathrm{T}},$$

constructed from input and output measurements, becomes singular no matter what the order of persistency of excitation of the input $u(k)$. To see this, note that a column of the matrix

$$\begin{bmatrix} U_{k-1,n,N} \\ Y_{k,n+1,N} \end{bmatrix}$$

is given by

$$v(k) = \begin{bmatrix} u(k-n) & u(k-n+1) & \cdots & u(k-1) & y(k-n) \\ & y(k-n+1) & \cdots & y(k) \end{bmatrix}^{\mathrm{T}}.$$

Constructing the vector

$$m^{\mathrm{T}} = \begin{bmatrix} b_n & b_{n-1} & \cdots & b_1 & a_n & a_{n-1} & \cdots & a_1 & -1 \end{bmatrix}$$

allows us to write the data-generating model (10.11) as $m^{\mathrm{T}} v(k) = 0$. Therefore,

$$\lim_{N \to \infty} \frac{1}{N} m^{\mathrm{T}} \begin{bmatrix} U_{k-1,n,N} \\ Y_{k,n+1,N} \end{bmatrix} \begin{bmatrix} U_{k-1,n+1,N} \\ Y_{k,n,N} \end{bmatrix}^{\mathrm{T}} m = 0.$$

Since $m \neq 0$, the matrix is indeed singular. The consequence is that the parameters of a one-step-ahead predictor of the form

$$\widehat{y}(k+1|k) = \theta_1 y(k) + \theta_1 y(k-1) + \cdots + \theta_{n+1} y(k-n)$$
$$+ \theta_{n+2} u(k-1) + \theta_{n+3} u(k-2) + \cdots + \theta_{2n+1} u(k-n)$$

cannot be determined uniquely. The reason for this lack of uniqueness is that the degree of the polynomial $1 + \theta_1 q^{-1} + \cdots + \theta_{n+1} q^{-n-1}$ exceeds that of the auto-regressive part (the part that involves $y(k)$) in the signal-generating system.

10.2.4.2 Persistency of excitation in subspace identification methods

Now we turn our attention to the role of persistency of excitation in subspace identification methods. In Chapter 9 it was explained that the key step in various subspace identification schemes is the estimation of the column space of the extended observability matrix $\mathcal{O}_s$. It was pointed out that, to obtain a consistent estimate of this subspace, under various assumptions on the additive-noise perturbations, the input sequence used for identification was required to satisfy certain rank conditions. These rank conditions are related to the notion of persistency of excitation.

We first focus on subspace identification in the case of an additive white perturbation to the output, as described in Section 9.3. The subspace algorithm for this case is the MOESP method based on Theorem 9.2 on page 310. The MOESP method requires the input to satisfy the rank condition (9.18) on page 302, which is repeated below for convenience:

$$\text{rank}\left(\begin{bmatrix} X_{0,N} \\ U_{0,s,N} \end{bmatrix}\right) = n + sm.$$

This condition involves the state sequence and is therefore difficult to verify on the basis of recorded input and output data. The following lemma shows that, if the input is persistently exciting of sufficient order, this rank condition is satisfied.

Lemma 10.4 *(Jansson, 1997) Given the state-space system*

$$x(k+1) = Ax(k) + Bu(k),$$
$$y(k) = Cx(k) + Du(k),$$

if the input $u(k)$ is persistently exciting of order $n + s$, then

$$\lim_{N \to \infty} \frac{1}{N} \begin{bmatrix} X_{0,N} \\ U_{0,s,N} \end{bmatrix} \begin{bmatrix} X_{0,N}^{\mathrm{T}} & U_{0,s,N}^{\mathrm{T}} \end{bmatrix} > 0. \tag{10.12}$$

This lemma can be proven using a multivariable extension of Lemma 10.3 (Jansson, 1997).

We conclude that, in the case of white measurement noise, the MOESP subspace method yields a consistent estimate of the column space of the extended observability matrix, provided that the input is persistently exciting of sufficient order.

We now look at the case of a nonwhite additive perturbation to the output. This case is dealt with by the PI-MOESP, PO-MOESP, and N4SID methods in Sections 9.5 and 9.6. In this case, persistency of excitation of the input signal is not sufficient to recover the column space of the extended observability matrix. Below we explain this statement by looking at the PO-MOESP and N4SID methods. The exposition that follows is largely based on the work of Jansson (1997) and Jansson and Wahlberg (1998).

In Section 9.6 it was argued that, as a precondition for consistently estimating the column space of the extended observability matrix $\mathcal{O}_s$, the following two rank conditions need to be satisfied:

$$\text{rank}\left(\lim_{N \to \infty} \frac{1}{\sqrt{N}} \begin{bmatrix} X_{0,N} \\ U_{0,2s,N} \end{bmatrix} \right) = n + 2ms,$$

(10.13)

$$\text{rank}\left(\lim_{N \to \infty} \frac{1}{N} \begin{bmatrix} X_{s,N} \\ U_{s,s,N} \end{bmatrix} \begin{bmatrix} Y_{0,s,N}^{\mathrm{T}} & U_{0,s,N}^{\mathrm{T}} & U_{s,s,N}^{\mathrm{T}} \end{bmatrix} \right) = n + sm.$$

(10.14)

Using the result of Lemma 10.4, it easy to see that condition (10.13) is satisfied if the input is persistently exciting of order $n + 2s$. To examine the rank condition (10.14), we split the system

$$x(k+1) = Ax(k) + Bu(k) + Ke(k),$$ (10.15)
$$y(k) = Cx(k) + Du(k) + e(k),$$ (10.16)

into a deterministic and a stochastic part using the superposition principle for linear systems. The deterministic part is given by

$$x^{\mathrm{d}}(k) = Ax^{\mathrm{d}}(k) + Bu(k),$$
$$y^{\mathrm{d}}(k) = Cx^{\mathrm{d}}(k) + Du(k),$$

and the stochastic part by

$$x^{\mathrm{s}}(k) = Ax^{\mathrm{s}}(k) + Ke(k),$$
$$y^{\mathrm{s}}(k) = Cx^{\mathrm{s}}(k) + e(k).$$

Of course, we have $x(k) = x^{\mathrm{d}}(k) + x^{\mathrm{s}}(k)$ and $y(k) = y^{\mathrm{d}}(k) + y^{\mathrm{s}}(k)$. Now the rank condition (10.14) can be written as

$$
\begin{bmatrix} X_{s,N} \\ U_{s,s,N} \end{bmatrix} \begin{bmatrix} Y_{0,s,N}^{\mathrm{T}} & U_{0,s,N}^{\mathrm{T}} & U_{s,s,N}^{\mathrm{T}} \end{bmatrix}
$$

$$
= \begin{bmatrix} X_{s,N}^{\mathrm{d}} \\ U_{s,s,N} \end{bmatrix} \begin{bmatrix} \left(Y_{0,s,N}^{\mathrm{d}} \right)^{\mathrm{T}} & U_{0,s,N}^{\mathrm{T}} & U_{s,s,N}^{\mathrm{T}} \end{bmatrix}
$$

$$
+ \begin{bmatrix} X_{s,N}^{\mathrm{s}} \\ 0 \end{bmatrix} \begin{bmatrix} \left(Y_{0,s,N}^{\mathrm{s}} \right)^{\mathrm{T}} & 0 & 0 \end{bmatrix}, \tag{10.17}
$$

where $X_{s,N}^{\mathrm{d}}$ and $X_{s,N}^{\mathrm{s}}$ denote the deterministic part and the stochastic part of the state, respectively; and $Y_{0,s,N}^{\mathrm{d}}$ and $Y_{0,s,N}^{\mathrm{s}}$ denote the deterministic part and the stochastic part of the output, respectively.

For the deterministic part of the rank condition, we have the following result.

Lemma 10.5 *For the minimal state-space system (10.15)–(10.16) with an input that is persistently exciting of order $n + 2s$, we have*

$$
\mathrm{rank}\left(\lim_{N \to \infty} \frac{1}{N} \begin{bmatrix} X_{s,N}^{\mathrm{d}} \\ U_{s,s,N} \end{bmatrix} \begin{bmatrix} \left(Y_{0,s,N}^{\mathrm{d}} \right)^{\mathrm{T}} & U_{0,s,N}^{\mathrm{T}} & U_{s,s,N}^{\mathrm{T}} \end{bmatrix} \right) = n + sm.
$$

Proof We can write the state sequence $X_{s,N}^{\mathrm{d}}$ as

$$
X_{s,N}^{\mathrm{d}} = A^{s} X_{0,N}^{\mathrm{d}} + \mathcal{C}_{s}^{\mathrm{r}} U_{0,s,N},
$$

where $\mathcal{C}_{s}^{\mathrm{r}}$ denotes the reversed controllability matrix

$$
\mathcal{C}_{s}^{\mathrm{r}} = \begin{bmatrix} A^{s-1}B & A^{s-2}B & \cdots & B \end{bmatrix}.
$$

Since we can write $Y_{0,s,N}^{\mathrm{d}} = \mathcal{O}_{s} X_{0,N}^{\mathrm{d}} + \mathcal{T}_{s} U_{0,s,N}$, we have

$$
\lim_{N \to \infty} \frac{1}{N} \begin{bmatrix} X_{s,N}^{\mathrm{d}} \\ U_{s,s,N} \end{bmatrix} \begin{bmatrix} \left(Y_{0,s,N}^{\mathrm{d}} \right)^{\mathrm{T}} & U_{0,s,N}^{\mathrm{T}} & U_{s,s,N}^{\mathrm{T}} \end{bmatrix}
$$

$$
= \lim_{N \to \infty} \frac{1}{N} \begin{bmatrix} A^{s} & \mathcal{C}_{s}^{\mathrm{r}} & 0 \\ 0 & 0 & I_{ms} \end{bmatrix} \underbrace{\begin{bmatrix} X_{0,N}^{\mathrm{d}} \\ U_{0,s,N} \\ U_{s,s,N} \end{bmatrix} \begin{bmatrix} \left(X_{0,N}^{\mathrm{d}} \right)^{\mathrm{T}} & U_{0,s,N}^{\mathrm{T}} & U_{s,s,N}^{\mathrm{T}} \end{bmatrix}}
$$

$$
\times \begin{bmatrix} \mathcal{O}_{s}^{\mathrm{T}} & 0 & 0 \\ \mathcal{T}_{s}^{\mathrm{T}} & I_{ms} & 0 \\ 0 & 0 & I_{ms} \end{bmatrix}.
$$

The matrix

$$
\begin{bmatrix} A^{s} & \mathcal{C}_{s}^{\mathrm{r}} & 0 \\ 0 & 0 & I_{ms} \end{bmatrix}
$$

has full row rank if the system is reachable. Since the system is assumed to be minimal, the matrix

$$\begin{bmatrix} \mathcal{O}_s^{\mathrm{T}} & 0 & 0 \\ \mathcal{T}_s^{\mathrm{T}} & I_{ms} & 0 \\ 0 & 0 & I_{ms} \end{bmatrix}$$

has full row rank. From Lemma 10.4 it follows that the limit of the underbraced matrix is also of full rank. The proof is concluded by a twofold application of Sylvester's inequality (Lemma 2.1 on page 16).

$\square$

The lemma shows that the deterministic part of Equation (10.17) is of full rank. However, there exist some exceptional cases in which, due to adding of the stochastic term

$$\begin{bmatrix} X_{s,N}^{\mathrm{s}} \\ 0 \end{bmatrix} \begin{bmatrix} \left(Y_{0,s,N}^{\mathrm{s}} \right)^{\mathrm{T}} & 0 & 0 \end{bmatrix},$$

the rank of the sum of the deterministic and stochastic parts in Equation (10.17) will drop below $n+sm$. Jansson has constructed an example with a persistently exciting input of sufficient order in which this happens (Jansson, 1997; Jansson and Wahlberg, 1998). In general this is very unlikely if $sm \gg n$, because then the stochastic term in (10.17) contains a lot of zeros, and hence the deterministic part will dominate the sum of the two. Jansson and Wahlberg (1998) derived sufficient conditions on the dimension parameter s for the case in which the input is a noise sequence generated by filtering a white-noise sequence with an ARX model. Bauer and Jansson (2000) showed that, if the input is persistently exciting of sufficient order, the set of ARX transfer functions for which the matrix (10.17) is of full rank is open and dense in the set of all stable transfer functions. This is a mathematical way of saying that it hardly ever happens that the matrix (10.17) drops rank when the input is persistently exciting of at least order $n+2s$. In conclusion, although we cannot prove that the relevant condition (10.14) on page 364 is satisfied, if the input is persistently exciting of order $n + 2s$, it will almost always hold in practice if $sm \gg n$.

10.2.5 *Types of input sequence*

Different types of input sequences are used as test sequences for system identification with different purposes. For example, in previous sections the use of elementary inputs, such as the step input, has been discussed as a means for gathering information needed to design a system-identification experiment. Other test sequences, such as an impulse or a harmonic signal, may also supply useful information in setting up an identification experiment. Some useful input sequences are depicted in Figure 10.6. In addition to the type of input, constraints on its spectral contents such as were outlined in the analysis of the persistency of excitation need to be considered in selecting the input sequence.

In this section, we illustrate by means of Example 10.7 that the optimality of solving parameter-estimation problems results in specific requirements on the input sequence. In addition to these requirements coming from numerical arguments, physical constraints in the applicability of the input sequence on the system to be identified may further constrain the class of input sequences. For example, when conducting helicopter-identification experiments employing a well-trained test pilot, one makes use of typical input sequences, such as the doublet input depicted in Figure 10.6. To generate more complex input patterns, such as the frequency sweep, an automatic rudder-actuation system would be required.

Example 10.7 (Minimum-variance parameter estimation and requirements on the input) Consider a dynamic system modeled by the FIR model

$$y(k) = h_1 u(k) + h_2 u(k-1) + h_3 u(k-2) + v(k),$$

with $u(k)$ and $v(k)$ independent stochastic processes. We want to determine constraints on the auto-correlation function $R_u(\tau) = E[u(k)u(k-\tau)]$ with $R_u(0) \leq 1$, such that the parameters θ_1, θ_2, and θ_3 of the model

$$\widehat{y}(k) = \theta_1 u(k) + \theta_2 u(k-1) + \theta_3 u(k-2)$$

can be optimally estimated in the sense that the determinant of their covariance matrix is maximal. The estimated parameter vector $\widehat{\theta}_N$ is obtained as

$$\widehat{\theta}_N = \arg\min \frac{1}{N} \sum_{k=1}^{N} \left(y(k) - \begin{bmatrix} u(k) & u(k-1) & u(k-2) \end{bmatrix} \begin{bmatrix} \theta_1 \\ \theta_2 \\ \theta_3 \end{bmatrix} \right)^2.$$

The covariance matrix of the estimated parameter vector $\widehat{\theta}_N$ is given by

$$E\left[\left(\widehat{\theta}_N - \begin{bmatrix} h_1 \\ h_2 \\ h_3 \end{bmatrix} \right) \left(\widehat{\theta}_N - \begin{bmatrix} h_1 \\ h_2 \\ h_3 \end{bmatrix} \right)^{\mathrm{T}} \right] = \frac{\sigma_v^2}{N} \begin{bmatrix} R_u(0) & R_u(1) & R_u(2) \\ R_u(1) & R_u(0) & R_u(1) \\ R_u(2) & R_u(1) & R_u(0) \end{bmatrix}^{-1}.$$

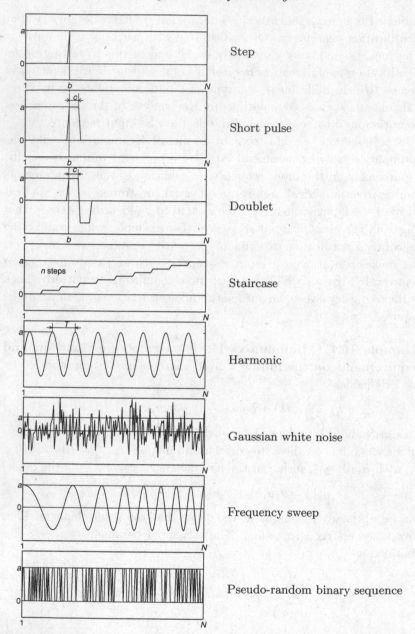

Fig. 10.6. Different types of standard input sequences used in system identification.

The determinant (see Section 2.4) of the matrix on the right-hand side, denoted by $\det(R^u)$, equals

$$\det(R^u) = R_u(0)\Big(R_u(0)^2 - R_u(1)^2\Big)$$
$$- R_u(1)\Big(R_u(0)R_u(1) - R_u(1)R_u(2)\Big)$$
$$+ R_u(2)\Big(R_u(1)^2 - R_u(0)R_u(2)\Big)$$
$$= R_u(0)^3 - R_u(0)R_u(1)^2 - R_u(0)R_u(1)^2 + R_u(1)^2 R_u(2)$$
$$+ R_u(1)^2 R_u(2) - R_u(0)R_u(2)^2$$
$$= R_u(0)^3 - 2R_u(1)^2\Big(R_u(0) - R_u(2)\Big) - R_u(0)R_u(2)^2.$$

Since $R_u(0) \geq R_u(i)$ for $i \neq 0$ and $R_u(0) \geq 0$, this quantity is maximized for $R_u(1) = R_u(2) = 0$. This property of the auto-correlation function of $u(k)$ can be satisfied if $u(k)$ is a zero-mean white-noise sequence.

10.3 Data pre-processing

After performing an identification experiment and after collecting the data, the next step is to polish the data to make them suitable for system identification. This step is referred to as the data pre-processing step. Below, we discuss several different pre-processing procedures.

10.3.1 Decimation

When the sampling frequency is too high with respect to the actual bandwidth (of interest) of the system under investigation, one may resample the data by selecting every jth sample (for $j \in \mathbb{N}$) from the original data sequences. If the original sampling frequency was ω_S (rad/s), the so-called down-sampled or decimated data sequence is sampled with a frequency of ω_S/j (rad/s). Prior to down-sampling a *digital* anti-aliasing filter with a cut-off frequency of $\omega_S/(2j)$ (rad/s) must be applied to prevent aliasing.

Too high a sampling frequency is indicated by the following:

(i) high-frequency disturbance in the data, above the frequency band of interest; and
(ii) the poles of an estimated discrete-time model cluster around the point $z = 1$ in the complex plane.

Decimation of the data with an accompanying low-pass anti-aliasing filter may take care of both phenomena. The following example illustrates the clustering of the poles.

Example 10.8 (Discrete-time system with poles around $z = 1$)
Consider the autonomous continuous-time state-space model

$$\frac{dx(t)}{dt} = \begin{bmatrix} -\lambda_1 & 0 \\ 0 & -\lambda_2 \end{bmatrix} x(t),$$

$$y(t) = \begin{bmatrix} c_1 & c_2 \end{bmatrix} x(t).$$

A discrete-time model that describes the sampled output sequence with a sampling frequency $1/\Delta T$ (Hz) is given by

$$x(k\,\Delta T + \Delta T) = \begin{bmatrix} e^{-\lambda_1 \Delta T} & 0 \\ 0 & e^{-\lambda_2 \Delta T} \end{bmatrix} x(k\Delta T),$$

$$y(k\Delta T) = \begin{bmatrix} c_1 & c_2 \end{bmatrix} x(k\Delta T),$$

where we have used the fact that, for a continuous-time system, $x(t) = e^{A(t-t_0)}x(t_0)$ for all $t \geq t_0$ (Rugh, 1996). In the limit for $\Delta T \to 0$, both poles of this discrete-time system approach $z = 1$.

10.3.2 Detrending the data

As discussed in Section 3.4.1, a linear model to be identified generally merely describes the underlying physical system in a restricted region of the system's operating range. For example, the operating range of an aircraft may be specified in terms of Mach number (velocity) and altitude. Over the operating range the dynamic behavior of the system is nonlinear. A linear model, resulting from linearizing the nonlinear dynamics at an equilibrium point in the selected operation range, may be used to approximate the system in a part of the operating region close to the point of linearization. Recall from Section 3.4.1 the notation $\overline{u}$, $\overline{y}$, and $\overline{x}$ used to represent the input, output, and state at the equilibrium point, and recall the notation $\widetilde{x}(k)$ used to represent the deviation of the state sequence from the equilibrium value $\overline{x}$. The linearized state-space model can be denoted by

$$\widetilde{x}(k+1) = A\widetilde{x}(k) + B(u(k) - \overline{u}), \tag{10.18}$$
$$y(k) = C\widetilde{x}(k) + \overline{y}. \tag{10.19}$$

There are two simple ways to deal with the presence of the (unknown) offsets $\overline{u}$ and $\overline{y}$ in the identification methods considered in the preceding chapters.

(i) Subtracting estimates of the offset from the input and output sequences, before applying the identification method. A widely used estimate of the offset is the sample mean of the measured sequences:

$$\bar{y} = \frac{1}{N} \sum_{k=1}^{N} y(k), \qquad \bar{u} = \frac{1}{N} \sum_{k=1}^{N} u(k).$$

(ii) Taking the constant offset terms into account in the identification procedure. This is illustrated below for the output-error and prediction-error methods. In these methods the offset terms can be included as unknown parameters of the state-space model (10.18)–(10.19) that need to be estimated. Let θ_n denote the vector containing the coefficients used to parameterize the matrices A and C, then the output of the model can be written as (see Theorem 7.1 on page 227)

$$\hat{y}(k, \theta) = C(\theta_n) A(\theta_n)^k x(0)$$
$$+ \sum_{\tau=0}^{k-1} C(\theta_n) A(\theta_n)^{k-1-\tau} [B \quad B\bar{u}] \begin{bmatrix} u(k) \\ 1 \end{bmatrix} + \bar{y},$$

or, equivalently,

$$\hat{y}(k, \theta) = \phi(k, \theta_n)\theta_\ell,$$

where

$$\theta_\ell = \begin{bmatrix} x(0) \\ \text{vec}(B) \\ B\bar{u} \\ \bar{y} \end{bmatrix}$$

and

$$\phi(k, \theta_n) = \left[C(\theta_n) A(\theta_n)^k, \sum_{\tau=0}^{k-1} u^{\mathrm{T}}(\tau) \otimes C(\theta_n) A(\theta_n)^{k-1-\tau}, \ldots \right.$$
$$\left. \sum_{\tau=0}^{k-1} C(\theta_n) A(\theta_n)^{k-1-\tau}, I_\ell \right].$$

From the estimates of the parameters θ_n and θ_ℓ we can then determine estimates of $\bar{u}$ and $\bar{y}$.

The data may contain not only an offset but also more general trends, such as linear drifts or periodic (seasonal) trends. The removal of such trends may be done by subtracting the best trend fitted through the input–output data in a least-squares sense.

10.3.3 Pre-filtering the data

Distortions in the data, for example trends and noise, that reside in a frequency band that can be separated from the frequency band (of interest) of the system can be removed by filtering the data with a high-pass or low-pass filter. Applying such filtering to the input and output sequences has some implications for the identification procedure.

Consider the SISO LTI system given by the following input–output model:

$$y(k) = G(q)u(k) + v(k).$$

On applying a filter $F(q)$ to the input and output sequences, we obtain their filtered versions $u_F(k)$ and $y_F(k)$ as

$$u_F(k) = F(q)u(k),$$
$$y_F(k) = F(q)y(k).$$

We can write the following input–output relationship between the filtered sequences:

$$y_F(k) = G(q)u_F(k) + v_F(k),$$

where $v_F(k) = F(q)v(k)$. We observe that the deterministic transfer function between $u_F(k)$ and $y_F(k)$ is $G(q)$. So seemingly the same information on $G(q)$ can be obtained using the filtered input–output sequences. This is not completely true, since the presence of the filter $F(q)$ does influence the accuracy of the estimate of $G(q)$. We illustrate this point by studying the bias in estimating an output-error model structure in which the number of data points approaches infinity. From Section 8.3.2 we know that the one-step-ahead prediction of the output equals

$$\widehat{y}(k, \theta) = \widehat{G}(q, \theta)u(k).$$

The bias analysis of Section 8.4 yields the following expression for the parameter estimate:

$$\widehat{\theta} = \arg\min \frac{1}{2\pi} \int_{-\pi}^{\pi} |G_0(e^{j\omega}) - \widehat{G}(e^{j\omega}, \theta)|^2 \Phi^{u_F}(\omega) + \Phi^{v_F}(\omega)d\omega$$

$$= \arg\min \frac{1}{2\pi} \int_{-\pi}^{\pi} |G_0(e^{j\omega}) - \widehat{G}(e^{j\omega}, \theta)|^2 |F(e^{j\omega})|^2 \Phi^u(\omega)$$
$$+ |F(e^{j\omega})|^2 \Phi^v(\omega)d\omega.$$

From this expression we can make two observations.

(i) The contribution of the disturbance to the cost function is directly affected by the filter $F(q)$. When the perturbation has a significant effect on the value of the cost function and the frequency band of the perturbation is known, the filter $F(q)$ may be selected to reduce this effect.

(ii) The presence of $|F(e^{j\omega})|^2$ as an additional weighting term for the bias term $|G_0(e^{j\omega}) - \widehat{G}(e^{j\omega})|^2$ offers the possibility of being able to emphasize certain frequency regions where the mismatch between G and $\widehat{G}$ should be small.

10.3.4 Concatenating data sequences

The experiment design is often characterized by conducting various experimental tests. Thus it may be important to combine or *concatenate* data sequences obtained from separate experiments. In estimating the parameters of the model, such a concatenation of data sets cannot be treated as a single data set and special care needs to be taken regarding the numerical algorithms used. The consequences for estimating an ARX model using input and output data sets from different experiments are treated in Exercise 10.4 on page 394.

10.4 Selection of the model structure

We have discussed identification methods to solve the problem of estimating a set of parameters (or system matrices) for a given model structure. However, in practice the model structure is often not known a priori and needs to be determined from the data as well. In this section we discuss the determination of the model structure using measured data.

10.4.1 Delay estimation

In physical applications, it is often the case that it takes some time before the action on an input (or steering) variable of the system becomes apparent on an output (or response) variable. The time that it takes for the output to react to the input is referred to as the *time delay* between the input and output variable. For example, in controlling the temperature in a room with a thermostat, it will take a few seconds before the temperature starts to rise after one's having changed the set-point temperature. Therefore, it is wise to extend the transfer from input to output with a delay. Assuming that the delay is an integer multiple of the sampling period, a SISO system in state-space form reads as

$$x(k+1) = Ax(k) + Bu(k-d),$$
$$y(k) = Cx(k).$$

Since $q^{-d}u(k)$ can be represented as the output $u_{\text{delay}}(k)$ of the state-space system,

$$z(k+1) = \begin{bmatrix} 0 & 1 & 0 & \cdots & 0 \\ 0 & 0 & 1 & & 0 \\ \vdots & & & \ddots & \vdots \\ 0 & 0 & & \cdots & 1 \\ 0 & 0 & & \cdots & 0 \end{bmatrix} z(k) + \begin{bmatrix} 0 \\ 0 \\ \vdots \\ 0 \\ 1 \end{bmatrix} u(k),$$

$$u_{\text{delay}}(k) = \begin{bmatrix} 1 & 0 & 0 & \cdots & 0 \end{bmatrix} z(k),$$

we can combine the two state-space models into one augmented system description as follows:

$$\begin{bmatrix} x(k+1) \\ z(k+1) \end{bmatrix} = \begin{bmatrix} A & B\begin{bmatrix} 1 & 0 & 0 & \cdots & 0 \end{bmatrix} \\ \hline \begin{matrix} 0 \\ 0 \\ \vdots \\ 0 \\ 0 \end{matrix} & \begin{matrix} 0 & 1 & 0 & \cdots & 0 \\ 0 & 0 & 1 & & 0 \\ \vdots & & & \ddots & \vdots \\ 0 & 0 & & \cdots & 1 \\ 0 & 0 & & \cdots & 0 \end{matrix} \end{bmatrix} \begin{bmatrix} x(k) \\ z(k) \end{bmatrix} + \begin{bmatrix} 0 \\ \hline 0 \\ 0 \\ \vdots \\ 0 \\ 1 \end{bmatrix} u(k),$$

$$y(k) = \begin{bmatrix} C & \begin{bmatrix} 1 & 0 & 0 & \cdots & 0 \end{bmatrix} \end{bmatrix} \begin{bmatrix} x(k) \\ z(k) \end{bmatrix}.$$

Therefore, a delay results in an increase of the order of the system. However, since the augmented part introduces zero eigenvalues, the augmented model is likely to represent a *stiff* system, that is, the eigenvalues of the system are likely to be far apart in the complex plane $\mathbb{C}$. Identifying such systems is difficult.

An alternative and equally valid route for SISO systems is to shift the input sequence d samples and then use the shifted input signal $u(k-d)$ together with the output signal $y(k)$ to identify a delay-free model; that is, we use the data set

$$\left\{ u(k-d), y(k) \right\}_{k=d}^{N-1}.$$

For MIMO systems with different delays in the relationships between the various inputs and outputs, the input–output transfer function may explicitly be denoted by

$$
\begin{bmatrix} y_1(k) \\ y_2(k) \\ \vdots \\ y_\ell(k) \end{bmatrix} = \begin{bmatrix} q^{-d_{1,1}}G_{1,1}(q) & q^{-d_{1,2}}G_{1,2}(q) & \cdots & q^{-d_{1,m}}G_{1,m}(q) \\ q^{-d_{2,1}}G_{2,1}(q) & q^{-d_{2,2}}G_{2,2}(q) & \cdots & q^{-d_{2,m}}G_{2,m}(q) \\ \vdots & \vdots & \ddots & \vdots \\ q^{-d_{\ell,1}}G_{\ell,1}(q) & q^{-d_{\ell,2}}G_{\ell,2}(q) & \cdots & q^{-d_{\ell,m}}G_{\ell,m}(q) \end{bmatrix} \begin{bmatrix} u_1(k) \\ u_2(k) \\ \vdots \\ u_m(k) \end{bmatrix}.
$$

To represent the delays in a standard state-space model, the following specific case different from the general case defined above is considered:

$$
\begin{bmatrix} y_1(k) \\ y_2(k) \\ \vdots \\ y_\ell(k) \end{bmatrix} = \begin{bmatrix} q^{-d_1^y} & & & 0 \\ & q^{-d_2^y} & & \\ & & \ddots & \\ 0 & & & q^{-d_\ell^y} \end{bmatrix} \begin{bmatrix} G_{1,1}(q) & G_{1,2}(q) & \cdots & G_{1,m}(q) \\ G_{2,1}(q) & G_{2,2}(q) & \cdots & G_{2,m}(q) \\ \vdots & \vdots & \ddots & \vdots \\ G_{\ell,1}(q) & G_{\ell,2}(q) & \cdots & G_{\ell,m}(q) \end{bmatrix}
$$

$$
\times \begin{bmatrix} q^{-d_1^u} & & & 0 \\ & q^{-d_2^u} & & \\ & & \ddots & \\ 0 & & & q^{-d_m^u} \end{bmatrix} \begin{bmatrix} u_1(k) \\ u_2(k) \\ \vdots \\ u_m(k) \end{bmatrix}.
$$

On the basis of this assumption regarding the delays, a MIMO state-space model without delay may be identified on the basis of "shifted" input and output data sequences:

$$
\left\{ \begin{bmatrix} u_1(k - d_1^u) \\ u_2(k - d_2^u) \\ \vdots \\ u_m(k - d_m^u) \end{bmatrix}, \begin{bmatrix} y_1(k + d_1^y) \\ y_2(k + d_2^y) \\ \vdots \\ y_\ell(k + d_\ell^y) \end{bmatrix} \right\}_{k=\max\{d_i^u\}}^{N-1-\max\{d_j^y\}}
$$

for $i = 1, 2, \ldots, m$ and $j = 1, 2, \ldots, \ell$.

One useful procedure for estimating the delay is based on the estimation of a FIR model. A SISO FIR model of order m is denoted by

$$
y(k) = h_0 u(k) + h_1 u(k - 1) + \cdots + h_m u(k - m). \tag{10.20}
$$

The estimation of the unknown coefficients h_i, $i = 0, 1, \ldots, m$ can simply be done by solving a linear least-squares problem. The use of this model structure in delay estimation is illustrated in the following example.

Example 10.9 (Delay estimation by estimating a FIR model)
Consider a continuous-time first-order system with a delay of 0.4 s and a time constant of 2 s. The transfer function of this system is given by

$$G(s) = e^{-s\tau_\mathrm{d}} \frac{1}{1+\tau s},$$

where the term $e^{-s\tau_\mathrm{d}}$ represents the delay, $\tau_\mathrm{d} = 0.4\,\mathrm{s}$ and $\tau = 2$ s. Discretizing the model with a zeroth-order hold on the input (Dorf and Bishop, 1998) and a sampling period $\Delta T = 0.1$ s yields the following model:

$$
\begin{aligned}
y(k) &= \frac{q^{-4}\left(1 - e^{-0.05}\right)}{q - e^{-0.05}} u(k) \\
&= \frac{0.048\,77}{1 - 0.951\,23 q^{-1}} q^{-5} u(k) \\
&= \frac{0.048\,77}{1 - 0.951\,23 q^{-1}} u(k-5).
\end{aligned}
$$

From this derivation we see that $u(k)$ has a delay of five samples. We take $u(k)$ equal to a zero-mean white-noise sequence with variance $\sigma_u^2 = 1$, and generate an output sequence of 120 samples. Using the generated input and output data, a FIR model of the form (10.20) with $m = 50$ is estimated. The estimates of the impulse-response parameters h_i are plotted in Figure 10.7. The figure clearly shows that a delay of five samples is present.

Next, $u(k)$ is taken equal to a doublet of 120 samples with $b = 30$ samples and $c = 30$ samples (see Figure 10.6 on page 368). The estimated FIR model with $m = 50$ is plotted in Figure 10.7. Now we observe that, though the estimate is biased, as is explained by solving Exercise 10.3 on page 393, the delay of five samples is still detectable.

10.4.2 *Model-structure selection in ARMAX model estimation*

In Chapters 7 and 8, the problem addressed was that of how to estimate the parameters in a pre-defined model parameterization. In Section 8.3 we discussed a number of SISO parametric input–output models and the estimation of their parameters. Having an input–output data set, we first have to select one of these particular model parameterizations and subsequently we have to select *which parameters* of this model parameterization need to be estimated. The first selection is targeted toward

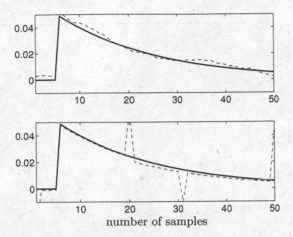

number of samples

Fig. 10.7. Top: estimated impulse-response coefficients of a FIR model of order 50 using a white-noise input sequence (broken line), together with the true impulse response (solid line). Bottom: estimated impulse-response coefficients of a FIR model based on a doublet input sequence (broken line), together with the true impulse response (solid line).

finding an appropriate way of modeling the noise (or perturbations) in the data, since the deterministic part of an ARX, ARMAX, OE, or BJ model can be chosen equal. This selection is far from trivial since prior (qualitative) information about the system to be identified is often available only for its deterministic part.

Prediction-error methods optimize a cost function given by

$$\min_{\theta} J_N(\theta) = \min_{\theta} \frac{1}{N} \sum_{k=0}^{N-1} \|y(k) - \widehat{y}(k|k-1, \theta)\|_2^2,$$

where $y(k)$ is the measured output and $\widehat{y}(k|k-1, \theta)$ is the output of the model corresponding to the parameters θ. Therefore, discriminating one model with respect to another model could be based on the (smallest) value of the cost function $J_N(\theta)$. However, when working with sequences of finite data length, focusing on the value of the cost function $J_N(\theta)$ alone can be misleading. This is illustrated in the following example.

Example 10.10 (Selecting the number of parameters on the basis of the value of the cost function) Consider the data generated by the system

$$y(k) = \frac{0.048\,77q^{-5}}{1 - 0.951\,23q^{-1}} u(k) + e(k),$$

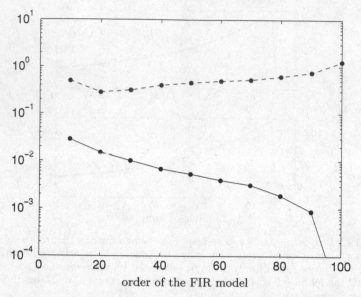

order of the FIR model

Fig. 10.8. The value of the cost function $J_N(\theta)$ for different orders of the FIR model (*solid line*) and the normalized error on the parameters (*broken line*) for Example 10.10.

where $u(k)$ and $e(k)$ are taken to be independent zero-mean white-noise sequences with standard deviations 1 and 0.1, respectively. Using a generated data set of 200 input and output samples, a FIR model of the form (10.20) is estimated for various orders m. The value of the cost function is given as

$$J_N(\theta) = \frac{1}{N-m} \sum_{k=m}^{N-1} (y(k) - \widehat{y}(k,\theta))^2,$$

with $N = 200$ and $\widehat{y}(k,\theta)$ the output predicted by the estimated FIR model. The values of $J_N(\theta)$ for $m = 10, 20, \ldots, 100$ are plotted in Figure 10.8. We observe that $J_N(\theta)$ is *monotonically decreasing*. Let us also inspect the error on the estimated parameter vector. For this, let θ denote the infinite-impulse response of the signal-generating system and let $\widehat{\theta}^m$ denote the parameter vector of the FIR model (10.20) for finite m, estimated using the available measurements. The normalized error on the estimated parameter vector is represented by

$$\frac{\|\theta(1:m) - \widehat{\theta}^m\|_2^2}{\|\theta(1:m)\|_2^2}.$$

This quantity is also shown in Figure 10.8 for various values of m. The figure clearly shows that this error is increasing with m. Thus, although for increasing m the cost function is decreasing, the error on the estimated parameters indicates that increasing m will not lead to a better model.

The example suggests that we should penalize the complexity of the model *in addition* to the value of the prediction error. The complexity is of the model expressed by the number of parameters used to parameterize the model. One often-used modification of the cost function is Akaike's information criterion (Akaike, 1981):

$$J_N^{\text{AIC}}(\theta) = J_N(\theta) + \frac{\dim(\theta)}{N}, \qquad (10.21)$$

where $\dim(\theta)$ is the dimension of the parameter vector θ. This criterion expresses the compromise between accuracy, in the prediction error sense, and model complexity. Another approach to avoid the estimation of too-complex models is by means of cross-validation. This is discussed at the end of the chapter in Section 10.5.3.

Besides the issue of model complexity, we also have to deal with the numerical complexity of solving the optimization problem. To reduce the numerical complexity, one generally starts with the estimation of simple models. Here the simplicity corresponds to the unknown parameters appearing *linearly* in the prediction error. This linearity holds both for FIR and for ARX types of model. For these models the optimization problem reduces to a linear least-squares problem.

For the selection both of the structure and of the initial parameter estimates of a SISO ARMAX model, Ljung (1999) has proposed that one should start the selection process by identifying a *high-order* ARX model. From the estimated parameters of this high-order ARX model, we attempt to retrieve the orders of the various polynomials in a more complex model such as an ARMAX model (Equation (8.22) on page 269). Using this information, the more complex optimization of the non-quadratic prediction-error cost function $J_N(\theta)$ that corresponds to the ARMAX model is solved.

The use of an ARX model in retrieving information about the parameterization and selection of the orders of the various polynomials in other parametric transfer-function models is illustrated in the following example.

Example 10.11 (Selecting the structural indices of an ARMAX model) Consider the following data-generating ARMAX model:

$$y(k) = \frac{q^{-1}}{1 - 0.95q^{-1}}u(k) + \frac{1 - 0.8q^{-1}}{1 - 0.95q^{-1}}e(k).$$

The input sequence $u(k)$ and the noise sequence $e(k)$ are taken to be independent, zero-mean, white-noise sequences of unit variance, with a length of 1000 samples.

We can use a Taylor-series expansion to express $1/(1 - 0.8q^{-1})$ as

$$\frac{1}{1 - 0.8q^{-1}} = 1 + 0.8q^{-1} + 0.64q^{-2} + 0.512q^{-3} + 0.4096q^{-4} + \cdots.$$

With this expression the ARMAX model of the given dynamic system can by approximated as

$$y(k) \approx \frac{q^{-1}}{1 - 0.95q^{-1}} \frac{1 + 0.8q^{-1} + 0.64q^{-2} + \cdots + 0.8^p q^{-p}}{1 + 0.8q^{-1} + 0.64q^{-2} + \cdots + 0.8^p q^{-p}}u(k)$$
$$+ \frac{1}{(1 - 0.95q^{-1})(1 + 0.8q^{-1} + 0.64q^{-2} + \cdots + 0.8^p q^{-p})}e(k).$$

for some choice of p. In this way, we have approximated the ARMAX model by a high-order ARX model. It should be remarked that such an approximation is always possible, provided that the polynomial $C(q)$ of the ARMAX model (Equation (8.22) on page 269) has all its roots within the unit circle. For this high-order ARX model, two observations can be made.

(i) On increasing the order of the denominator of the ARX model (increasing p), the noise part of the ARMAX model will be better approximated by the noise part of the high-order ARX model.

(ii) The deterministic transfer from $u(k)$ to $y(k)$ of the ARX model approximates the true deterministic transfer of the ARMAX model and in addition contains a large number of pole–zero cancellations.

Let us now evaluate the consequences of using a high-order ARX model to identify the given ARMAX model. We inspect the value of the prediction-error cost function $J_N(\theta_{\mathrm{ARX}})$ of an estimated $(p+1)$th-order ARX model using 1000 input–output samples generated with the given ARMAX model. The cost function is plotted in against p Figure 10.9. We observe clearly that, on increasing the value of p, the value of the cost function of the ARX model approaches the optimal value, which equals the standard deviation of $e(k)$.

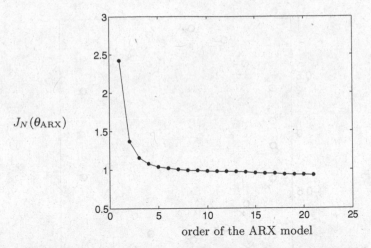

$J_N(\theta_{\mathrm{ARX}})$

order of the ARX model

Fig. 10.9. The value of $J_N(\theta_{\mathrm{ARX}})$ versus the order of the ARX model estimated using the input–output data of the ARMAX signal-generating system in Example 10.11.

Figure 10.10 shows the poles and zeros of the deterministic part of the identified 21st-order ARX model. We can make the following observations.

(i) The near pole–zero cancellations in the deterministic transfer function of the 21st-order ARX model are clearly visible. The noncanceling pole, which is equal to 0.95, is an accurate approximation of the true pole of the deterministic part of the ARMAX model. This is not the case if p is taken equal to zero; the estimated pole of the deterministic part of a first-order ARX model equals 0.8734.

(ii) The polynomial that represents the poles and zeros that approximately cancel out in the deterministic part of the high-order ARX model can be taken as an initial estimate of the polynomial $C(q)^{-1}$ in the ARMAX model. This information may be used to generate initial estimates for the coefficients of the polynomial $C(q)$ in the estimation of an ARMAX model.

The example highlights the fact that, starting with a simple ARX model estimation, it is possible to retrieve information on the order of the polynomials of an ARMAX model, as well as initial estimates of the parameters. An extension of the above recipe is found by solving Exercise 10.2 on page 392.

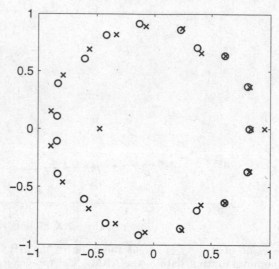

Fig. 10.10. The complex plane showing the poles (crosses) and zeros (circles) of the estimated 21st-order ARX model of Example 10.11. One zero outside the range of the figure is not shown.

Nevertheless, in practical circumstances inspecting a pole–zero cancellation and estimating high-order ARX models is generally much more cumbersome than has been illustrated in the above example. Therefore, we need to develop more-powerful estimation techniques that require only the solution of linear least-squares problems and provide consistent estimates for much more general noise-model representations. The subspace identification methods described in Chapter 9 are good candidates to satisfy this need.

10.4.3 Model-structure selection in subspace identification

Subspace identification methods aim at directly estimating a state-space model in the innovation form from input and output data. These methods do not require an a priori parameterization of the system matrices A, B, C, and D. More importantly, the estimates of the system matrices are obtained by solving a series of linear least-squares problems and performing one SVD. For these steps, well-conditioned reliable numerical solutions are available in most common software packages, such as Matlab (MathWorks, 2000b).

In subspace identification methods, model-structure selection boils down to the determination of the order of the state-space model. The size of the block Hankel matrices processed by subspace algorithms is

determined by an *upper bound* on the order of the state-space model.
In Chapter 9 (Equation (9.4) on page 295) this upper bound is denoted
by the integer s. Let n be the order of the state-space model, then the
constraint on s used in the derivation of the subspace identification algorithms is $s > n$.

The choice of the dimensioning parameter s will influence the accuracy
of the state-space model identified. Attempts have been made to analyze
the (information-) theoretical and experimental selection of the parameter s for a given input–output data set (Bauer *et al.*, 1998). Practical
experience has shown that the dimensioning parameter s can be fixed
at a value of two to three times a rough estimate of the order of the
underlying system.

For a particular choice of the dimensioning parameter s (and a particular value of the time delay between input and output), the *only*
model-structure index that needs to be determined in subspace identification is the order of the state-space model. The fact that the order
is the only structural model index leads to a feasible and systematic
evaluation, as well as a visual (two-dimensional) inspection of a scalar
model-quality measure versus the model order. Two such scalar quality measures can be used *in combination* for selecting the model order:
first, the singular values obtained in calculating the approximation of
the column space of $\mathcal{O}_s$; and second, the value of the prediction-error
cost function $J_N(A, B, C, D, K)$ or a scaled variant can be evaluated for
state-space models of various orders.

A scaled version of the cost function $J_N(\theta)$ that is often used for
assessing the quality of a model is the variance accounted for (VAF).
The VAF is defined as

$$
\text{VAF}(y(k), \widehat{y}(k,\theta)) = \max\left(0, \left(1 - \frac{\frac{1}{N}\sum_{k=1}^{N} \|y(k) - \widehat{y}(k,\theta)\|_2^2}{\frac{1}{N}\sum_{k=1}^{N} \|y(k)\|_2^2} \right) \cdot 100\% \right).
$$

$$(10.22)$$

The VAF has a value between 0% and 100%; the higher the VAF, the
lower the prediction error and the better the model.

Below, we illustrate the combination of the singular values obtained
from subspace identification and the VAF in selecting an appropriate
model order for the acoustical-duct example treated earlier in
Example 8.3.

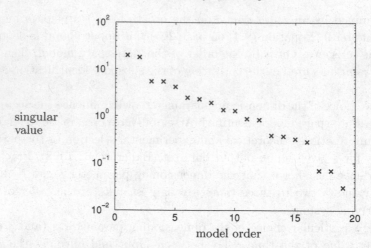

Fig. 10.11. The first 20 singular values calculated with the PO-MOESP subspace identification method for $s = 40$ in Example 10.12.

Example 10.12 (Model-order selection with subspace methods)
Consider the acoustical duct of Example 8.3, depicted in Figure 8.2. We use the transfer-function model between the speaker M and microphone to generate 4000 samples of input and output measurements $u(k)$ and $y(k)$, with $u(k)$ taken as a zero-mean white-noise sequence with unit variance. On the basis of these input and output measurements, we estimate various low-order state-space models using the PO-MOESP subspace method treated in Section 9.6.1. The dimension parameter s in this algorithm is taken equal to 40.

As discussed in Section 9.3, the singular values calculated in the subspace method can be used to determine the order of the system. To determine the order, we look for a gap between a set of dominant singular values that correspond to the dynamics of the system and a set of small singular values due to the noise. The singular values obtained by applying the PO-MOESP method to the data of the acoustical duct are plotted in Figure 10.11. From the largest 20 singular values plotted in this figure, it becomes clear that detecting a clear gap between the ordered singular values is not trivial. To make a better decision on the model order, we inspect the VAF of the estimated output obtained with the models of order 2–12 in Figure 10.12. The VAF values show that, although the simulated model was of order 20, accurate prediction is possible with an eighth-order state-space model identified with a subspace method.

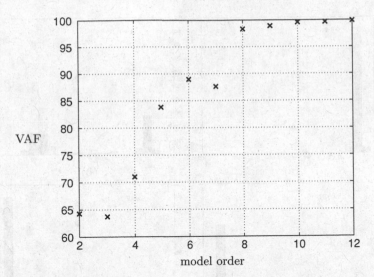

Fig. 10.12. The VAF of the estimated output obtained by models having order $n = 2$–12 determined with the PO-MOESP subspace identification method for $s = 40$ in Example 10.12.

10.4.3.1 Combination of subspace identification and prediction-error methods

Subspace identification methods are able to provide an initial estimate of the model structure and the parameters (system matrices) of the innovation state-space model without constraining the parameter set. This property, in combination with the numerically well-conditioned way by which these estimates can be obtained, leads to the proposal of using these estimates as initial estimates for prediction-error methods (Haverkamp, 2000; Bergboer *et al.*, 2002). The following example illustrates the usefulness of this combination.

Example 10.13 (Combining subspace identification and prediction-error methods) Consider the acoustical duct of Example 8.3 on page 278. We perform 100 identification experiments using different input and output realizations from this acoustical duct. We use 4000 data points and we estimate low-order models with an order equal to six. We compare the VAF values of the estimated outputs obtained from the following four types of models:

(i) a sixth-order ARX model estimated by solving a linear least squares problem;

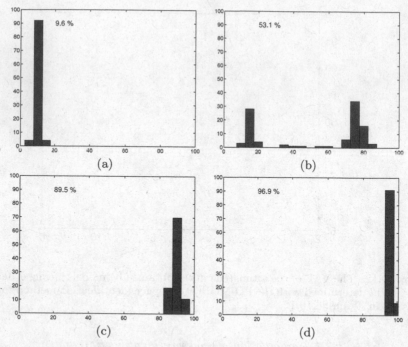

Fig. 10.13. Distributions of the VAF values for 100 identification experiments on the acoustical duct of Example 10.13: (a) ARX models, (b) OE models, (c) state-space models obtained from subspace identification, and (d) state-space models obtained from prediction-error optimization. The percentages indicate the mean VAF value.

(ii) a sixth-order OE model estimated by iteratively solving a nonlinear least-squares problem using the oe function from the Matlab system-identification toolbox (MathWorks, 2000a);

(iii) a sixth-order state-space model identified with the PO-MOESP algorithm described in Subsection 9.6.1 using $s = 40$; and

(iv) a sixth-order state-space model estimated using a prediction-error method with the matrices (A, C) parameterized using the output normal form presented in Subsection 7.3.1 (as an initial starting point the state-space model identified by the PO-MOESP algorithm is used).

Figure 10.13 shows the VAF values of the different models for the different identification experiments. The ARX model structure is not suitable, as is indicated by the very low VAF values. The OE models have better VAF values. These models use the estimates of the ARX model as initial estimates. However, the high number of VAF values in the region 10%–30% indicates that the algorithm gets stuck at local minima. A similar observation was made in Example 8.4 on page 281. We also see that the combination of subspace identification and prediction-error optimization does yield accurate *low-order* models. In fact, the models obtained directly by subspace identification are already quite good.

10.5 Model validation

As discussed in the previous section, selecting the best model among a number of possibilities can be done on the basis of a criterion value of the prediction-error cost function $J_N(\theta)$ or the modified version $J_N^{\mathrm{AIC}}(\theta)$ in Equation (10.21) on page 379. However, once a model has been selected we can further *validate* this model on the basis of a more detailed inspection. One class of inspection tests is called residual tests. These tests are briefly discussed in this section.

In the context of innovation state-space models the optimal model means that the variance of the one-step-ahead prediction error is minimal. For a parameter estimate $\widehat{\theta}$ of θ the one-step-ahead prediction error $\widehat{\epsilon}(k,\widehat{\theta})$ can be estimated via the Kalman-filter equations

$$\widehat{x}(k+1,\widehat{\theta}) = \Big(A(\widehat{\theta}) - K(\widehat{\theta})C(\widehat{\theta})\Big)\widehat{x}(k,\widehat{\theta})) + \Big(B(\widehat{\theta}) - K(\widehat{\theta})D(\widehat{\theta})\Big)u(k)$$
$$+ K(\widehat{\theta})y(k),$$
$$\widehat{\epsilon}(k,\widehat{\theta}) = y(k) - C(\widehat{\theta})\widehat{x}(k,\widehat{\theta}) - D(\widehat{\theta})u(k).$$

If the system to be identified can be represented by a model in the model class of innovation state-space models, this prediction-error sequence has a number of additional properties when the identification experiment is conducted in *open-loop* mode.

(i) The sequence $\widehat{\epsilon}(k,\widehat{\theta})$ is a zero-mean white-noise sequence.
(ii) The sequence $\widehat{\epsilon}(k,\widehat{\theta})$ is statistically independent from the input sequence $u(k)$.

If the identification is performed in *closed-loop* mode, only the first property can be verified and the second property will hold only for past inputs, that is, $E[\widehat{\epsilon}(k,\widehat{\theta})u(k-\tau)] = 0$ for $\tau > 0$. The verification of both properties of the prediction-error sequence is done via the estimation of the auto-correlation function and the cross-correlation function (see Section 4.3.1).

10.5.1 The auto-correlation test

If the length of the estimated prediction-error sequence equals N, then an estimate of its auto-correlation matrix can be obtained as

$$\widehat{R}_\epsilon(\tau) = \frac{1}{N-\tau} \sum_{k=\tau+1}^{N} \widehat{\epsilon}(k,\widehat{\theta})\widehat{\epsilon}(k-\tau,\widehat{\theta})^{\mathrm{T}}.$$

For a scalar sequence $\widehat{\epsilon}(k,\widehat{\theta}) \in \mathbb{R}$, the central-limit theorem indicates that, when $\widehat{\epsilon}(k,\widehat{\theta})$ is

(i) a zero-mean white-noise sequence,
(ii) Gaussian- and symmetrically distributed, and
(iii) independent from the input sequence $u(j)$ for all $k,j \in \mathbb{Z}$,

the vector

$$r_\epsilon = \frac{\sqrt{N}}{\widehat{R}_\epsilon(0)} \begin{bmatrix} \widehat{R}_\epsilon(1) \\ \widehat{R}_\epsilon(2) \\ \vdots \\ \widehat{R}_\epsilon(M) \end{bmatrix}$$

is asymptotically Gaussian-distributed with mean zero and a covariance matrix equal to the identity matrix I_M. This information allows us to evaluate model quality by checking whether the estimated auto-correlation function represents a unit pulse (within a certain confidence interval based on the probability distribution).

10.5.2 The cross-correlation test

Using the estimated one-step-ahead prediction-error sequence $\widehat{\epsilon}(k,\widehat{\theta})$ and the input sequence $u(k)$, we can develop a test similar to the auto-correlation test to check the independence between the one-step-ahead prediction error and the input. The cross-correlation matrix is given by

$$\widehat{R}_{\epsilon u}(\tau) = \frac{1}{N-\tau} \sum_{k=\tau+1}^{N} \widehat{\epsilon}(k,\widehat{\theta})u(k-\tau,\widehat{\theta})^{\mathrm{T}}.$$

For scalar sequences $\widehat{\epsilon}(k,\widehat{\theta}) \in \mathbb{R}$ and $u(k) \in \mathbb{R}$, the central-limit theorem indicates that, when $\widehat{\epsilon}(k,\widehat{\theta})$ is

(i) a zero-mean white-noise sequence,
(ii) Gaussian and symmetrically distributed, and
(iii) independent from the input sequence $u(j)$ for all $k,j \in \mathbb{Z}$,

the vector

$$R_{\epsilon u} = \frac{\sqrt{N}}{\widehat{R}_{\epsilon}(0)} \begin{bmatrix} \widehat{R}_{\epsilon u}(1) \\ \widehat{R}_{\epsilon u}(2) \\ \vdots \\ \widehat{R}_{\epsilon u}(M) \end{bmatrix}$$

is asymptotically Gaussian-distributed with mean zero and a covariance matrix given by

$$\widehat{R}^u = \frac{1}{N-M} \sum_{k=M}^{N} \begin{bmatrix} u_k \\ u_{k-1} \\ \vdots \\ u_{k-M+1} \end{bmatrix} \begin{bmatrix} u_k & u_{k-1} & \cdots & u_{k-M+1} \end{bmatrix}.$$

This information allows us to check with some probability whether the estimated cross-correlation function equals zero.

Example 10.14 (Residual test) The ARMAX model of Example 10.11 on page 380 is used to generate 1000 measurements of a white-noise input sequence $u(k)$ and its corresponding output sequence $y(k)$. On the basis of the data generated, we estimate three types of first-order parametric models: an ARX, an OE, and an ARMAX model. The one-step-ahead prediction errors of these models are evaluated using their auto-correlations and the cross-correlations with the input sequence, as discussed above. Figure 10.14 displays these correlation functions for a maximal lag τ equal to 20. From this figure we observe that the residual of the first-order ARMAX model complies with the theoretical assumptions of the residua; that is, the auto-correlation function approximates a unit pulse and the cross-correlation function is approximately zero.

For the estimated first-order OE model, only the latter property of the one-step-ahead prediction error holds. This is because the OE model attempts to model only the deterministic part (that is, the transfer between the measurable input and output) of the data-generating system. This type of residual test with an OE model indicates that we have found the right deterministic part of the model.

Both the auto-correlation and the cross-correlation of the one-step-ahead prediction error obtained with the first-order ARX model indicate that the model is still not correct. A possible model refinement would

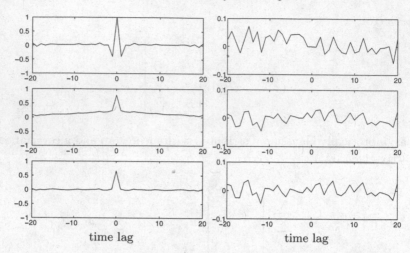

Fig. 10.14. The auto-correlation function (left) of the one-step-ahead prediction error and the cross-correlation function (right) between one-step-ahead prediction error and input for estimated first-order models. From top to bottom: ARX, OE, and ARMAX models.

be to increase the order of the polynomials *or* to select a different type of model, as was illustrated in Example 10.11.

10.5.3 The cross-validation test

The validation tests discussed in the previous two subsections are performed on the same data set as was used for identifying the model. In cross-validation, a data set different from the one used for estimating the parameters is used. This can be made possible by, for example, dividing the data batch into two parts: the first 2/3 of the total number of samples is used for the parameter estimation, while the final 1/3 is used for evaluating the quality of the model by computing the value of the prediction-error cost function and performing the correlation tests. Cross-validation overcomes the chance of "over-fitting" by selecting too large a number of parameters to be estimated, as illustrated in Example 10.10 on page 377.

10.6 Summary

This chapter has highlighted the fact that parameter estimation is only a tiny part of the task of identifying an unknown dynamical system. The many choices and subtasks that need to be addressed prior to being able to define and solve a parameter-estimation problem are described by the system-identification cycle which is summarized in Figure 10.1 on page 347.

After designing and carrying out an identification experiment, the collected input and output measurements can be polished in the *data pre-processing* step by simple digital-processing schemes to remove outliers, trends, and undesired noise. These polished data are used to determine a model. Prior to estimating model parameters, the type of model and which parameters are to be estimated need to be determined. This *structure selection* may request a vast amount of time and energy in the overall identification task. The class of subspace identification methods of Chapter 9 has the potential to simplify greatly the task of structure selection. This is because a single and general MIMO state-space model is considered. The generality of this type of model stems from the fact that it covers a wide class of parametric models, such as ARX, ARMAX, OE, and Box–Jenkins models (see Chapter 8). An important part in judging the outcome of the system-identification cycle is *model validation*. Several strategies for validating a model on the basis of (newly) measured input–output data sets are briefly listed.

The introduction to systematically setting up, modifying, and analyzing a system-identification experiment given here has left a number of important practical tricks involved in getting accurate models untouched. Such tricks are often problem-specific and can be learned by applying system identification in practice.

The authors have developed a Matlab toolbox that implements the subspace and prediction-error methods for the identification of state-space models described in this book. It is highly recommended that you experiment with the software using artificially generated data or your own data set. To get you started, a comprehensive software guide is distributed together with the toolbox. This guide includes illustrative examples on how to use the software and it bridges the gap between the theory presented in this book and the Matlab programs.

Exercises

10.1   Let T be the sampling period in seconds by which the input and output signals of a continuous time system are sampled. The filtering of these signals is done in an analog manner and introduces a pure delay with transfer function

$$G(s) = e^{-s\tau_d}, \quad \tau_d < T < 1,$$

with s the complex variable in the Laplace transform. A discrete-time approximation of this (infinite-dimensional) system is to be determined by solving the following steps.

(a) *Rational approximation by a continuous-time transfer function.* Let the Taylor-series expansion of $e^{-s\tau_d}$ be given as

$$e^{-s\tau_d} = 1 - s\tau_d + \frac{1}{2!}(s\tau_d)^2 - \frac{1}{3!}(s\tau_d)^3 + O(s^4),$$

with $O(s^j)$ denoting a polynomial in s with lowest power s^j. Determine the coefficient a in the first-order rational (Padé (Hayes, 1996)) approximation of $G(s)$, given by

$$G(s) = \frac{a - s}{a + s} + O(s^3).$$

(b) *Bilinear transformation of the complex plane.* Using a similar Taylor-series expansion to that given in part (a), show that the relationship between the complex variable s and the variable z given by

$$z = e^{sT}$$

can be approximated as

$$z = \frac{1 + (T/2)s}{1 - (T/2)s} + O(s^3).$$

(c) Let the transformation from the variable s to the variable z be given as

$$z = \frac{1 + (T/2)s}{1 - (T/2)s}.$$

Determine the inverse transformation from z to s.

(d) Using the approximation of the variable s derived in part (c), determine the coefficient α in the discrete-time transfer-function approximation of the delay $G(s)$:

$$\mathcal{Z}(G(s)) \approx \frac{1 + \alpha z}{z + \alpha}.$$

(e) Is the approximation of part (d) of minimum phase?

10.2 Let the system that generates the output sequence $y(k) \in \mathbb{R}$ from the input sequence $u(k) \in \mathbb{R}$ be given by

$$y(k) = \frac{B(q)}{A(q)}u(k) + \frac{1}{C(q)}e(k),$$

with $e(k)$ a zero-mean white-noise sequence that is independent from $u(k)$.

(a) Show that, if $A(q)$ has all its roots within the unit circle, we have

$$\frac{1}{A(q)} = \sum_{i=1}^{\infty} h_i q^{-i}, \quad |h_i| < \infty.$$

(b) Using the relationship derived in part (a), show that the output can be accurately approximated by the model with the following special ARX structure:

$$C(q)y(k) = \left(\sum_{i=1}^{n} \gamma_i q^{-i} \right) u(k) + e(k). \qquad \text{(E10.1)}$$

10.3 Let the system that generates the output sequence $y(k) \in \mathbb{R}$ from the input sequence $u(k) \in \mathbb{R}$ be given by

$$y(k) = \left(\sum_{i=1}^{\infty} h_i q^{-i} \right) u(k) + e(k),$$

with $u(k)$ and $e(k)$ independent zero-mean white-noise sequences. This system is modeled using a FIR model that predicts the output as

$$\widehat{y}(k, \alpha_i) = \left(\sum_{i=1}^{n} \alpha_i q^{-i} \right) u(k).$$

The coefficients α_i of the FIR model are estimated as

$$\widehat{\alpha}_i = \arg \min E\left[\left(y(k) - \widehat{y}(k, \alpha_i) \right)^2 \right], \quad i = 1, 2, \dots, n.$$

(a) Show that the estimates $\widehat{\alpha}_i$, $i = 1, 2, \dots, n$ are unbiased; that is, show that they satisfy

$$E[\widehat{\alpha}_i - h_i] = 0,$$

provided that

$$\sum_{i=n+1}^{\infty} |h_i|^2 < \infty.$$

(b) Show that

$$\widehat{\alpha}_i = \frac{E[y(k)u(k-i)]}{E[u(k)^2]}.$$

(c) Show that the estimates of α_i are biased in the case that $u(k)$ differs from being zero-mean white noise.

10.4   Let two batches of input–output data collected from an unknown dynamical system be denoted by

$$\{u(k),y(k)\}_{k=N_{b,1}}^{N_{e,1}} \quad \text{and} \quad \{u(k),y(k)\}_{k=N_{b,2}}^{N_{e,2}}$$

Determine the matrix A and the vector b, when using the two batches simultaneously to estimate the parameters a_j and b_j $(j = 1, \ldots, n)$ of an nth-order ARX model via the least-squares problem

$$\min_{a_j,b_j} \left\| A \begin{bmatrix} a_1 \\ \vdots \\ a_n \\ b_1 \\ \vdots \\ b_n \end{bmatrix} - b \right\|_2^2.$$

10.5   Show how pre-filtering can be used to make the weighting term of the model error for the ARX model with the number of measurements $N \to \infty$ equal to the weighting term for the OE model. You can assume that the measured data from the true system are noise-free.

References

Akaike, H. (1981). Modern development of statistical methods. In Eykhoff, P. (ed.), *Trends and Progress in System Identification*. Elmsford, New York: Pergamon Press, pp. 169–184.

Anderson, B. D. O. and Moore, J. B. (1979). *Optimal Filtering*. Englewood Cliffs, New Jersey: Prentice Hall.

Åström, K. J. and Wittenmark, B. (1984). *Computer Controlled Systems*. Englewood Cliffs, New Jersey: Prentice Hall.

Bauer, D. (1998). 'Some asymptotic theory for the estimation of linear systems using maximum likelihood methods or subspace algorithms.' PhD thesis, Technische Universität Wien, Vienna.

Bauer, D., Deistler, M., and Scherrer, W. (1998). User choices in subspace algorithms. In *Proceedings of the 37th IEEE Conference on Decision and Control*, Tampa, Florida. Piscataway, New Jersey: IEEE Press, pp. 731–736.

Bauer, D. and Jansson, M. (2000). Analysis of the asymptotic properties of the MOESP type of subspace algorithms. *Automatica*, **36**, 497–509.

Bergboer, N., Verdult, V., and Verhaegen, M. (2002). An efficient implementation of maximum likelihood identification of LTI state-space models by local gradient search. In *Proceedings of the 41st IEEE Conference on Decision and Control*, Las Vegas, Nevada. Piscataway, New Jersey: IEEE Press, pp. 616–621.

Box, G. E. P. and Jenkins, G. M. (1970). *Time Series Analysis, Forecasting and Control*. San Francisco, California: Holden-Day.

Boyd, S. and Baratt, C. (1991). *Linear Controller Design, Limits of Performance*. Englewood Cliffs, New Jersey: Prentice-Hall.

Brewer, J. W. (1978). Kronecker products and matrix calculus in system theory. *IEEE Transactions on Circuits and Systems*, **25**(9), 772–781.

Broersen, P. M. T. (1995). A comparison of transfer function estimators. *IEEE Transactions on Instrumentation and Measurement*, **44**(3), 657–661.

Bruls, J., Chou, C. T., Haverkamp, B., and Verhaegen, M. (1999). Linear and non-linear system identification using separable least-squares. *European Journal of Control*, **5**(1), 116–128.

Bühler, W. K. (1981). *Gauss: A Biographical Study*. Berlin: Springer-Verlag.

Chiuso, A. and Picci, G. (2005). Consistency analysis of certain closed-loop subspace identification methods. *Automatica*, **41**(3), 377–391.

Chou, C. T. and Verhaegen, M. (1997). Subspace algorithms for the identification of multivariable dynamic errors-in-variables models. *Automatica*, **33**(10), 1857–1869.

Clarke, D. W., Mohtadi, C., and Tuffs, P. S. (1987). Generalized predictive control part I: the basic algorithm. *Automatica*, **23**(2), 137–148.

David, B. (2001). 'Parameter estimation in nonlinear dynamical systems with correlated noise.' PhD thesis, Université Catholique de Louvain, Louvain-La-Neuve, Belgium.

David, B. and Bastin, G. (2001). An estimator of the inverse covariance matrix and its application to ML parameter estimation in dynamical systems. *Automatica*, **37**(1), 99–106.

De Bruyne, F. and Gevers, M. (1994). Identification for control: can the optimal restricted complexity model always be indentified? In *Proceedings of the 33rd IEEE Conference on Decision and Control*, Orlando, Florida. Piscataway, New Jersey: IEEE Press, pp. 3912–3917.

Dorf, R. and Bishop, R. (1998). *Modern Control Systems* (8th edn.). New York: Addison-Wesley.

Duncan, D. and Horn, S. (1972). Linear dynamic recursive estimation from the viewpoint of regression analysis. *Journal of the American Statistical Association*, 67, 815–821.

Eker, J. and Malmborg, J. (1999). Design and implementation of a hybrid control strategy. *IEEE Control Systems Magazine*, **19**(4), 12–21.

Friedlander, B., Kailath, T., and Ljung, L. (1976). Scattering theory and linear least-squares estimation. Part II: discrete-time problems. *Journal of the Franklin Institute*, **301**, 71–82.

Garcia, C. E., Prett, D. M., and Morari, M. (1989). Model predictive control: theory and practice – a survey. *Automatica*, **25**(3), 335–348.

Gevers, M. (1993). Towards a joint design of identification and control? In Trentelman, H. L. and Willems, J. C. (eds.), *Essays on Control: Perspectives in the Theory and its Applications*. Boston, Massachusetts: Birkhäuser, pp. 111–151.

Golub, G. H. and Pereyra, V. (1973). The differentiation of pseudo-inverses and nonlinear least squares problems whose variables separate. *SIAM Journal of Numerical Analysis*, **10**(2), 413–432.

Golub, G. H. and Van Loan, C. F. (1996). *Matrix Computations* (3rd edn.). Baltimore, Maryland: The Johns Hopkins University Press.

Gomez, C. (1999). *Engineering and Scientific Computation with Scilab*. Boston, Massachusetts: Birkhäuser.

Grimmett, G. R. and Stirzaker, D. R. (1983). *Probability and Random Processes*. Oxford: Oxford University Press.

Hanzon, B. and Ober, R. J. (1997). Overlapping block-balanced canonical forms and parametrizations: the stable SISO case. *SIAM Journal of Control and Optimization*, **35**(1), 228–242.

(1998). Overlapping block-balanced canonical forms for various classes of linear systems. *Linear Algebra and its Applications*, **281**, 171–225.

Hanzon, B., Peeters, R., and Olivi, M. (1999). Balanced parametrizations of discrete-time stable all-pass systems and the tangential Schur algorithm. In *Proceedings of the European Control Conference 1999*, Karlsruhe. Duisburg: Universität Duisburg (CD Info: http://www.uni-duisburg.de/euca/ecc99/proceedi.htm).

Hanzon, B. and Peeters, R. L. M. (2000). Balanced parametrizations of stable SISO all-pass systems in discrete time. *Mathematics of Control, Signals, and Systems*, **13**(3), 240–276.

Haverkamp, B. (2000). 'Subspace method identification, theory and practice.' PhD thesis, Delft University of Technology, Delft, The Netherlands.

Hayes, M. H. (1996). *Statistical Digital Signal Processing and Modeling*. New York: John Wiley and Sons.

Ho, B. L. and Kalman, R. E. (1966). Effective construction of linear, state-variable models from input/output functions. *Regelungstechnik*, **14**(12), 545–548.

Isermann, R. (1993). Fault diagnosis of machines via parameter estimation and knowledge processing: tutorial paper. *Automatica*, **29**(4), 815–835.

Jansson, M. (1997). 'On subspace methods in system identification and sensor array signal processing.' PhD thesis, Royal Institute of Technology (KTH), Stockholm, Sweden.

   (2003). Subspace identification and ARX modeling. In *Preprints of the IFAC Symposium on System Identification (SYSID)*, Rotterdam, The Netherlands. Oxford: Elsevier Science Ltd, pp. 1625–1630.

Jansson, M. and Wahlberg, B. (1998). On consistency of subspace methods for system identification. *Automatica*, **34**(12), 1507–1519.

Johansson, R. (1993). *System Modeling and Identification*. Englewood-Cliffs, New Jersey: Prentice-Hall.

Kailath, T. (1968). An innovation approach to least-squares estimation. *IEEE Transactions on Automatic Control*, **16**(6), 646–660.

   (1980). *Linear Systems*. Englewood Cliffs, New Jersey: Prentice-Hall.

Kailath, T., Sayed, A. H., and Hassibi, B. (2000). *Linear Estimation*. Upper Saddle River, New Jersey: Prentice-Hall.

Kalman, R. E. (1960). A new approach to linear filtering and prediction problems. *Transactions of the ASME, Journal of Basic Engineering*, **82**, 34–45.

Kamen, E. W. (1990). *Introduction to Signals and Systems* (2nd edn.). New York: Macmillan Publishing Company.

Katayama, T. (2005). *Subspace Methods for System Identification*. London: Springer-Verlag.

Kourouklis, S. and Paige, C. C. (1981). A constrained least squares approach to the general Gauss–Markov linear model. *Journal of the American Statistical Association*, **76**, 620–625.

Kung, S. (1978). A new identification and model reduction algorithm via singular value decompositions. In *Proceedings of the 12th Asilomar Conference on Circuits, Systems and Computers*, Pacific Grove, California. Piscataway, New Jersey: IEEE Press, pp. 705–714.

Kwakernaak, H. (1993). Robust control and H_∞ optimization – tutorial paper. *Automatica*, **29**(2), 255–273.

Kwakernaak, H. and Sivan, R. (1991). *Modern Signals and Systems*. Englewood Cliffs, New Jersey: Prentice-Hall.

Larimore, W. E. (1990). Canonical variate analysis in identification, filtering and adaptive control. In *Proceedings of the 29th IEEE Conference on Decision and Control*, Honolulu, Hawaii. Piscataway, New Jersey, IEEE Press, pp. 596–604.

Lee, L. H. and Poolla, K. (1999). Identification of linear parameter-varying systems using nonlinear programming. *Journal of Dynamic System Measurement and Control*, **121**(1), 71–78.

Leon-Gracia, A. (1994). *Probability and Random Processes for Electrical Engineering* (2nd edn.). Reading, Massachusetts: Addison-Wesley.

Ljung, L. (1978). Convergence analysis of parametric identification methods. *IEEE Transactions on Automatic Control*, **23**(5), 770–783.

(1999). *System Identification: Theory for the User* (2nd edn.). Upper Saddle River, New Jersey: Prentice-Hall.

Ljung, L. and Glad, T. (1994). *Modelling of Dynamic Systems*. Englewood Cliffs, New Jersey: Prentice-Hall.

Luenberger, D. (1964). Observing the state of a linear system. *IEEE Transactions of Military Electronics*, **8**, 74–80.

MathWorks (2000a). *System Identification Toolbox User's Guide*. Natick, Massachusetts: MathWorks.

(2000b). *Using Matlab*. Natick, Massachusetts: MathWorks.

(2000c). *Using the Control Systems Toolbox*. Natick, Massachusetts: MathWorks.

McKelvey, T. (1995). 'Identification of state-space model from time and frequency data.' PhD thesis, Linköping University, Linköping, Sweden.

McKelvey, T. and Helmersson, A. (1997). System identification using overparametrized model class – improving the optimization algorithm. In *Proceedings of the 36th IEEE Conference on Decision and Control*, San Diego. Piscataway, New Jersey: IEEE Press, pp. 2984–2989.

Moré, J. J. (1978). The Levenberg–Marquardt algorithm: implementation and theory. In Watson, G. A. (ed.), *Numerical Analysis*, volume 630 of *Lecture Notes in Mathematics*. Berlin: Springer-Verlag, pp. 106–116.

Oppenheim, A. V. and Willsky, A. S. (1997). *Signals and Systems* (2nd edn.). Upper Saddle River, New Jersey: Prentice-Hall.

Paige, C. C. (1979). Fast numerically stable computations for generalized linear least squares problems. *SIAM Journal on Numerical Analysis*, **16**(1), 165–171.

(1985). Covariance matrix representation in linear filtering. In Datta, B. N. (ed.), *Linear Algebra and Its Role in Systems Theory*. Providence, Rhode Island: AMS Publications, pp. 309–321.

Paige, C. C. and Saunders, M. A. (1977). Least squares estimation of discrete linear dynamic systems using orthogonal transformations. *SIAM Journal on Numerical Analysis*, **14**, 180–193.

Papoulis, A. (1991). *Probability, Random Variables, and Stochastic Processes* (3rd edn.). New York: McGraw-Hill.

Peternell, K., Scherrer, W., and Deistler, M. (1996). Statistical analysis of novel subspace identification methods. *Signal Processing*, **52**(2), 161–177.

Powell, J. D., Franklin, G. F., and Emami-Naeini, A. (1994). *Feedback Control of Dynamic Systems* (3rd edn.). Reading, Massachusetts: Addison-Wesley.

Qin, S. J. and Ljung, L. (2003). Closed-loop subspace identification with innovation estimation. In *Preprints of the IFAC Symposium on System Identification (SYSID)*, Rotterdam, The Netherlands. Oxford: Elsevier Science Ltd, pp. 887–892.

Rao, S. S. (1986). *Mechanical Vibrations*. Reading, Massachusetts: Addison-Wesley.

Rudin, W. (1986). *Real and Complex Analysis* (3rd edn.). New York: McGraw-Hill.

Rugh, W. J. (1996). *Linear System Theory* (2nd edn.). Upper Saddle River, New Jersey: Prentice-Hall.

Schoukens, J., Dobrowiecki, T., and Pintelon, R. (1998). Parametric identification of linear systems in the presence of nonlinear distortions. A frequency domain approach. *IEEE Transactions of Automatic Control*, **43**(2), 176–190.

Sjöberg, J., Zhang, Q., Ljung, L., Benveniste, A., Delyon, B., Glorennec, P.-Y., Hjalmarsson, H., and Juditsky, A. (1995). Nonlinear black-box modeling in system identification: a unified overview. *Automatica*, **31**(12), 1691–1724.

Skogestad, S. and Postlethwaite, I. (1996). *Multivariable Feedback Control: Analysis and Design*. Chichester: John Wiley and Sons.

Söderström, S. and Stoica, P. (1983). *Instrumental Variable Methods for System Identification*. New York: Springer-Verlag.

(1989). *System Identification*. Englewood Cliffs, New Jersey: Prentice-Hall.

Soeterboek, R. (1992). *Predictive Control: A Unified Approach*. New York: Prentice-Hall.

Sorid, D. and Moore, S. K. (2000). The virtual surgeon. *IEEE Spectrum*, **37**(7), 26–31.

Strang, G. (1988). *Linear Algebra and its Applications* (3rd edn.). San Diego: Harcourt Brace Jovanovich.

Tatematsu, K., Hamada, D., Uchida, K., Wakao, S., and Onuki, T. (2000). New approaches with sensorless drives. *IEEE Industry Applications Magazine*, **6**(4), 44–50.

van den Hof, P. M. J. and Schrama, R. J. P. (1994). Identification and control – closed loop issues. In *Preprints of the IFAC Symposium on System Identification*. Copenhagen, Denmark. Oxford: Elsevier Science Ltd, pp. 1–13.

Van Overschee, P. and De Moor, B. (1994). N4SID: subspace algorithms for the identification of combined deterministic and stochastic systems. *Automatica*, **30**(1), 75–93.

(1996a). *Closed-loop subspace identification*. Technical Report ESAT-SISTA/TR 1996-52I, KU Leuven, Leuven, Belgium.

(1996b). *Subspace Identification for Linear Systems; Theory, Implementation, Applications*. Dordrecht: Kluwer Academic Publishers.

Verdult, V. (2002). 'Nonlinear system identification: a state-space approach.' PhD thesis, University of Twente, Faculty of Applied Physics, Enschede, The Netherlands.

Verdult, V., Kanev, S., Breeman, J., and Verhaegen, M. (2003). Estimating multiple sensor and actuator scaling faults using subspace identification. In *IFAC Symposium on Fault Detection, Supervision and Safety of Technical Processes (Safeprocess) 2003*, Washington DC. Oxford: Elsevier Science Ltd, pp. 387–392.

Verdult, V. and Verhaegen, M. (2002). Subspace identification of multivariable linear parameter-varying systems. *Automatica*, **38**(5), 805–814.

Verhaegen, M. (1985). 'A new class of algorithms in linear system theory.' PhD thesis, KU Leuven, Leuven, Belgium.

(1993). Subspace model identification part 3. Analysis of the ordinary output-error state-space model identification algorithm. *International Journal of Control*, **56**(3), 555–586.

(1994). Identification of the deterministic part of MIMO state space models given in innovations form from input–output data. *Automatica*, **30**(1), 61–74.

Verhaegen, M. and Dewilde, P. (1992a). Subspace model identification part 1. The output-error state-space model identification class of algorithms. *International Journal of Control*, **56**(5), 1187–1210.

(1992b). Subspace model identification part 2. Analysis of the elementary output-error state-space model identification algorithm. *International Journal of Control*, **56**(5), 1211–1241.

Verhaegen, M. and Van Dooren, P. (1986). Numerical aspects of different Kalman filter implementations. *IEEE Transactions on Automatic Control*, **31**(10), 907–917.

Verhaegen, M., Verdult, V., and Bergboer, N. (2003). *Filtering and System Identification: Matlab Software*. Delft: Delft Center for Systems and Control.

Viberg, M. (1995). Subspace-based methods for the identification of linear time-invariant systems. *Automatica*, **31**(12), 1835–1851.

Wahlberg, B. and Ljung, L. (1986). Design variables for bias distribution in transfer function estimation. *IEEE Transactions on Automatic Control*, **31**(2), 134–144.

Zhou, K., Doyle, J. C., and Glover, K. (1996). *Robust and Optimal Control*. Upper Saddle River, New Jersey: Prentice-Hall.

Index

absolutely summable, 46
accuracy
  least-squares method, 108
  parameter, 211, 242, 248, 264–265,
    275, 355, 379
  subspace method, 310, 383
  transfer function, 4, 372
adjugate, 20
Akaike's information criterion, 379, 387
aliasing, 349–350, 369
amplitude, 46, 73, 352
ARMAX, 3, 5, 266–271, 348, 355
  prediction error, 276, 389
  structure selection, 376–382
ARX, 3, 269–271, 279, 355, 373, 385,
  389
  persistency of excitation, 359–363,
    366
  prediction error, 277, 285
  structure selection, 377, 379–382

bandwidth, 179, 350–352, 354, 369
basis, 12, 16, 18, 27, 241
Bayes' rule, 93
bias
  frequency-response function, 191,
    199
  linear estimation, 110–120, 376
  output-error method, 5, 211, 243,
    244, 245, 246, 372
  prediction-error method, 5, 255, 264,
    265, 275–286
  state estimation, 133, 135, 136, 137,
    144, 148
  subspace method, 310, 312, 313, 314,
    315, 318, 319, 320, 329, 338
bijective, 214, 218
Box–Jenkins model, 271–275, 280,
  282–283, 377

canonical form, 5, 73–75, 78, 216, 218,
  229, 268
cascade connection, 80
Cauchy–Schwartz inequality, 11
causality, 56, 260, 337
Cayley–Hamilton theorem, 21, 66
characteristic polynomial, 21
Cholesky factorization, 26, 222, 224,
  247
closed-loop, *see* feedback
completion of squares, 30–31, 114, 115,
  118, 139
concatenation, 373
consistent
  frequency-response function, 191, 198,
    199
  output-error method, 244
  prediction-error method, 256, 285,
    338, 382
  subspace method, 319, 329, 336, 338,
    363, 364
control design, 9, 208, 255, 256, 275,
  346, 348
controllability, 64–65, 69
  matrix, 65, 75, 317, 365
controller, 169, 208, 256, 278, 283, 284,
  285, 337, 338
convolution, 47, 49, 53, 72, 107, 181,
  183, 185
correlation, 98, 105, 246, 390
  auto-, 101, 103–104, 106, 107, 193,
    358, 367, 388, 389
  cross-, 101, 201, 388–390
cost function, 210, 245, 264, 275,
  293
  least-squares, 5, 29–31, 37, 108, 118,
    235, 246
  output-error, 210–211, 227–231, 243,
    244, 247

Printed in the United States
By Bookmasters